Undergraduate Texts in Mathematics

Editors
F. W. Gehring
P. R. Halmos

Advisory Board
C. DePrima
I. Herstein

Undergraduate Texts in Mathematics

Apostol: Introduction to Analytic Number Theory.

Armstrong: Basic Topology.

Bak/Newman: Complex Analysis.

Banchoff/Wermer: Linear Algebra Through Geometry.

Childs: A Concrete Introduction to Higher Algebra.

Chung: Elementary Probability Theory with Stochastic Processes.

Croom: Basic Concepts of Algebraic Topology.

Curtis: Linear Algebra: An Introductory Approach.

Dixmier: General Topology.

Driver: Why Math?

Ebbinghaus/Flum/Thomas Mathematical Logic.

Fischer: Intermediate Real Analysis.

Fleming: Functions of Several Variables. Second edition.

Foulds: Optimization Techniques: An Introduction.

Foulds: Combination Optimization for Undergraduates.

Franklin: Methods of Mathematical Economics.

Halmos: Finite-Dimensional Vector Spaces. Second edition.

Halmos: Naive Set Theory.

Iooss/Joseph: Elementary Stability and Bifurcation Theory.

Jänich: Topology.

Kemeny/Snell: Finite Markov Chains.

Klambauer: Aspects of Calculus.

Lang: Undergraduate Analysis.

Lang: A First Course in Calculus. Fifth Edition.

Lang: Calculus of One Variable. Fifth Edition.

Lang: Introduction to Linear Algebra. Second Edition.

Lax/Burstein/Lax: Calculus with Applications and Computing, Volume 1. Corrected Second Printing.

LeCuyer: College Mathematics with APL.

Lidl/Pilz: Applied Abstract Algebra.

Macki/Strauss: Introduction to Optimal Control Theory.

Malitz: Introduction to Mathematical Logic.

Marsden/Weinstein: Calculus I, II, III. Second edition.

Jerome Malitz

Introduction to Mathematical Logic

Set Theory
Computable Functions
Model Theory

Springer-Verlag
New York Berlin Heidelberg
London Paris Tokyo

J. Malitz
Department of Mathematics
University of Colorado
Boulder, CO 80309
U.S.A.

Editorial Board

F. W. Gehring
Department of Mathematics
University of Michigan
Ann Arbor, MI 48104
U.S.A.

P. R. Halmos
Department of Mathematics
Santa Clara University
Santa Clara, CA 95053
U.S.A.

AMS Subject Classification: 02-01, 04-01

With 2 Figures

Library of Congress Cataloging in Publication Data

Malitz, J.
　Introduction to mathematical logic.

　Bibliography: p.
　Includes index.
　1. Logic, Symbolic and mathematical.　I. Title.
QA9.M265　　511'.3　　78-13588

All rights reserved.

No part of this book may be translated or reproduced
in any form without written permission from Springer-Verlag.

© 1979 by Springer-Verlag New York Inc.

Printed in the United States of America

9 8 7 6 5 4 3 2 (Second Corrected Printing, 1987)

ISBN 0-387-90346-1　Springer-Verlag New York
ISBN 3-540-90346-1　Springer-Verlag Berlin Heidelberg

For Sue, Jed, and Seth

Contents

Preface		ix
Glossary of Symbols		xi

Part I: An Introduction to Set Theory

1.1	Introduction	1
1.2	Sets	1
1.3	Relations and Functions	6
1.4	Pairings	9
1.5	The Power Set	14
1.6	The Cantor–Bernstein Theorem	17
1.7	Algebraic and Transcendental Numbers	20
1.8	Orderings	21
1.9	The Axiom of Choice	27
1.10	Transfinite Numbers	31
1.11	Paradise Lost, Paradox Found (Axioms for Set Theory)	43
1.12	Declarations of Independence	51

Part II: An Introduction to Computability Theory

2.1	Introduction	59
2.2	Turing Machines	60

2.3	Demonstrating Computability without an Explicit Description of a Turing Machine	68
2.4	Machines for Composition, Recursion, and the "Least Operator"	79
2.5	Of Men and Machines	89
2.6	Non-computable Functions	90
2.7	Universal Machines	95
2.8	Machine Enumerability	100
2.9	An Alternate Definition of Computable Function	105
2.10	An Idealized Language	110
2.11	Definability in Arithmetic	118
2.12	The Decision Problem for Arithmetic	120
2.13	Axiomatizing Arithmetic	124
2.14	Some Directions in Current Research	129

Part III: An Introduction to Model Theory

3.1	Introduction	135
3.2	The First Order Predicate Calculus	136
3.3	Structures	138
3.4	Satisfaction and Truth	142
3.5	Normal Forms	150
3.6	The Compactness Theorem	158
3.7	Proof of the Compactness Theorem	162
3.8	The Löwenheim–Skolem Theorem	167
3.9	The Prefix Problem	174
3.10	Interpolation and Definability	180
3.11	Herbrand's Theorem	185
3.12	Axiomatizing the Validities of L	187
3.13	Some Recent Trends in Model Theory	190

Subject Index   195

Preface

This book is intended as an undergraduate senior level or beginning graduate level text for mathematical logic. There are virtually no prerequisites, although a familiarity with notions encountered in a beginning course in abstract algebra such as groups, rings, and fields will be useful in providing some motivation for the topics in Part III.

An attempt has been made to develop the beginning of each part slowly and then to gradually quicken the pace and the complexity of the material. Each part ends with a brief introduction to selected topics of current interest.

The text is divided into three parts: one dealing with set theory, another with computable function theory, and the last with model theory. Part III relies heavily on the notation, concepts and results discussed in Part I and to some extent on Part II. Parts I and II are independent of each other, and each provides enough material for a one semester course.

The exercises cover a wide range of difficulty with an emphasis on more routine problems in the earlier sections of each part in order to familiarize the reader with the new notions and methods. The more difficult exercises are accompanied by hints. In some cases significant theorems are developed step by step with hints in the problems. Such theorems are not used later in the sequence.

The part dealing with set theory is intended to provide a notational and conceptual framework for areas of mathematics outside of logic as well as to introduce the student to those topics that are of particular interest to those working in the foundations of set theory.

We hope that the part of the text devoted to computable functions will be of interest to those who intend to work with real world computers.

We believe that the notation, methodology, and results of elementary logic should be a part of a general mathematics program and are of value in a wide variety of disciplines within mathematics and outside of mathematics.

Boulder, Colorado J. MALITZ
March 1979

Glossary of Symbols

$\{\cdots\}$	2	f^{-1}	7
I	2	$f\upharpoonright C$	7
N	2	$f \circ g$	7
N$^+$	2	$\sim$	9
Q	2	$\prec, \preccurlyeq$	15
Q$^+$	2	$(\mathbf{R},<), (\mathbf{Q},<), (\mathbf{I},<), <_A$	22
R	2		
R$^+$	2	$(\mathbf{N},<)$	23
$\{x:\cdots\}$	2	Ord α	33
$\varnothing$	2	$\cong$	34
$\in$	2	Card	36
$\subseteq, \supseteq, \subset, \supset$	2, 3	$c(x)$	37
$\cup$	3	ZF	38
$\bigcup X$	4	ZFC	46 IC 48
$\bigcup_{i \in I} A_i$	4	$\vDash$	51
$\cap$	4	$M(t)$	61
$\bigcap X$	4	Sum	63
$B-A$	4	$C_{k,d}$	63
$P(X)$	5	$P_{k,t}$	63
$A \times B$	6	Pred	63
$[B]_k$	6	Prod$_n$	69
Dom R	7	Mult	70
Ran zR	7	Pow	71
1–1	7	Diff'	71
$f:A \to B$	7	$m \dotminus n$	71
$^A B$	7	$\exists x \leqslant y$	75
$f[C]$	7	$\forall x \leqslant y$	75
		$P(\bar{n}, x)$	75
		Prime	75

xi

Glossary of Symbols

Prim	75	In	101
Exp′	76	Halt	101
Max	77	∀	103
$M\|t \leadsto s$	80	∃	103
$\dot n$	80	Rec	107
compress	80	Rem	108
$M_1 \downarrow M$	82	L	111 (see also 136)
		≈	111
$\begin{smallmatrix} & M & & M \\ \swarrow & , & \searrow \\ M_1 & & & M_1 \end{smallmatrix}$	82	∨, ∧, ¬	111
		∀, ∃	111
		⊕, ⊙	111
		[,]	111
$\begin{smallmatrix} & M & \\ \swarrow & & \searrow \\ M_1 & & M_2 \end{smallmatrix}$	82	Trm	111
		$t\langle z \rangle$	112
$\rightarrow M, M \leftarrow$	82	Fm	113
		⊨	114
		⊢	124
copy k	83	Cons$_\mathfrak{S}$	129
		Prf$_\mathfrak{S}$	129
shift right	84	P	133
		NP	133
shift left	84	τ	136
		Fm$_s$	137
erase	84	$\mathfrak{A} \subseteq \mathfrak{B}$	140
# k	91	$\mathfrak{B} \upharpoonright s$	140
TS	95	$\equiv_g, \equiv$	140
STP	95	$z\binom{u}{a}$	142
decode	95	$t^\mathfrak{A}\langle z \rangle$	142
		$\mathfrak{A} \models \varphi\langle z \rangle$	142
code	98	Th $\mathfrak{A}$	144
Exp	98	≡	144
RR	98	Mod Σ	144
RC	98	$\mathfrak{D}(\mathfrak{A})$	147
NP	99	Π_F	166
NS	99	∝	167
NST	99	$\mathfrak{D}^c \mathfrak{A}$	169
T	99	$\bigcup_{\alpha < k} \mathfrak{A}_\alpha$	170
STP	99	$S(\mathfrak{B}, X)$	177
Row	100	$\mathfrak{A} \propto_\Gamma \mathfrak{B}$	178
Mach	101	∀∃-formula	178
		Th$_{\forall\exists} K$	178

PART I
An Introduction to Set Theory

1.1 Introduction

Through the centuries mathematicians and philosophers have wondered if size comparisons between infinite collections of objects can be made in a meaningful way. Does it make sense to ask if there are as many even numbers as odd numbers? What does it mean to say that one infinite collection has greater magnitude than another? Can one speak of different sizes of infinity?

Before the last three decades of the nineteenth century, mathematicians and philosophers generally agreed that such notions are not meaningful. But then in the early 1870s, a German mathematician, Georg Cantor (1845–1918), in a remarkable series of papers, formulated a theory in which size comparisons between infinite collections could be made. This theory became known as set theory. As with many radical departures from traditional approaches, his ideas were at first violently attacked but now have come to be regarded as a useful and basic part of modern mathematics. This chapter is an introduction to set theory.

1.2 Sets

We use the term *set* to refer to any collection of objects. The objects composing a set will be referred to as the *members* or *elements* of the set. There are various ways to denote sets. One approach is to list the elements of the set in some way and enclose this list in braces. For example, using this convention, the set consisting of the numbers 1, 2, and 3 is denoted by $\{1,2,3\}$. A set is completely determined by its members, and so the order in which we list the elements is immaterial. Thus $\{1,2,3\} = \{2,3,1\} = \{3,2,1\} = \{1,3,2\} = \{2,1,3\} = \{3,1,2\}$.

A set may have so many members belonging to it that it is impractical or impossible to use the above method of notation, and so other notational devices must be used. For example, instead of using the method described above to denote the set of all positive integers less than or equal to 10^{10}, we might use $\{1,2,3,\ldots,10^{10}\}$ to denote this set. The three dots indicate that some members of the set being described have not been listed explicitly. Of course, in using this notational device it is important to include enough members of the list before and after the three dots so that the reader will know which elements belong to the set and which do not. For example, the set of even integers between -100 and 100 inclusive should not be denoted by $\{-100,\ldots,100\}$ but by something like $\{-100,-98,\ldots,-4,-2,0,2,4,\ldots,98,100\}$ or by $\{0,2,-2,4,-4,\ldots,98,-98,100,-100\}$. Again, the order in which the elements are listed is arbitrary as long as the reader understands which elements of the set have not been mentioned explicitly.

The method for denoting sets using the three dots abbreviation can also be used for infinite sets. For example, the set of even integers can be denoted by $\{0,2,-2,4,-4,\ldots\}$ or by $\{\ldots,-6,-4,-2,0,2,4,6,\ldots\}$. We will use $\mathbf{N}$ to denote the set $\{0,1,2,3,\ldots\}$ of natural numbers, while $\mathbf{N}^+$ will denote $\{1,2,3,\ldots\}$. $\mathbf{I}$ will denote the set of integers $\{0,1,-1,2,-2,3,-3,\ldots\}$. $\mathbf{Q}$ will denote the set of rationals, $\mathbf{Q}^+$ the set of positive rationals, $\mathbf{R}$ the set of real numbers, and $\mathbf{R}^+$ the set of positive reals.

If a set consists of exactly those objects satisfying a certain condition, say P, we may denote it by $\{x:P(x)\}$, which is read: "the set of all x such that $P(x)$ is true." For example, $\{x: 3 \leqslant x \leqslant 8$ and x is a rational number$\}$ is the set of rationals between 3 and 8 inclusive. Notice that x merely represents a typical object in the set under consideration, and any letter will serve just as well in place of x. Thus $\{1,2,3\} = \{x:x$ is an integer and $1 \leqslant x \leqslant 3\} = \{y:y$ is an integer and $1 \leqslant y \leqslant 3\} = \{x:x$ is an integer and $0 < x < 4\}$. Notice that the last two conditions are different but define the same set.

We consider as a set the collection which has no members. We call this set the null set and denote it by $\varnothing$, rather than $\{\ \}$.

A set may contain other sets as elements. For example, the set $\{1,\{2,3\}\}$ is the set whose elements are the number 1 and the set $\{2,3\}$. It is important to understand that this set has only two elements, namely 1 and $\{2,3\}$. 2 is an element of $\{2,3\}$, but 2 is not an element of $\{1,\{2,3\}\}$.

We write $x \in A$ when x is an element of A, and $x \notin A$ otherwise.

Let A be a set. We say that a set B is a *subset* of A if each element of B is an element of A. If B is a subset of A we write $B \subseteq A$ or $A \supseteq B$. If $B \subseteq A$

1.2 Sets

and, in addition, $A \neq B$, we write $B \subset A$ or $A \supset B$ and say that B is a proper subset of A. So $\{1,\{2,3\}\} \subset \{1,\{2,3\},4\}$ but $\{1,\{2,3\}\} \not\subset \{1,\{2\},\{3\}\}$.

Notice that $A \subseteq A$ and $\emptyset \subseteq A$ for every set A (since $\emptyset$ has no elements, it is true that every element of $\emptyset$ is an element of A). Another trivial observation is that if $A \subseteq B$ and $B \subseteq C$, then $A \subseteq C$.

We note that if A and B are sets such that $A \subseteq B$ and $B \subseteq A$, then $A = B$. For if $x \in A$, then since $A \subseteq B$, $x \in B$. Similarly, if $y \in B$ we have $y \in A$. Thus A and B contain precisely the same elements and so are equal. This will be used frequently in what follows; two sets A and B will be shown to be equal by proving both $A \subseteq B$ and $B \subseteq A$.

Next we consider ways of combining sets to get new sets.

The *union* of A and B, denoted by $A \cup B$, is the set whose elements belong either to A or to B. In other words $A \cup B = \{x : x \in A \text{ or } x \in B\}$. (In mathematics we use the word "or" in the inclusive sense. So when we say that an object is in A or in B we include the case where the object is in both A and B.) For example

$$\{1,2\} \cup \{3,4\} = \{1,2,3,4\},$$
$$\{a,b,c\} \cup \{a,c,d\} = \{a,b,c,d\},$$
$$\{x : x \text{ is an even integer}\} \cup \{x : x \text{ is an odd integer}\}$$
$$= \{x : x \text{ is an integer}\}.$$

Some of the elementary properties of the union operation are summarized below.

Theorem 2.1.

i. $A \subseteq B$ implies that $A \cup B = B$.
ii. $A \cup B = B \cup A$.
iii. $A \cup (B \cup C) = (A \cup B) \cup C$.

The proof of the theorem is very easy, and we leave all parts but iii as exercises.

To prove part iii, first suppose that $x \in A \cup (B \cup C)$. Then either $x \in A$ or $x \in B \cup C$. If $x \in A$, then $x \in A \cup B$, and so $x \in (A \cup B) \cup C$. If $x \in B \cup C$, then $x \in B$ or $x \in C$. If $x \in B$, then $x \in A \cup B$, and so $x \in (A \cup B) \cup C$. If $x \in C$, then $x \in (A \cup B) \cup C$. Hence we have shown that whenever $x \in A \cup (B \cup C)$, then $x \in (A \cup B) \cup C$, in other words, we have shown that $A \cup (B \cup C) \subseteq (A \cup B) \cup C$. In the same way one proves that $A \cup (B \cup C) \supseteq (A \cup B) \cup C$ (the reader should check this). Hence $A \cup (B \cup C) = (A \cup B) \cup C$ as claimed.

Because of part iii, no confusion can arise if parentheses are omitted from $(A \cup B) \cup C$ and we write $A \cup B \cup C$.

It should be clear what is meant by $A_1 \cup A_2 \cup \ldots \cup A_n$, namely, $\{x : x \in A_1 \text{ or } x \in A_2 \text{ or } \ldots \text{ or } x \in A_n\}$. An alternative notation for this set is

$\bigcup \{A_i : i \in \mathbf{N}^+ \text{ and } i \leq n\}$. In general, if X is a non-empty set of sets, then $\bigcup X = \{y : \text{there is a } Y \in X \text{ such that } y \in Y\}$. This is called the *union over* X. So if $X = \{A_1, A_2, \ldots, A_n\}$, then $\bigcup X = A_1 \cup A_2 \cup \ldots \cup A_n$. For example, if $A_i = \{x : x = i/n \text{ for some } n \in \mathbf{N}^+\}$ (so that $A_5 = \{5/1, 5/2, 5/3, 5/4, \ldots\}$), then $\bigcup \{A_i : i \in \mathbf{N}^+\} = \mathbf{Q}^+$. Instead of writing $\bigcup \{A_i : i \in I\}$ we may write $\bigcup_{i \in I} A_i$.

The *intersection* of A and B, $A \cap B$, is the set whose elements are simultaneously elements of A and of B. In other words $A \cap B = \{x : x \in A \text{ and } x \in B\}$. For example $\{1, 3, 9\} \cap \{1, 5, 9\} = \{1, 9\}$ and $\{x : x \in \mathbf{R}^+ \text{ and } x < 5\} \cap \{x : x \in \mathbf{Q} \text{ and } x \geq 3\} = \{x : x \in \mathbf{Q} \text{ and } 3 \leq x < 5\}$.

Theorem 2.1. (Cont.)

i'. $A \subseteq B$ implies that $A \cap B = A$.
ii'. $A \cap B = B \cap A$.
iii'. $A \cap (B \cap C) = (A \cap B) \cap C$.

The proofs are very easy and left as exercises.

As in the case of the union operation, the intersection operation generalizes to the intersection over a set of sets. Letting X be a non-empty set of sets, we define the *intersection* over X, $\bigcap X$, to be $\{y : \text{for all } Y \in X, y \in Y\}$. So if $X = \{A_1, A_2, \ldots, A_n\}$, then $\bigcap X = A_1 \cap \ldots \cap A_n$. (As before, we use iii' to justify our omission of parentheses in $A_1 \cap \ldots \cap A_n$.) As another example let $A_n = \{x : x \in \mathbf{R} \text{ and } |x| < 1/n\}$. Let $X = \{A_n : n \in \mathbf{N}^+\}$. Then $\bigcap X = \{0\}$.

We say that A and B are *disjoint* if $A \cap B = \emptyset$. Similarly, X is a set of *pairwise disjoint sets* if for all $A, B \in X$, either $A = B$ or $A \cap B = \emptyset$.

We next state some easily proved facts relating union and intersection. The proofs are left for the exercises.

Theorem 2.1. (Cont.)

iv. $A \cap (B \cup C) = (A \cap B) \cup (A \cap C)$, and more generally $\left(\bigcup X\right) \cap \left(\bigcup Y\right) = \bigcup (A \cap B : A \in X \text{ and } B \in Y)$.
iv'. $A \cup (B \cap C) = (A \cup B) \cap (A \cup C)$, and more generally $\left(\bigcap X\right) \cup \left(\bigcap Y\right) = \bigcap (\{A \cup B : A \in X \text{ and } B \in Y\})$.

The *difference of A from B*, denoted $B - A$, is the set of elements in B but not in A; in other words we define $B - A = \{x : x \in B \text{ and } x \notin A\}$. For example, $\mathbf{Q}^+ - \{x : x \in \mathbf{R} \text{ and } x \leq 3\}$ is the set of positive rationals greater than 3. As another example, $\{1, 4, 9\} - \{3, 4, 8\} = \{1, 9\}$. $B - A$ is also called the *complement of A in B*.

We next state several relations between the above notions.

Theorem 2.1. (Cont.)

v. $A \subseteq B$ implies $B - (B - A) = A$.
vi. $C \supseteq B \supseteq A$ implies $C - A \supseteq C - B$.
vii. $C - (A \cup B) = (C - A) \cap (C - B)$, and more generally
$$C - \left(\bigcup X\right) = \bigcap \{C - A : A \in X\}.$$
viii. $C - (A \cap B) = (C - A) \cup (C - B)$, and more generally
$$C - \left(\bigcap X\right) = \bigcup \{C - A : A \in X\}.$$

We prove vii, leaving the proof of the other clauses for the exercises. Here and throughout the text we use 'iff' to abbreviate 'if and only if'.

$$x \in C - \left(\bigcup X\right) \quad \text{iff}$$
$$x \in C \text{ and } x \notin A \text{ for all } A \in X \quad \text{iff}$$
$$x \in C - A \text{ for all } A \in X \quad \text{iff}$$
$$x \in \bigcap \{C - A : A \in X\}.$$

In other words, $C - \left(\bigcup X\right)$ and $\bigcap \{C - A : A \in X\}$ have the same members and so are identical, as claimed in vii.

Clauses vii and viii are called De Morgan's rules.

We next define the *power set*, $P(X)$, of a set X. This is the set of all subsets of X, i.e., $P(X)$ is defined as $\{Y : Y \subseteq X\}$.

For example, if $X = \{1, 2, 3\}$, then $P(X) = \{\phi, \{1\}, \{2\}, \{3\}, \{1, 2\}, \{1, 3\}, \{2, 3\}, \{1, 2, 3\}\}$. Clearly, we always have $\phi \in P(X)$ and $X \in P(X)$. Elementary properties of the power set operation will be found in Exercise 7 below.

EXERCISES FOR §1.2

1. How many elements are there in each of the following sets?
$$\{1, 2, \phi\}, \quad \{1, \{1, \phi\}\}, \quad \{\phi\}, \quad \{1\}, \quad \{\{1\}\}.$$

2. Which of the following are true?
$$\emptyset \in \emptyset, \quad \emptyset \subseteq \emptyset, \quad \{1\} \in \{1, 2\}, \quad 1 \in \{\{1\}, 2\}.$$

3. Show that
   (a) if $A \subseteq C$ and $B \subseteq C$, then $A \cup B \subseteq C$, and
   (b) if $C \subseteq A$ and $C \subseteq B$, then $C \subseteq A \cap B$.

4. Supply the missing proofs for Theorem 2.1.

5. List the elements of $P(\{1, 2, 3, 4\})$.

6. List the elements of $P(P(P(\emptyset)))$.

7. Show that
   (a) $A \supseteq B$ implies $P(A) \supseteq P(B)$;
   (b) $P(A \cup B) \supseteq P(A) \cup P(B)$, and more generally
   $$P\left(\bigcup X\right) \supseteq \bigcup \{P(A) : A \in X\};$$
   (c) $P(A \cap B) \subseteq P(A) \cap P(B)$, and more generally
   $$P\left(\bigcap X\right) \subseteq \bigcap \{P(A) : A \in X\};$$
   When does equality hold in (b) and in (c)?

1.3 Relations and Functions

The aim of this section is to supply definitions of 'relation', 'function' and related notions in enough generality to be of service throughout the book. These notions ultimately rest on that of the ordered pair (a,b). Although 'ordered pair' can be defined in terms of the membership relation, as can all the notions of the classical mathematics, we will not do this until later. For the time being we shall take the ordered pair (a,b) to be an undefined notion with the property that $(a,b) = (c,d)$ if and only if $a = c$ and $b = d$. For example $(3,8) \neq (8,3)$ (although $\{3,8\} = \{8,3\}$). Similarly, the only property of n-tuples that we shall use is that $(a_1, \ldots, a_n) = (b_1, \ldots, b_n)$ iff $a_i = b_i$ for all $i \leq n$.

The *Cartesian product* of A and B, written $A \times B$, is $\{(x,y) : x \in A$ and $y \in B\}$. More generally, we define $A_0 \times A_1 \times \ldots \times A_n$ to be $\{(a_0, \ldots, a_n) : a_i \in A_i$ for each $i \in \{0, \ldots, n\}\}$. For example, $(1, \frac{3}{4}) \in \mathbf{N} \times \mathbf{Q}$, but $(\frac{3}{4}, 1) \notin \mathbf{N} \times \mathbf{Q}$.

If for each $i,j \in \{0, \ldots, k-1\}$ we have $B = A_i = A_j$, then we abbreviate $A_0 \times \ldots \times A_{k-1}$ by $[B]_k$. For example, $[\mathbf{R}]_n$ is Euclidean n-space.

A *binary relation* is a set of ordered pairs. For example $\{(x,y) : x < y$ and $x \in \mathbf{N}, y \in \mathbf{N}\}$ is a binary relation. So is $\{(3,4),(1,1)\}$, as well as the circle in Euclidean 2-space of radius 3 with center $(4,\pi)$, namely $\{(x,y) : (x-4)^2 + (y-\pi)^2 = 3^2\}$.

The *domain* of a binary relation R, sometimes written Dom R, is $\{x$: there is a y such that $(x,y) \in R\}$; the *range* of R, Ran R, is $\{y$: for some x, $(x,y) \in R\}$. The *field* of R is Dom $R \cup$ Ran R. In the first of the three examples above we have Dom $R = \mathbf{N}$, Ran $R = \mathbf{N}$; in the second Dom $R = \{3,1\}$, Ran $R = \{4,1\}$; and in the third Dom $R = \{x : 1 \leq x \leq 7\}$, Ran $R = \{y : \pi - 3 \leq y \leq \pi + 3\}$. One frequently writes xRy instead of $(x,y) \in R$, and $x\not R y$ if $(x,y) \notin R$.

More generally, a *k-relation* is a set of ordered k-tuples (so a 2-relation is a binary relation). As an example of a 3-relation we have $\{(x,y,z) : (x,y,z) \in [\mathbf{N}]_3$ and z is the least common multiple of x and $y\}$. Another example is $\{(x,y,z) : (x,y,z) \in [\mathbf{R}]_3$ and $x + y = z\}$. We do not define the domain or range of a k-relation when $k \neq 2$.

The set of all primes is an example of a 1-relation, as is the set of all multiples of π.

1.3 Relations and Functions

A *function* f is a 2-relation such that for every x there is at most one y for which $(x,y) \in f$. In other words, if $(x,y) \in f$ and $(x,z) \in f$, then $y=z$. When f is a function, one usually writes $f(x) = y$ instead of $(x,y) \in f$, and says that y is the value of f at x.

For example, $\{(1,3),(3,1),(\pi,1)\}$ is a function, but $\{(1,3),(3,1),(1,\pi)\}$ is not. $\{(x,y): x=y^3 \text{ and } x \in \mathbf{N} \text{ and } y \in \mathbf{N}\}$ is a function, but $\{(x,y): x=y^2 \text{ and } x \in \mathbf{N} \text{ and } y \in \mathbf{I}\}$ is not.

A function f is *one to one*, abbreviated 1-1, if $\{(y,x): f(x) = y\}$ is a function, i.e., if when $f(x) = y$ and $f(z) = y$ we have $x = z$. In our examples of functions above, the second is 1-1 but the first is not.

We say that a function f is *on* A if $\text{Dom} f = A$; *into* B if $\text{Ran} f \subseteq B$; *onto* B if $\text{Ran} f = B$. If f is a function on A into B, we may write $f:A \to B$. The notation $f:A \xrightarrow{1\text{-}1} B$ adds the condition that f is 1-1, while $f:A \xrightarrow[\text{onto}]{} B$ adds the condition that f is onto B. The set of all functions on A into B is denoted by ${}^A B$, i.e., ${}^A B = \{f : f:A \to B\}$.

By $f[C]$ we mean $\{y : \text{for some } x \in C, f(x) = y\}$. Notice that no restriction is placed on C; C need not be included in $\text{Dom} f$. For example, if $f = \{(x,y): y = x^2 \text{ and } x \in \mathbf{N}\}$ and $C = \{x: x < \pi \text{ and } x \in \mathbf{R}\}$, then $f[C] = \{0,1,4,9\}$.

Define $f^{-1}[Y]$ to be $\{x : f(x) \in Y\}$. $f^{-1}[Y]$ is defined even if $Y \not\subseteq \text{Ran} f$. So if $f(x) = 3x+2$ for all $x \in \mathbf{R}^+$, then $f^{-1}[\{y: 0 \leq y \leq 11\}] = \{x: 0 < x \leq 3\}$. As another example, if $f(x) = x^2$ for each $x \in \mathbf{R}$, then $f^{-1}[\{y\}] = \{-\sqrt{y}, \sqrt{y}\}$ for each $y \in \mathbf{R}^+$. If $f:A \xrightarrow[\text{onto}]{1\text{-}1} B$, then the set $\{(b,a): f(a) = b\}$ is a 1-1 function on B onto A which we call f *inverse*, written f^{-1}. Notice that $f(f^{-1}(b)) = b$ and $f^{-1}(f(a)) = a$ for all $a \in A$ and all $b \in B$.

The *restriction of f to C*, abbreviated $f \restriction C$, is the function g with domain $C \cap \text{Dom} f$ such that for each $x \in C \cap \text{Dom} f$ we have $g(x) = f(x)$. In other words $g = \{(x,y): x \in C \cap \text{Dom} f \text{ and } y = f(x)\}$.

Notice that C is arbitrary and need not be a subset of $\text{Dom} f$. For example, if $f = \{(x,y): y = x^2 \text{ and } x \in \mathbf{N}\}$ and $C = \{x: x < \pi \text{ and } x \in \mathbf{R}\}$, then $f \restriction C = \{(0,0),(1,1),(2,4),(3,9)\}$.

If $g = f \restriction C$ and $C \subseteq \text{Dom} f$, then we say that f is an extension of g.

Let $f \in {}^B C$ and let $g \in {}^A B$. The *composite of f and g*, written $f \circ g$, is that element of ${}^A C$ defined by $(f \circ g)(x) = f(g(x))$ for all $x \in A$.

Theorem 3.1. *Let $f \in {}^B C$, $g \in {}^A B$. Then*

i. *if f and g are 1-1, then so is $f \circ g$.*
ii. *if f is onto C and g is onto B, then $f \circ g$ is onto C.*

PROOF OF i. Suppose f and g are 1-1, and $(f \circ g)(a) = (f \circ g)(b)$. Then $f(g(a)) = f(g(b))$. Since f is 1-1, $g(a) = g(b)$. Since g is 1-1, $a = b$. □

We leave the proof of part ii as an exercise (see Exercise 15).

EXERCISES FOR §3.
1. What are the elements of $\{1,3\} \times \{1,\pi,4\}$?
2. If A has m elements and B has n elements, how many elements does $A \times B$ have?
3. Prove that if $A_i \subseteq B_i$ for each $i \in \{1,2,\ldots,k\}$ then $A_1 \times \ldots \times A_k \subseteq B_1 \times \ldots \times B_k$.
4. Show that the relation $<$ on $\mathbf{Q}$ is not a set of the form $A \times B$.
5. If the following statement is true, prove it; if not, give a counter example:
$$\text{If } A \supseteq B \cup C \text{ then } (A \times A) - (B \times C) = (A - B) \times (A - C).$$
6. Prove or disprove the following statement:
$$(A_1 \times A_2) \cup (B_1 \times B_2) = (A_1 \cup B_1) \times (A_2 \cup B_2).$$
7. Prove or disprove the following statement:
$$(A_1 \times A_2) \cap (B_1 \times B_2) = (A_1 \cap B_1) \times (A_2 \cap B_2).$$
8. For each relation R below, find Dom R, Ran R, and the field of R:
   (a) $R = \{(1,4),(\pi,3),(\pi,1),(1,\pi)\}$.
   (b) $R = \{(x,y): |x| + |y| = 1\}$.
   (c) $R = \{p : p \in [R]_2 \text{ and } |p - (1,0)| + |p - (1,0)| = 3\}$
   $\left[\text{where } |(x_1,y_1) - (x_2,y_2)| = \sqrt{(x_1 - x_2)^2 + (y_1 - y_2)^2} \right]$.
9. Which of the following are functions; which of the functions are 1-1?
   (a) $\{(x,y): x > 0, x^2 + y^2 = 1, \text{ and } x \in \mathbf{R}, y \in \mathbf{R}\}$.
   (b) $\{(x,y): y > 0, x^2 + y^2 = 1, \text{ and } x \in \mathbf{R}, y \in \mathbf{R}\}$.
   (c) $\{(x,y): x > 0, y > 0, x^2 + y^2 = 1 \text{ and } x \in \mathbf{R}, y \in \mathbf{R}\}$
   (d) $\{(x,y,z): x,y,z \in \mathbf{N}^+ \text{ and } z = 2^x 3^y\}$.
   (e) $\{(x,y,z): x,y,z \in \mathbf{N}^+ \text{ and } z = 2x + 3y\}$.
10. Prove:
    (a) $f\left[\bigcup X\right] = \bigcup \{f[A]: A \in X\}$.
    (b) $f\left[\bigcap X\right] \subseteq \bigcap \{f[A]: A \in X\}$.
    Show that equality need not hold in (b) by describing sets A and B and a function f such that $f[A \cap B] \neq f[A] \cap f[B]$.
11. Show that
    (a) $(f \upharpoonright C) \upharpoonright D = f \upharpoonright (C \cap D)$.
    (b) $\bigcap \{f \upharpoonright C : C \in K\} = f \upharpoonright \bigcap \{C : C \in K\}$.
12. Suppose A has n elements and B has m elements. How many elements are there in $^A B$? Give a proof.
13. Prove:
    (a) $f^{-1}\left[\bigcup X\right] = \bigcup \{f^{-1}[A]: A \in X\}$.
    (b) $f^{-1}\left[\bigcap X\right] \subseteq \bigcap \{f^{-1}[A]: A \in X\}$.
    (c) $f^{-1}[A - B] \supseteq f^{-1}[A] - f^{-1}[B]$.
    Show that equality need not hold in (b) or in (c).
14. Find functions f and g such that Ran$f = $ Ran$g = $ Dom$f = $ Dom$g$ and $f \circ g \neq g \circ f$.
15. Show that if $f: B \underset{\text{onto}}{\to} C$ and $g: A \underset{\text{onto}}{\to} B$, then $f \circ g: A \underset{\text{onto}}{\to} C$.

1.4 Pairings

Suppose that A and B are sets and f is a 1-1 function from A onto B. f can be thought of as an association or pairing of the elements of A with those of B such that each element x of A has a unique associate $f(x)$ in B and, conversely, each element y in B has a unique associate $f^{-1}(y)$ in A. Clearly, if A has ten elements, then so does B; if A has one million elements, then so does B. Indeed, the existence of such a pairing f assures us that if A has n elements, where $n \in \mathbf{N}$, then B also has n elements. But what if A is infinite? It seems natural to use the existence of such an f to assert that A and B have equal magnitude or are equinumerous even when both sets are infinite. It is this simple idea that underlies the theory of infinite sets. We now discuss this idea in more detail and consider some of the surprising consequences.

Let A and B be arbitrary sets. A *pairing* between A and B is a 1-1 function on A onto B.

EXAMPLE 4.1. Let $A = \{1,2,3\}$, $B = \{4,5,6\}$. There are several pairings of A with B. One such is $\{(1,4),(2,5),(3,6)\}$; another is $\{(1,6),(2,4),(3,5)\}$.

EXAMPLE 4.2. Let $A = \{1,3,5,7,9,11,\ldots\}$, $B = \{0,2,4,6,8,\ldots\}$. One pairing between A and B is defined by $P(n) = n-1$ for all odd $n \in \mathbf{N}^+$, i.e., $P = \{(1,0),(3,2),(5,4),(7,6),\ldots\}$. Clearly P is 1-1 and onto.

EXAMPLE 4.3. Let $A = \mathbf{N}$, $B = \{0, -1, -2, -3, \ldots\}$. One pairing of A and B is given by $P(n) = -n$ for all $n \in \mathbf{N}$, i.e., $P = \{(0,0),(1,-1),(2,-2),\ldots\}$.

EXAMPLE 4.4. Let $A = \mathbf{Q}^+ = \{p/q \mid p,q \in \mathbf{N}^+, p/q \text{ in lowest terms}\}$. Let $B = \mathbf{N}^+$. Consider the function P from A into B defined by $P(p/q) = p+q$. This is not a pairing between A and B, since P is neither 1-1 nor onto (verify this). This does not show that no pairing exists between A and B, but only that this particular function is not a pairing. We shall return to the question of whether or not there is a pairing between this A and B later in this section.

EXAMPLE 4.5. Let $A = \{1,2,3\}$, $B = \{1,2\}$. The reader can easily list all functions from A into B and check that no pairing is possible between A and B.

Definition. Let A and B be sets. We say that A is *equinumerous* to B if there is a pairing between A and B. If A and B are equinumerous we write $A \sim B$.

This definition fits well with our intuition, and yet some of its consequences at first glance seem bizarre. For example, a set A may properly contain a set B and still be equinumerous to B. Of course this can not happen if A is finite, but it can happen when A is infinite.

To illustrate, let $B=\{n^2:n\in\mathbf{N}\}$. Then $\mathbf{N}\supset B$. Yet $\{(x,x^2):x\in\mathbf{N}\}$ is a pairing between $\mathbf{N}$ and B, so $\mathbf{N}$ and B are equinumerous. At first this seems so strange that one balks at accepting the above definition of equinumerous. In fact, for centuries this very pairing was used as evidence that it is nonsensical to compare magnitudes of infinite sets.

It was Cantor who in 1874 took the bold position that the above definition of equinumerous is the "correct" mathematical definition in spite of such examples. Moreover, he proceeded to show that this definition leads to a significant and beautiful mathematical theory.

Why did mathematicians come to accept Cantor's views? There are several compelling reasons. First of all, Cantor's ideas were found to be applicable to many problems in several branches of mathematics outside of set theory, problems that did not seem on the surface to be concerned with comparisons of infinite magnitudes. In §1.7 we give a proof of Cantor's of a striking fact about the real numbers which shed light on a major problem of the 19th century. Furthermore, set theory was found to provide a uniform notational and conceptual framework within which all of mathematics can be expressed. We touched on this aspect briefly in §3, but much more will be said in the next chapter. Later we will give an axiomatization of set theory which provides an axiomatic framework for all of classical mathematics. Esthetically, Cantor's proofs are among the most beautiful in mathematics, and there is no question that mathematics has been considerably enriched by his theories and methods.

Definition. A set A is *countable* if $A\sim B$ for some $B\subseteq\mathbf{N}$. Sometimes the term *denumerable* is used for A when $A\sim\mathbf{N}$. If for some $n\in\mathbf{N}$, $A\sim\{0,1,\ldots,n-1\}$, then A is *finite*. If no such n exists, then A is *infinite*.

We now give more examples of sets that are equinumerous with $\mathbf{N}$.

EXAMPLE 4.6. A pairing P between $\mathbf{N}$ and $\mathbf{I}$ is given by

$\mathbf{N}$: 0   1   2   3   4   5   6   ...

$\mathbf{I}$: 0   1 -1 2 -2 3 -3 ...

That is, $P=\{(0,0),(1,1),(2,-1),\ldots\}$. P can also be described by the following equations:

$$P(n)=-\tfrac{1}{2}n \quad \text{if } n \text{ is even,}$$

$$P(n)=\tfrac{1}{2}(n+1) \quad \text{if } n \text{ is odd.}$$

EXAMPLE 4.7. Let $A=\mathbf{N}$ and $B=\mathbf{N}^+$. Define $P:\mathbf{N}\to\mathbf{N}^+$ by $P(n)=n+1$ for all $n\in\mathbf{N}$. Since P is 1-1 and onto, $\mathbf{N}\sim\mathbf{N}^+$.

We have used the word 'equinumerous' for the relation '$\sim$' so as to imply an analogy between 'equinumerous' as the word is used informally

1.4 Pairings

when referring to finite sets having the same number of elements and the technical meaning given to the word in the present context. This analogy would be weak indeed without the following.

Theorem 4.8. *For all sets A, B, and C:*

i. $A \sim A$.
ii. *If* $A \sim B$, *then* $B \sim A$.
iii. *If* $A \sim B$ *and* $B \sim C$, *then* $A \sim C$.

PROOF: i. Define $P: A \to A$ by $P(a) = a$ for all $a \in A$. Then P is a pairing between A and A, so $A \sim A$.

ii. Suppose $A \sim B$. Then there is a P such that $P: A \xrightarrow[\text{onto}]{1\text{-}1} B$. Clearly $P^{-1}: B \xrightarrow[\text{onto}]{1\text{-}1} A$, and so $B \sim A$.

iii. Suppose $P: A \xrightarrow[\text{onto}]{1\text{-}1} B$ and $Q: B \xrightarrow[\text{onto}]{1\text{-}1} C$. It is immediate from Theorem 3.1 that $Q \circ P$ is a pairing between A and C. □

We have already observed that $\{n^2 : n \in \mathbf{N}\}$ is a proper subset of $\mathbf{N}$ which is equinumerous to $\mathbf{N}$. This is a specific instance of the following more general statement.

Theorem 4.9. *If $A \subseteq \mathbf{N}$ and A is infinite, then $\mathbf{N} \sim A$.*

PROOF: For each non-empty subset Y of $\mathbf{N}$, let $f(Y)$ be the smallest element of Y [so for all $x \in Y$ we have $f(Y) \leq x$]. Now define

$$P(0) = f(A),$$
$$P(1) = f(A - \{P(0)\}),$$
$$P(2) = f(A - \{P(0), P(1)\}),$$
$$P(3) = f(A - \{P(0), P(1), P(2)\}),$$

and so on. In general we have $P(n) = f(A - \{P(j) : j < n\})$. Notice that for each n, $P(n)$ is defined, since $A - \{P(j) : j < n\}$ is non-empty [in fact, since A is infinite and $\{P(j) : j < n\}$ is finite, $A - \{P(j) : j < n\}$ is infinite]. Clearly, if $m > n$, then $P(n) \notin (A - \{P(i) : i < m\})$, and so $P(m) > P(n)$. Hence P is 1-1. We need only show that P is onto. If not, let m^* be the least element of $A - P$. Let $X = \{n : P(n) < m^*\}$. X is finite, since P is 1-1. Hence there is an n^* such that $n^* > n$ for every $n \in X$. Then $P(n^*) = f(A - \{P(j) : j < n^*\}) = m^*$. So $m^* \in \operatorname{Ran} P$—a contradiction. □

As an application we see that the set of primes is equinumerous to $\mathbf{N}$.

Corollary 4.10. *A countable set is finite or denumerable.*

Theorem 4.11. *The following statements are equivalent:*

i. $f: \mathbf{N} \xrightarrow[\text{onto}]{} A$ *for some f.*
ii. *Either A is finite and non-empty, or $\mathbf{N} \sim A$.*

PROOF THAT i IMPLIES ii. Let $f:\mathbf{N} \underset{\text{onto}}{\to} A$. By the preceding theorem, it is enough to find a pairing P between A and a subset of $\mathbf{N}$. For each $a \in A$ let $P(a)$ be the least $n \in \mathbf{N}$ such that $f(n) = a$. Clearly P is 1-1 and onto a subset of N as needed. □

PROOF THAT ii IMPLIES i. If $\mathbf{N} \sim A$ then there is a pairing P between $\mathbf{N}$ and A which we can take as the f of clause i). If A is finite then there is an n and a g such that $g:\{0,1,\ldots,n-1\} \underset{\text{onto}}{\overset{\text{1-1}}{\to}} A$. Extend g to $f:\mathbf{N} \to A$ by defining $f(x) = g(x)$ if $x \in \{0,1,\ldots,n-1\}$ and $f(m) = g(0)$ if $m \geq n$. □

Recall that a number $n \in \mathbf{N}^+$ is a prime if $n > 1$ and whenever $k \cdot l = n$ with $k, l \in \mathbf{N}^+$ then either $k = 1$ or $l = 1$. For example, $6 = 3 \cdot 2$ and so is not prime. But $2, 5, 7, 11, 13, 17, 19, 23$ are primes. One of the most basic theorems of arithmetic is the Prime Factorization Theorem. We shall use it frequently and so we state it below for easy reference. A proof may be found in almost any text on elementary algebra.

Theorem (Prime Factorization Theorem). *For each $x \in \mathbf{N}^+$, $x > 1$, there is exactly one sequence $k, n_1, n_2, \ldots, n_k$ of elements of $\mathbf{N}^+$ and one increasing sequence $p_1, p_2, \ldots, p_k$ of primes such that $x = p_1^{n_1} p_2^{n_2} \ldots p_k^{n_k}$.*

Recall that $[\mathbf{N}]_k = \{(n_1, \ldots, n_k) : n_i \in \mathbf{N} \text{ for } i = 1, 2, \ldots, k\}$.

Theorem 4.12. $\bigcup_{k \in \mathbf{N}^+} [\mathbf{N}]_k \sim \mathbf{N}$.

PROOF: Let $x \in \bigcup_{k \in \mathbf{N}^+} [\mathbf{N}]_k$, say $x = (n_1, \ldots, n_m) \in [\mathbf{N}]_m$. Define $f(x) = 2^{n_1+1} \cdot 3^{n_2+1} \cdot \ldots \cdot p_m^{n_m+1}$. Then $f : \bigcup_{k \in \mathbf{N}^+} [\mathbf{N}]_k \to \mathbf{N}$, and by the prime factorization theorem f is 1-1. Now apply Corollary 4.10.

Of course with f as above, $f \upharpoonright [\mathbf{N}]_k$ is 1-1 on $[\mathbf{N}]_k$ into $\mathbf{N}$, so by Corollary 4.10 $[\mathbf{N}]_k \sim \mathbf{N}$ for each $k \in \mathbf{N}^+$. (Another proof of this is found in Exercise 8.)

The existence of a pairing between $\mathbf{N} \times \mathbf{N}$ and $\mathbf{N}$ (different from those above) can be seen pictorially as follows:

$(0,0) \to (0,1) \quad (0,2) \to (0,3) \ldots$

↙ ↗ ↙

$(1,0) \quad (1,1) \quad (1,2) \quad (1,3) \ldots$

↓ ↗ ↙

$(2,0) \quad (2,1) \quad (2,2) \quad (2,3) \ldots$

↙

$(3,0) \quad (3,1) \quad (3,2) \quad (3,3) \ldots$

↓

1.4 Pairings

The arrows indicate the ordering of $\mathbf{N} \times \mathbf{N}$ imposed by the pairing with $\mathbf{N}$:

$$0 = P(0,0), \quad 1 = P(0,1), \quad 2 = P(1,0),$$
$$3 = P(2,0), \quad 4 = P(1,1), \quad 5 = P(0,2),\ldots.$$

Theorem 4.13. *Suppose that A_n is countable for each $n \in \mathbf{N}$. Then $\bigcup_{n \in \mathbf{N}} A_n$ is countable.*

PROOF: The hypothesis asserts that for each $n \in \mathbf{N}$ there is a function $f: A_n \xrightarrow{1\text{-}1} \mathbf{N}$. For each n we choose one such f; call it f_n. (In general there are infinitely many such f's; cf. Exercise 1 of this section.) For each $x \in \bigcup_{n \in \mathbf{N}} A_n$, let $k(x)$ be the smallest j such that $x \in A_j$. Define $F(x) = 2^{k(x)} 3^{f_{k(x)}(x)}$. Clearly $F: \bigcup_{n \in \mathbf{N}} A_n \to \mathbf{N}$. We claim that F is 1-1. For suppose that $F(x) = 2^{k(x)} 3^{f_{k(x)}(x)}$, $F(y) = 2^{k(y)} 3^{f_{k(y)}(y)}$, and $F(x) = F(y)$. By the Prime Factorization Theorem, $k(x) = k(y)$. Hence $f_{k(x)}(x) = f_{k(x)}(y)$, and since $f_{k(x)}$ is 1-1, we have $x = y$. Therefore, $F: \bigcup_{n \in \mathbf{N}} A_n \xrightarrow{1\text{-}1} \mathbf{N}$, i.e., $\bigcup_{n \in \mathbf{N}} A_n$ is countable.

EXAMPLE 4.14. We show that $\mathbf{Q}^+ \sim \mathbf{N}$. Let $A_0 = \{\frac{1}{1}, \frac{2}{1}, \frac{3}{1}, \frac{4}{1}, \ldots\}$, $A_1 = \{\frac{1}{2}, \frac{2}{2}, \frac{3}{2}, \frac{4}{2}, \ldots\}$, and in general $A_n = \{m/n : m \in \mathbf{N}^+\}$. Then $\mathbf{Q}^+ = \bigcup_{i=0}^{\infty} A_i$. Clearly $f_n: A_n \xrightarrow[\text{onto}]{1\text{-}1} \mathbf{N}^+$, where $f_n(m/n) = m$ for each $m \in \mathbf{N}^+$. Now apply Theorem 4.13.

EXAMPLE 4.15. We show that $\mathbf{Q} \sim \mathbf{N}$. Let $A_0 = \{0\}$, $A_1 = \mathbf{Q}^+$, and $A_2 = \{-x : x \in \mathbf{Q}^+\}$. Clearly $\mathbf{Q} = A_0 \cup A_1 \cup A_2$. Moreover, it follows from Example 4.14 that each A_i is countable. Hence, by Theorem 4.13, $\mathbf{Q}$ is countable.

So far all the examples discussed in this section have involved sets that are either finite or equinumerous to $\mathbf{N}$. Are there any infinite sets that are not equinumerous to $\mathbf{N}$, or is there only one "size of infinity"? If all infinite sets are equinumerous, then there is little else to say about this notion. However, this is not the case, as we shall see in the next section.

EXERCISES FOR §4

1. Show that if A is countable and not empty, then there are infinitely many f's such that $f: A \xrightarrow{1\text{-}1} \mathbf{N}$.

2. Prove that $\mathbf{Q}^+ \sim \mathbf{N}$ by making use of a table as in the second proof that $\mathbf{N} \times \mathbf{N} \sim \mathbf{N}$.

3. Let $A \sim B$, and suppose that A has 20 elements. How many pairings are there between A and B? Justify your answer.

In Exercises 4 through 6 let A, B, C, and D be sets such that $A \sim C$ and $B \sim D$.

4. (a) Give an example to show that $A \cap B$ need not be equinumerous to $C \cap D$.
   (b) Give an example to show that $A \cup B$ need not be equinumerous to $C \cup D$.

5. (a) Suppose $A \cap B = C \cap D = \phi$. Prove that $A \cup B \sim C \cup D$.
   (b) Suppose $A \cup B \sim C \cup D$. Must $A \cap B \sim C \cap D$?

6. Suppose A, B, C, and D are finite sets. If $A \cap B \sim C \cap D$, prove that $A \cup B \sim C \cup D$.

7. Let P_1 be a pairing between A and C, and let P_2 be a pairing between B and D. Suppose $A \cap B = C \cap D$ and $P_1(x) = P_2(x)$ for all $x \in A \cap B$. Prove that $A \cup B \sim C \cup D$.

8. (a) Define $f: \mathbf{N} \times \mathbf{N} \to \mathbf{N}$ by $f(m,n) = (2^m(2n+1)) - 1$. Prove that f is 1-1 and onto.
   (b) Let $g: [\mathbf{N}]_k \xrightarrow[\text{onto}]{1\text{-}1} \mathbf{N}$. Define $h(n_1, n_2, \ldots, n_k, n_{k+1}) = (2^{g(n_1, n_2, \ldots, n_k)}(2n_{k+1}+1)) - 1$. Show that $h: [\mathbf{N}]_{k+1} \xrightarrow[\text{onto}]{1\text{-}1} \mathbf{N}$.

9. Let $f: \mathbf{N} \xrightarrow[\text{onto}]{} A$, and suppose that $f^{-1}[a]$ is finite for each $a \in A$. Show that $\mathbf{N} \sim A$.

10. Prove that $P(A) \sim {}^A\{0,1\}$ for any A. (*Hint*: For B a subset of A, define
    $$f_B(x) = 1 \quad \text{if } x \in B,$$
    $$f_B(x) = 0 \quad \text{if } x \in A - B.$$
    Now let F be the function on $P(A)$ defined by $F(B) = f_B$. Show that $F: P(A) \xrightarrow[\text{onto}]{1\text{-}1} {}^A\{0,1\}$.)

11. Prove that $A \sim B$ implies ${}^AX \sim {}^BX$, ${}^XA \sim {}^XB$, and $P(A) \sim (B)$.

12. Prove that $A \times B \sim B \times A$ and that $A \times (B \times C) \sim (A \times B) \times C$.

1.5 The Power Set

In this section we show that there are infinite sets that are not equinumerous to **N**. It is then quite natural to ask if these sets have greater magnitude than **N**, and we shall consider several that do. However, in order to give a cohesive and general discussion of size comparisons one needs an assumption that we have not explicitly mentioned so far and that is considerably more subtle in content. This new principle, the axiom of choice, is discussed in §1.9, and there we shall again take up the problem of size comparisons. (Covert use of this axiom was made in Theorem 4.13.)

Theorem 5.1. *Let A be an arbitrary set, and suppose $f: A \to P(A)$. Then f is not onto $P(A)$, i.e., there is a $B \in P(A)$ such that $B \notin \operatorname{Ran} f$.*

PROOF: Let $B = \{x : x \in A \text{ and } x \notin f(x)\}$. Then $B \subseteq A$ and so $B \in P(A)$. Suppose $B \in \operatorname{Ran} f$. Then $B = f(a)$ for some $a \in A$. We ask whether or not $a \in B$. If $a \in B$, then, by the definition of B, $a \notin f(a)$. But $B = f(a)$ so $a \notin B$, a contradiction. On the other hand, if $a \notin B$, then $a \notin f(a)$, so, again by the definition of B, we conclude that $a \in B$, contradiction. Since both assumptions $a \in B$ and $a \notin B$ lead to contradictions, our assumption that $B \in \text{Ran} f$ is erroneous, i.e., $B \notin \text{Ran} f$, as we needed to show. □

Corollary 5.2. *Let A be an arbitrary set. Then $A \not\sim P(A)$.*

PROOF: Suppose for some set A we have $A \sim P(A)$. Then there is a pairing $P: A \underset{\text{onto}}{\overset{1\text{-}1}{\to}} P(A)$, contradicting Theorem 5.1. □

The argument used in the proof of Theorem 5.1 is known as a "diagonal argument." To see why, consider the special case $\mathbf{N} = A$. Again let $f: \mathbf{N} \to P(\mathbf{N})$. Now consider the following table:

	0	1	2	3	4	...
$f(0)$	a_{00}	a_{01}	a_{02}	a_{03}	a_{04}	
$f(1)$	a_{10}	a_{11}	a_{12}	a_{13}	a_{14}	
$f(2)$	a_{20}	a_{21}	a_{22}	a_{23}	a_{24}	
$f(3)$	a_{30}	a_{31}	a_{32}	a_{33}	a_{34}	
⋮						

Here a_{ij} is 0 if $j \notin f(i)$, and a_{ij} is 1 if $j \in f(i)$. For example, if $f(3)$ is the set of primes, then the fourth line of the table begins as follows: 0011010100010.... We define B as before, that is, $B = \{x : x \in \mathbf{N}, x \notin f(x)\} = \{m : m \in \mathbf{N} \text{ and } a_{m,m} = 0\}$. Then for each n, $B \neq f(n)$, since $n \in f(n)$ iff $a_{nn} = 1$ and $a_{nn} = 1$ iff $n \notin B$. Note how the elements of B are determined by the diagonal of the table, namely the entries a_{nn} for $n \in \mathbf{N}$.

Diagonal arguments play an important role in many proofs in this book. As with the proof of Theorem 5.1, the "diagonal" considered may not be immediately obvious but it is usually helpful to rewrite the proof in tabular form.

The power set of an infinite set is certainly infinite, since $P(A) \supset \{\{a\} : a \in A\}$, and so Corollary 5.2 implies that there are different sizes of infinity. Is there a reasonable way to compare different sizes of infinity? Can one speak of one infinite set having greater "magnitude" than another infinite set? The following definition fits well with our intuition and provides a basis for size comparisons of infinite sets.

We say that B is *at least as numerous* as A if A is equinumerous with a subset of B. If this is the case, we write $B \geqslant A$ or $A \leqslant B$. In other words, $B \geqslant A$ if there is a 1-1 function from A into B. If B is at least as numerous as A but not equinumerous to A, we say that B is *more numerous* than A (or A is *less numerous* than B) and we write $B > A$ (or $A < B$).

If A is an arbitrary set, then the function f defined by $f(a) = \{a\}$ for all $a \in A$ yields a 1-1 function from A into $P(A)$. Thus $A \leqslant P(A)$. By Corollary 5.2, A is not equinumerous with $P(A)$. Hence A is less numerous than $P(A)$, and we have proved:

Theorem 5.3. *Let A be an arbitrary set. Then $A \prec P(A)$.*

Our definitions of "more numerous" and "at least as numerous" allow us to make size comparisons between infinite sets. Do size comparisons between infinite sets obey the basic rules as size comparisons between finite sets? If so then our experience with finite sets will be of help when comparing infinite sets. For example, if any of the statements in the following theorem were false, it would be difficult to think of "at least as numerous" as a generalization of this concept for finite sets.

Theorem 5.4. *Let A, B, and C be arbitrary sets. Then:*
i. $A \subseteq B$ implies $A \leqslant B$.
ii. $A \leqslant A$.
iii. If $A \leqslant B$ and $B \leqslant C$, then $A \leqslant C$.
iv. If $A \leqslant B$ and $B \leqslant A$, then $A \sim B$.
v. Either $A \leqslant B$ or $B \leqslant A$.

The proofs of parts i and ii are immediate from the definitions. Part iii follows from Theorem 3.1i. We will prove iv in the next section. Part v is considerably harder and will not be proved until §1.9.

Note that Corollary 4.10 and Theorem 5.4v together imply that the magnitude of $\mathbf{N}$ is minimal among the infinite sets. In other words, we have

Corollary 5.5. *If A is inifinite, then $\mathbf{N} \leqslant A$.*

Here is a direct proof of Corollary 5.5 that does not make use of theorem 5.4. Let f be a function whose domain is $P(A) - \{\varnothing\}$ and such that $f(X) \in X$ for each $X \in P(A) - \{\phi\}$. Now define $g: \mathbf{N} \xrightarrow{1\text{-}1} A$ as follows:

$$g(0) = f(A),$$
$$g(1) = f(A - \{g(0)\}),$$
$$g(2) = f(A - \{g(0), g(1)\}),$$
$$g(3) = f(A - \{g(0), g(1), g(2)\}),$$

and so on. In other words $g(n+1) = f(A - \{g(m): m \leqslant n\})$. Notice that for each n, $g(n+1)$ is defined, since $A - \{g(m): m \leqslant n\} \neq \phi$. Since $g(n+1) \notin \{g(m): m \leqslant n\}$, it is clear that g is 1-1. Hence $g: \mathbf{N} \xrightarrow{1\text{-}1} A$.

EXERCISES FOR §1.5

1. If $A \sim B$, prove that $P(A) \sim P(B)$.
2. If $A \leqslant B$ and $B \sim C$, prove that $A \leqslant C$.
3. If $A \prec B$ and $B \sim C$, prove that $A \prec C$.
4. Show that if $\mathbf{N} \leqslant A \cup B$, then $\mathbf{N} \leqslant A$ or $\mathbf{N} \leqslant B$.

5. Prove that if A has n elements, where $n \in \mathbb{N}$, then $P(A)$ has 2^n elements. (Try a proof by induction on n, using Exercise 10 of §1.4.)

6. Prove:
   (a) $A \leqslant B \cap C$ implies $A \leqslant B$ and $A \leqslant C$.
   (b) $A \prec B \cap C$ implies $A \prec B$ and $A \prec C$.
   (c) $B \cup C \leqslant A$ implies $B \leqslant A$ and $C \leqslant A$.
   (d) $B \cup C \prec A$ implies $B \prec A$ and $C \prec A$.

1.6 The Cantor–Bernstein Theorem

In this section we shall prove of Theorem 5.4iv, which states that if $A \leqslant B$ and $B \leqslant A$, then $A \sim B$. This is called the Cantor-Bernstein Theorem. We begin by proving a related result.

Theorem 6.1. *If $A \supseteq B$, $B \supseteq C$, and $A \sim C$, then $A \sim B$.*

PROOF: By assumption there is an $f: A \xrightarrow[\text{onto}]{1\text{-}1} C$. Since f maps A onto C, $f[A - B] = \{f(x) : x \in A, \ x \notin B\} \subseteq C$. Let $A_0 = A - B$, $A_1 = f[A_0] = f[A - B]$, $A_2 = f[A_1]$, and so on. The situation is illustrated in Figure 1.1.

Let $D = \bigcup_{i=0}^{\infty} A_i$. We define a function f' from A into B as follows:

$$f'(a) = f(a) \quad \text{if } a \in D,$$
$$f'(a) = a \quad \text{if } a \notin D.$$

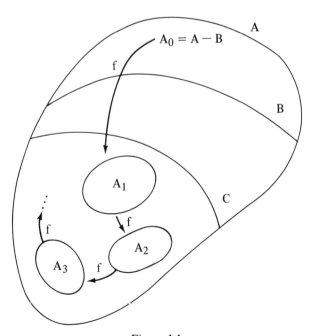

Figure 1.1

(A glance at the illustration may be helpful at this point.) We show that f' is a pairing between A and B.

To see that f' is 1-1 we take distinct elements a_1 and a_2 of A and consider four cases:

Case 1: $a_1 \in D$ and $a_2 \in D$. Then $f'(a_1) = f(a_1) \neq f(a_2) = f'(a_2)$, since f is 1-1.
Case 2: $a_1 \notin D$ and $a_2 \notin D$. Then of course $f'(a_1) \neq f'(a_2)$, since $f'(a_1) = a_1$ and $f'(a_2) = a_2$.
Case 3: $a_1 \in D$ and $a_2 \notin D$. Then $f'(a_1) = f(a_1) \in D$ and $f'(a_2) = a_2 \notin D$.
Case 4: $a_1 \notin D$ and $a_2 \in D$ is handled as in case 3.

We next show that f' is onto. Suppose $b \in B$. If $b \notin D$, then $f'(b) = b$, so $b \in \mathrm{Ran} f'$. Suppose $b \in D$. Then $b \in A_i$ for some $i > 0$ ($i \neq 0$, since $A_0 \cap B = \phi$). Since $A_i = f[A_{i-1}]$, $b = f(c)$ for some $c \in A_{i-1}$. Then $c \in D$, so $f'(c) = f(c) = b$. Thus $b \in \mathrm{Ran} f'$, and so f' is onto B. Hence $f' : A \xrightarrow[\text{onto}]{\text{1-1}} B$, so $A \sim B$ as the theorem claims. □

We can now prove

Theorem 6.2 (Cantor-Bernstein Theorem). *If $A \leqslant B$ and $B \leqslant A$, then $A \sim B$.*

PROOF: By assumption there are functions f and g such that $f : A \xrightarrow{\text{1-1}} B$, $g : B \xrightarrow{\text{1-1}} A$. Then $g \circ f : A \xrightarrow{\text{1-1}} A$. Let $C = g \circ f[A]$. So $A \sim C$ and $A \supseteq C$. Let $B' = g[B]$. If $x \in C$, then $x = g(f(a))$ for some $f(a) \in B$, so $x \in B'$. Thus $A \supseteq B' \supseteq C$ and $A \sim C$. By Theorem 6.1 we conclude that $A \sim B'$. Since $g : B \xrightarrow[\text{onto}]{\text{1-1}} g[B]$, $B \sim B'$, and so $A \sim B$, proving the Cantor-Bernstein theorem. □

As an application of the Cantor-Bernstein theorem we prove

Theorem 6.3. $\mathbf{R} \sim P(\mathbf{N})$.

PROOF: We first exhibit a function $f : P(\mathbf{N}) \xrightarrow{\text{1-1}} \mathbf{R}$. Let $X \in P(\mathbf{N})$, i.e., $X \subseteq \mathbf{N}$. Define $f(X)$ to be the real number $0.a_0 a_1 a_2 a_3 \ldots$, where $a_i = 0$ if $i \notin X$ and $a_i = 1$ if $i \in X$. To see that f is 1-1, we need only recall that if two decimal expansions are equal, then one of them ends in an infinite string of 9's. This shows that $P(\mathbf{N}) \leqslant \mathbf{R}$.

Next we show that $\mathbf{R} \leqslant P(\mathbf{N})$. First notice that $\mathbf{R} \sim \{r : -1 < r < 1\}$, as the pairing $f : x \to x/(1 + |x|)$ shows. Then $\{r : -1 < r < 1\} \sim \{r : 0 < r < 1\}$, as the pairing $g : x \to \frac{1}{2}(x + 1)$ shows. Hence it is enough to show that $A \leqslant P(\mathbf{N})$, where $A = \{r : 0 < r < 1\}$. Given $r \in A$, let $0.r_1 r_2 r_3 \ldots$ be its non-terminating decimal representation. Let $h(r) = \{r_1, 10 + r_2, 100 + r_3, 1000 + r_4, \ldots\}$. Clearly $h : A \xrightarrow{\text{1-1}} P(\mathbf{N})$, and so $\mathbf{R} \leqslant P(\mathbf{N})$. Since we have already shown $\mathbf{R} \leqslant P(\mathbf{N})$, an application of the Cantor-Bernstein theorem gives $A \sim P(\mathbf{N})$. □

1.6 The Cantor-Bernstein Theorem

We now have the machinary to prove

Theorem 6.4. *Let A, B, and C be arbitrary sets. Then*:
i. *If $A \leq B$ and $B \prec C$, then $A \prec C$.*
ii. *If $A \prec B$ and $B \leq C$, then $A \prec C$.*

We prove part i and leave ii as an exercise (see Exercise 5). Suppose $A \leq B$ and $B \prec C$. Then there are functions f and g with $f: A \xrightarrow{1\text{-}1} B$, $g: B \xrightarrow{1\text{-}1} C$. Thus $g \circ f: A \xrightarrow{1\text{-}1} C$, and so $A \leq C$. If $A \sim C$, we would have $C \supseteq g[B] \supseteq g \circ f[A]$ and $g: f[A] \sim A \sim C$. By Theorem 6.1 this would imply $C \sim g[B] \sim B$, a contradiction. Thus $A \prec C$.

Let $P_0(\mathbf{N}) = \mathbf{N}$, $P_1(\mathbf{N}) = P(\mathbf{N})$, $P_2(\mathbf{N}) = P(P(\mathbf{N})), \ldots$, and in general $P_{n+1}(\mathbf{N}) = P_n(\mathbf{N})$. Using part i of the above theorem and Corollary 5.3, we see that $P_n(\mathbf{N}) \prec P_m(\mathbf{N})$ for all $n, m \in \mathbf{N}$ with $n < m$. Hence there are infinitely many sizes of infinity. Are there sets more numerous than any of the P_n's? Indeed there are. For example, let $X = \{P_n(\mathbf{N}): n \in \mathbf{N}\}$. Then for any $B \in X$ we have $B \prec P(B)$ and $P(B) \subseteq \bigcup X$. Hence $B \prec P(B) \leq \bigcup X$, and so by Theorem 6.4ii we have $B \prec \bigcup X$.

Of course, there are sets that are more numerous than $\bigcup X$, such as $P(\bigcup X)$, $P(P(\bigcup X))$, and so on.

We have seen in Theorem 5.4 that there are no infinite sets less numerous than $\mathbf{N}$. Does the magnitude of $\mathbf{N}$ have an immediate successor, i.e., is there a set X such that $\mathbf{N} \prec X$ and for no Y do we have $\mathbf{N} \prec Y \prec X$? In particular, is $P(\mathbf{N})$ such an X? This problem, posed by Cantor, has been one of the most vexing in set theory, and many ramifications arising from this problem are still under active investigation. We shall say more about this famous problem in a later section.

EXERCISES FOR §1.6

1. Let a, b, c, d be real numbers with $a < b$ and $c < d$. Let $A = \{r: r \in \mathbf{R} \text{ and } a < r < b\}$, $B = \{r: r \in \mathbf{R} \text{ and } a \leq r \leq b\}$, and $C = \{r: r \in \mathbf{R} \text{ and } c < r < d\}$. Show that $A \sim B \sim C$.

2. Let $A = \{r: r \in \mathbf{R} \text{ and } 0 \leq r \leq 1\}$. Show that $A \times A \sim A$. [*Hint*: let r and s be reals with decimal expansions $0.r_1 r_2 r_3 \ldots$ and $0.s_1 s_2 s_3 \ldots$ respectively. Consider $f(r,s) = 0.r_1 s_1 r_2 s_2 r_3 s_3 \ldots$.]

3. Let $A = \{r: r \in \mathbf{R} \text{ and } -1 < r < 1\}$. Show that
   (a) $A \times A \sim \mathbf{R} \times \mathbf{R}$,
   (b) $\mathbf{R} \times \mathbf{R} \sim \mathbf{R}$.

4. Let $B = \{(x,y): (x-a)^2 + (y-b)^2 = c^2 \text{ and } x \in \mathbf{R}, y \in \mathbf{R}\}$, where a, b, and c are fixed members of $\mathbf{R}$ and $c > 0$. Suppose $\mathbf{R} \times \mathbf{R} \supseteq A \supseteq B$. Show that $\mathbf{R} \times \mathbf{R} \sim A$.

5. Prove Theorem 6.4ii.

6. Show that $^N\mathbf{R} \sim \mathbf{R}$. [*Hint*: $\mathbf{R} \leqslant {}^N\mathbf{R}$ is clear. Let $A = \{r : 0 \leqslant r < 1 \text{ and } r \in R\}$. Let $f \in {}^N A$; say $f(n) = 0.r_{n,1}r_{n,2}r_{n,3}\ldots$. Let $F(f) = 0.s_1 s_2 \ldots$, where $s_j = 0$ if j is not of the form $2^k 3^l$, and $s_j = r_{kl}$ otherwise. Show that $F : {}^N A \xrightarrow{1\text{-}1} \mathbf{R}$.]

7. Show that
   (a) $^N\mathbf{N} \sim \mathbf{R}$,
   (b) $^Q\mathbf{N} \sim \mathbf{R}$.
   (*Hint*: Show $\mathbf{R} \leqslant {}^N\mathbf{N} \leqslant {}^N\mathbf{R}$).

8. Let $\mathcal{C}$ be the class of continuous functions. Then $\mathcal{C} \sim \mathbf{R}$.

1.7 Algebraic and Transcendental Numbers

In §1.4 we mentioned the applicability of Cantor's methods to problems outside of set theory. In this section we shall present one of Cantor's most striking results, achieved by applying his methods to the study of real numbers.

By an *integral polynomial* in the variable x, we mean a polynomial $a_0 + a_1 x + \ldots + a_n x^n$, where the $a_i \in \mathbf{I}$ ($\mathbf{I}$ being the set of integers) and $a_n \neq 0$. A *real root* of an integral polynomial $a_0 + a_1 x + \ldots + a_n x^n = f(x)$ is a real number α such that $f(\alpha) = 0$. Of course, some integral polynomials have no real roots, for example, $1 + x^2$. A real number is said to be *algebraic* if it is the root of some integral polynomial. Otherwise, it is called *transcendental*. Are there transcendental numbers, or is every real number algebraic?

Although the above question is a natural one to ask, no answer was known until the middle of the nineteenth century. Then in 1844, the French mathematician Liouville produced the first examples of transcendental numbers. This was a considerable achievement, and to this day there are many simply stated but unanswered questions concerning the transcendence of various numbers. For example, while π and e were shown to be transcendental in 1882, it is still not known whether $e + \pi$ or $e \cdot \pi$ is transcendental (although one of them must be; see Exercise 3).

One is tempted to conclude that transcendental numbers are scarce. But in 1874 Cantor proved the remarkable result that most real numbers are transcendental. More precisely,

Theorem 7.1. *Let T be the set of transcendental numbers and A the set of algebraic numbers. Then $T \sim \mathbf{R}$ and $A \sim \mathbf{N}$.*

PROOF: We first show that there are countably many algebraic numbers.

Let $\mathbf{I}[x]$ be the set of all integral polynomials in the variable x. If $f(x) \in \mathbf{I}[x]$, [say $f(x) = a_0 + a_1 x + \ldots + a_n x^n$, $a_n \neq 0$], we call n the degree of $f(x)$. Recall that a polynomial of degree n with real coefficients can have at most n real roots, a fact usually proved in courses in elementary algebra.

We first prove that $\mathbf{N} \sim A$. Define f as follows:

$f(n) = y$ if $n = 2^{a_0} 3^{a_1} \cdots p_k^{a_k} p_{k+1}^{a_{k+1}}$ for some $a_0, a_1, \ldots, a_k, a_{k+1}$, and y is the a_{k+1} largest root of $a_0 + a_1 x + \ldots + a_n x^n$.
$f(n) = 1$ otherwise.

Clearly $f: \mathbf{N} \xrightarrow[\text{onto}]{1\text{-}1} A$, and so $\mathbf{N} \sim A$ by 4.11.

We now prove that T, the set of transcendental numbers, is equinumerous with $\mathbf{R}$. Clearly, $T \leqslant \mathbf{R}$, since $T \subseteq \mathbf{R}$. By the Cantor-Bernstein theorem, we need only show that $\mathbf{R} \leqslant T$. We know that $\mathbf{R} \sim \mathbf{R}_{(-1,1)}$, where $\mathbf{R}_{(-1,1)} = \{z : -1 < z < 1 \text{ and } z \in \mathbf{R}\}$ (see the proof of Theorem 6.3). So it is enough to show that $\mathbf{R}_{(-1,1)} \leqslant T$. Choose $t \in T$ so that $t > 1$ (we know that such a t exists, for otherwise $\{z : 1 < z \text{ and } z \in \mathbf{R}\} \subseteq A$, and this is impossible, since $\{z : 1 < z \text{ and } z \in \mathbf{R}\} \sim \mathbf{R}$ but $A \sim \mathbf{N}$). Notice that for each $n \in \mathbf{N}^+$, nt is transcendental, for if nt is a root of $a_0 + a_1 x + \ldots + a_k x^k$, then t is a root of $a_0 + (a_1 n) x + \ldots + (a_k n^k) x^k$. Let $h: A \xrightarrow{1\text{-}1} \mathbf{N}$. Define $H(z) = z$ if $z \in \mathbf{R}_{(-1,1)} \cap T$ and $H(z) = (h(z))t$ if $z \in \mathbf{R}_{(-1,1)} \cap A$. Then clearly $H: \mathbf{R}_{(-1,1)} \xrightarrow{1\text{-}1} T$, which completes the proof. □

Later we shall prove that if $B \leqslant C$ and C is infinite, then $B \cup C \sim C$. Since $A \cup T = R$, this gives an immediate proof that $T \sim R$.

EXERCISES FOR §1.7

1. Let A' be the set of real roots of polynomials with rational coefficients. Show that $A = A'$.

2. Let A^* be the set of real roots of polynomials whose coefficients are algebraic numbers. Show that $A^* = A$. (Requires a bit of field theory.)

3. It is known that e and π are transcendental. Show that either $e + \pi$ or $e \cdot \pi$ is transcendental by examining the polynomial $x^2 - (e + \pi)x + e \cdot \pi$ (and using some field theory). However, it is not known which of the two is transcendental. As a completely trivial problem in the same vein, show that either $e + \pi$ or $e - \pi$ is transcendental.

1.8 Orderings

In every branch of mathematics orderings of one sort or another are encountered. We have no intention of giving a comprehensive classification of the various orderings that arise, but instead we restrict our attention to those that arise most frequently and that are particularly important in logic.

A *partial ordering* is a binary relation R such that for every x, y, z

i. $x \mathcal{R} x$;
ii. $x R y$ implies $y \mathcal{R} x$;
iii. $x R y$ and $y R z$ implies $x R z$.

[Recall that xRy means $(x,y) \in R$ and $x\cancel{R}y$ means $(x,y) \notin R$.] We shall often use the symbol $<$ to denote a partial ordering.

An ordered pair (A, R) is a *partially ordered structure* if $A \neq \emptyset$ and R is a partial ordering with field $\subseteq A$.

EXAMPLE 8.1. For any X, $(P(X), \subset)$ is a partially ordered structure.

EXAMPLE 8.2. For $a, b \in \mathbf{N}^+$ we write $a|b$ if there is a $c \in \mathbf{N}$ such that $c \neq 1$ and $c \neq a$ and $ac = b$. Then $(\mathbf{N}, |)$ is a partially ordered structure.

EXAMPLE 8.3. Let A be the set of polynomials with real coefficients. Let $p, q \in A$. Write $p|q$ if for some $r \in A$, $p \cdot r = q$ and the degree of p and of r is positive. Then $|$ is a partial ordering with field A.

EXAMPLE 8.4. Let $\mathcal{F} = {}^\mathbf{R}\mathbf{R}$. Given $f, g \in \mathcal{F}$, define $f \triangle g$ if $f(x) \leqslant g(x)$ for each $x \in \mathbf{R}$ and $f(z) < g(z)$ for some $z \in \mathbf{R}$. Then $(\mathcal{F}, \triangle)$ is a partially ordered structure.

EXAMPLE 8.5. Let $\mathcal{C}$ be the set of continuous functions with domain $\{x : 0 \leqslant x \leqslant 1\}$. For $f, g \in \mathcal{C}$ define $f <^* g$ if $\int_0^1 f < \int_0^1 g$. Then $(\mathcal{C}, <^*)$ is a partially ordered structure.

A *linear ordering* is a partial ordering R that satisfies the additional requirement that any two elements of its field are comparable, i.e.,

iv. for all x, y in the field of R, either xRy or $x = y$ or yRx.

(A, R) is a *linearly ordered structure* if R is a linear ordering with field A and $A \neq \emptyset$.

Requirement iv forces the elements of A to be arranged in a chain. Examples of linearly ordered structures are $(\mathbf{R}, <)$, $(\mathbf{Q}, <)$, $(\mathbf{I}, <)$, and $(\mathbf{N}, <)$ (the relation $<$ is the usual "less than" relation on $\mathbf{R}$, restricted appropriately). Examples 8.1 through 8.5 are not linearly ordered structures.

A linear ordering R is *dense* if $\text{Dom}\, R$ has at least two elements and between any two elements of $\text{Dom}\, R$ there is another element of $\text{Dom}\, R$—i.e., for every x, y, if xRy, then there is a z such that xRz and zRy. (A, R) is a *densely ordered structure* if R is a dense linear ordering with field A. Both $\mathbf{Q}$ and $\mathbf{R}$ are densely ordered by the usual $<$, but $\mathbf{N}$ is not.

We often will ignore the distinction between the ordered structure (A, R) and the ordering R, referring to both as orderings of the appropriate kind.

Let (A, R) be an ordering with $x, y \in A$. y is a *successor* of x in (A, R) if xRy. y is the *immediate successor* of x if y is a successor of x and there is no $z \in A$ such that xRz and zRy. Predecessor and immediate predecessor are defined analogously. A linear ordering is *discrete* if every element having a successor has an immediate successor and every element having a

1.8 Orderings

predecessor has an immediate predecessor. $(\mathbf{N}, <)$ and $(\mathbf{I}, <)$ are discretely ordered, but $(\mathbf{Q}, <)$ and $(\mathbf{R}, <)$ are not.

If $B \neq \emptyset$ and $(B \times B) \cap R$ is a linear ordering, then B is called a *chain* or a *branch*. In Example 8.2 above, $\{3^n : n \in \mathbf{N}^+\}$ is a chain. A *minimal element* in a partially ordered structure (A, R) is an element $a \in A$ such that for no $b \in A$ do we have bRa. If there is no $b \in A$ such that aRb, then a is a *maximal element*. If for each $b \in A$ we have aRb or $a = b$, then a is the *least element*, and if for each $b \in A$ we have bRa or $a = b$, then a is the *greatest element*.

In Example 8.1, ϕ is the least element and X is the greatest element. In Example 8.5 there is no minimal element and no maximal element. If we let $A = P(\mathbf{N}) - \{\emptyset, \mathbf{N}\}$, then $(A, \subseteq)$ is a partially ordered structure with as many minimal elements and many maximal elements as there are elements in A (see Exercise 9).

A *well ordering* is a linear ordering R having the additional property that

v. Every non-empty subset of the field of R has a least element.

In other words, for every non-empty $X \subseteq$ the field of R, there is an element $x \in X$ such that xRy for all $y \in X - \{x\}$.

With the usual "less than" relation the natural numbers are well ordered. However, neither $\mathbf{I}$ nor $\mathbf{R}^+$ is well ordered. To see this let X be the set of elements less than 1 (of $\mathbf{I}$ and $\mathbf{R}^+$ respectively). Clearly X does not have a least element.

Next we describe a construction that provides many examples of well-ordered structures.

Let $(A, <_A)$ be a linearly ordered structure, and let $(B_a, <_a)$ be a linearly ordered structure for each $a \in A$. Define $<_A \Sigma \{(B_a, <_a) : a \in A\}$ to be the structure $(B, <_B)$, where $B = \bigcup \{(a, b) : a \in A$ and $b \in B_a\}$, and $(a, b) <_B (c, d)$ iff $a <_A c$ or $(a = c$ and $b <_a d)$.

Loosely speaking, $(B, <_B)$ is obtained by replacing each $a \in A$ with a copy of $(B_a, <_a)$. For example, if $A = \{0, 1, 2\}$, $B_i = \mathbf{N}$ for each $i \in A$, and $<_A$ and $<_i$ are the usual orderings on A and on B_i, then $<_A \Sigma \{(B_i, <_i) : i \in A\}$ can be viewed as the ordering obtained by stacking three copies of $(\mathbf{N}, <)$ one upon the other. It is easy to see that the resulting structure is well ordered. More generally we have the following.

Theorem 8.6.

i. $<_A \Sigma \{(B_a, <_a) : a \in A\}$ *is linearly ordered.*
ii. *If* $(A, <_A)$ *is well ordered and if* $(B_a, <_a)$ *is well ordered for each* $a \in A$, *then* $<_A \Sigma \{(B_a, <_a) : a \in A\}$ *is well ordered.*

PROOF: i. Let $(B, <_B) = <_A \Sigma \{(B_a, <_a) : a \in A\}$. Let $(a, b), (c, d), (e, f) \in B$. Clearly $(a, b) \not<_B (a, b)$, for otherwise we would have $a <_A a$ or $b <_a b$. Suppose $(a, b) <_B (c, d)$. Then either $a <_A c$, or $a = c$ and $b <_a d$. If $a <_A c$, then $c \not<_A a$, and so we can not have $(c,d)<_B(a,b)$. If $a=c$ and $b<_a d$, then $d \not<_a b$, and so again $(c,d) \not<_B(a,b)$. Now suppose that $(a,b)<_B(c,d)$ and $(c,d)<_B(e,f)$. Then either

i. $a<_A c<_A e$,
ii. $a<_A c=e$ and $d<_c f$,
iii. $a=c<_A e$ and $b<_a d$, or
iv. $a=c=e$ and $b<_a d$ and $d<_c f$.

In either case $a<_A e$, or $a=e$ and $b<_a f$; hence $(a,b)<_B(e,f)$. □

PROOF: ii. Let X be a non-empty subset of B. Let $X_1=\{a:(a,b)\in X\}$. $X_1 \neq \emptyset$, and $X_1 \subseteq A$. Since $(A, <_A)$ is well ordered, X_1 has a least element, say a^*. Let $X_2=\{b:(a^*,b)\in X\}$. Then $X_2 \neq \emptyset$ and $X_2 \subseteq B_{a^*}$. Since $(B_{a^*}, <_{a^*})$ is well ordered, X_2 has a least element, say b^*. Clearly (a^*,b^*) is the least element of X. □

An *initial segment* of a linearly ordered structure $(A, <)$ is a subset $X \subseteq A$ such that for each $x \in X$ and $a \in A$, if $a<x$ then $a \in X$.

For example, $\{x:x\in \mathbf{Q} \text{ and } x<\pi\}$ is an initial segment of $\mathbf{Q}$ but not of $\mathbf{R}$, and $\{x:x\in \mathbf{R} \text{ and } 0<x\leq 4\}$ is an initial segment of $\mathbf{R}^+$ but not of $\mathbf{R}$.

Our next theorem shows that given two well-ordered structures an initial segment of one of them is a copy of the other. For this we need some definitions and several easy lemmas.

Through the remainder of this section we let $(A_1, <_1)$ and $(A_2, <_2)$ be well-ordered structures.

A binary relation $S \subseteq A_1 \times A_2$ is *order preserving* [with respect to $(A_1, <_1)$ and $(A_2, <_2)$] if whenever $(x_1,y_1)\in S$ and $(x_2,y_2)\in S$, then $x_1<_1 x_2$ iff $y_1<_2 y_2$.

Clearly, an order preserving relation is a 1-1 function.

An order preserving relation S is an *initial pairing* if $\mathrm{Dom}\,S$ is an initial segment of $(A_1, <_1)$ and $\mathrm{Ran}\,S$ is an initial segment of $(A_2, <_2)$.

Lemma 8.7. *If S and S' are initial pairings and $\mathrm{Dom}\,S = \mathrm{Dom}\,S'$, then $S=S'$.*

PROOF: Assume that the hypotheses holds but that for some $x\in \mathrm{Dom}\,S$ we have $Sx \neq S'x$. Let x^* be the least such x (in the sense of $<_1$), say $S'x^* <_2 Sx^*$. Then $Sz=S'z<_2 S'x^*$ for each $z<_1 x^*$, and $Sx^* \leq_2 Sz$ for each $z \geq_1 x^*$, $z\in \mathrm{Dom}\,S$. Thus $S'x^* \notin \mathrm{Ran}\,S$, and so $\mathrm{Ran}\,S$ is not an initial segment of $(A_2, <_2)$. This contradicts the assumption that S is an initial pairing. □

Lemma 8.8. *If X is an initial segment of $(A_1, <_1)$ and S is an initial pairing, then $S \restriction X$ is an initial pairing.*

PROOF: We need only show that $\mathrm{Ran}(S \restriction X)$ is an initial segment of $(A_2, <_2)$. So suppose that $y<_2 y'$ and that $Sx'=y'$, where $x'\in X$. Since

1.8 Orderings

Ran S is an initial segment and $y' \in \text{Ran } S$, there is an $x \in \text{Dom } S$ such that $Sx = y$. Since S is order preserving, $x <_1 x'$ and so $x \in X$. Hence $\text{Ran}(S \restriction X)$ is an initial segment. □

Lemma 8.9. *If S and S' are initial pairings, then $S \supseteq S'$ or $S' \supseteq S$.*

PROOF: Suppose $\text{Dom } S \subseteq \text{Dom } S'$. By Lemma 8.8, $S' \restriction \text{Dom } S$ is an initial pairing, and so by Lemma 8.7 $S' \restriction \text{Dom } S = S$, i.e., $S \subseteq S'$. □

Lemma 8.10. *If $\mathcal{C}$ is a set of initial segments of $(A_1, <_1)$, then $\bigcup \mathcal{C}$ is an initial segment of $(A_1, <_1)$.*

PROOF: If $x_1 <_1 x_2$ and $x_1, x_2 \in A_1$ with $x_2 \in \bigcup \mathcal{C}$, then for some $X \in \mathcal{C}$, $x_2 \in X$. Since X is an initial segment, $x_1 \in X$ and so $x_1 \in \bigcup \mathcal{C}$. □

Theorem 8.11. *If $(A_1, <_1)$ and $(A_2, <_2)$ are well-ordered structures, then there is a unique order preserving function S such that either $\text{Dom } S = A_1$ and $\text{Ran } S$ is an initial segment of $(A_2, <_2)$, or $\text{Ran } S = A_2$ and $\text{Dom } S$ is an initial segment of $(A_1, <_1)$.*

PROOF: Let $\mathcal{K}$ be the set of all initial pairings. Let $S = \bigcup \mathcal{K}$.

We first show that S is order preserving. For suppose $(x_1, y_1) \in S$ and $(x_2, y_2) \in S$ and $x_1 <_1 x_2$. Then $(x_1, y_1) \in S_1$ and $(x_2, y_2) \in S_2$ for some $S_1, S_2 \in \mathcal{K}$. By Lemma 8.9, $S_1 \subseteq S_2$ or $S_2 \subseteq S_1$; let's say $S_1 \subseteq S_2$. Hence $y_1 <_2 y_2$, since S_2 is order preserving. Thus S is order preserving.

$\text{Dom } S = \bigcup \{\text{Dom } S' : S' \in \mathcal{K}\}$, and this is an initial segment of $(A_1, <_1)$ by Lemma 8.10. Similarly $\text{Ran } S$ is an initial segment of $(A_2, <_2)$. Thus S is an initial pairing.

Suppose $\text{Dom } S \subset A_1$ and $\text{Ran } S \subset A_2$. Let a be the least member (in the sense of $<_1$) of $A_1 - \text{Dom } S$, and let b be the least member (in the sense of $<_2$) of $A_2 - \text{Ran } S$. Then clearly $S \cup \{(a,b)\} \in \mathcal{K}$, and so $(a,b) \in S$. But then $a \in \text{Dom } S$—a contradiction. Hence $\text{Dom } S = A_1$ or $\text{Ran } S = A_2$.

The unicity of S follows from Lemma 8.7. □

Let $<_B$ be a linear ordering with field B, and let $<_C$ be a linear ordering with field C. We say that $<_B$ is an *initial segment* of $<_C$ if B is an initial segment of $<_C$ and $<_B = <_C \cap (B \times B)$. The following theorem will be useful in the next section.

Theorem 8.12. *Suppose $\mathcal{K}$ is a set of linear orderings (well orderings) such that whenever $<_B \in \mathcal{K}$ and $<_C \in \mathcal{K}$, one of them is an initial segment of the other. Then $\bigcup \mathcal{K}$ is a linear ordering (well ordering) and each $<_B \in \mathcal{K}$ is an initial segment of $\bigcup \mathcal{K}$.*

PROOF: $\bigcup \mathcal{K}$ is a binary relation; call it $<$. Suppose that $y < z$ and $z < y$. Then $y <_B z$ and $z <_C y$ for some $<_B, <_C \in \mathcal{K}$. We may assume that $<_B$ is an initial segment of $<_C$. But then $y<_C z$ and $z<_C y$, contrary to the assumption that $<_C$ is a linear ordering. Hence $y<z$ implies $z \not< y$. Similarly one shows that $x \not< x$, and that if $x<y$ and $y<z$ then $x<z$. Hence $<$ is a linear ordering.

Suppose that $a<b$ when $b \in B$, and B is the field of $<_B \in \mathcal{K}$. Then $a<_C b$ for some $<_C \in \mathcal{K}$. Either $<_C$ is an initial segment of $<_B$, or $<_B$ is an initial segment of $<_C$; and both alternatives imply that $a \in B$. Hence $<_B$ is an initial segment of $<$.

Finally, suppose each $<_B \in \mathcal{K}$ is well ordered and X is a non-empty subset of the field of $\mathcal{K}$. Let $a \in X$. Then $a \in C$, where C is the field of some $<_C \in \mathcal{K}$. Let d be the least element of $X \cap \{x : x \leqslant_C a\}$. Then d is the least element of X in the sense of $<$, since $<_C$ is an initial segment of $<$. Hence $<$ is a well ordering. □

Exercises for §1.8

1. Prove: If A is finite and $(A, <_A)$ is a linearly ordered structure, then $(A, <_A)$ is discretely ordered.

2. Prove: If $(A, <)$ is densely ordered, then A is infinite.

3. Prove: A well-ordered structure is not densely ordered.

4. (a) Find a well-ordered structure that is not discretely ordered.
   (b) Find a discretely ordered structure that is not well ordered.

5. (a) How many linear orderings are there on a set A of n elements?
   (b) Show that the set of linear orderings of $\mathbf{N}$ is equinumerous to $\mathbf{R}$.

6. Let (A, R) be a well-ordered structure, and let $B \cap A \neq \emptyset$. Show that $R \cap (B \times B)$ is a well ordering.

7. Let $<_\mathbf{Q}$ be the usual dense ordering of the rationals, and for each $r \in \mathbf{Q}$ let $B_r = \mathbf{I}$ and $<_r$ be the usual discrete ordering on $\mathbf{I}$. Show that $<_\mathbf{Q} \Sigma \{(B_r, <_r) : r \in \mathbf{Q}\}$ is a discretely ordered structure.

8. Let A be an uncountable subset of $\mathbf{R}$. Let $<$ be the usual order relation on $\mathbf{R}$. Then $(A \times A) \cap <$ is neither a discrete ordering nor a well ordering.

9. Let $A = P(\mathbf{N}) - \{\emptyset, \mathbf{N}\}$.
   (a) Show that $(A, \subset)$ is a partially ordered structure.
   (b) Show that the set of maximal elements and the set of minimal elements of $(A, \subset)$ are each equinumerous with $\mathbf{N}$.
   (c) Show that the set of branches of $(A, \subset)$ is equinumerous with $\mathbf{R}$.

10. Let $<_A$ be a binary relation with field A. Let B be a non-empty subset of A, and let $<_B = <_A \cap (B \times B)$. Give examples to show that the following are possible:
    (a) $<_A$ is a dense ordering, and $<_B$ is an infinite discrete ordering.
    (b) $<_A$ is a discrete ordering, and $<_B$ is a dense ordering.

1.9 The Axiom of Choice

Let A be a set of non-empty sets. We say that a function f is a *choice function* for A if the domain of f is A and $f(X) \in X$ for each $X \in A$. The terminology is quite descriptive, since f "chooses" an element in each $X \in A$, namely $f(X)$. There are some sets A for which the existence of a choice function is obvious. For example, if $A \subseteq P(\mathbf{N})$ and each member of A is non-empty, we can define $f(X)$ to be the least member of X.

On the other hand, suppose that $A = P(R) - \{\varnothing\}$. We can no longer define a choice function on A by defining $f(X)$ to be the least member of X. This is because R is not well ordered by the usual "less than" relation, and so X need not have a least member. In fact, in a sense which will be made clearer in a later section, the existence of a choice function on A must be taken on faith. This is a special case of the following principle, called the *axiom of choice*:

Every set of non-empty sets has a choice function.

We have already used the axiom of choice in proving that a countable union of countable sets is countable (Theorem 4.13), and in the direct proof of Corollary 5.5, in which it is shown that $\mathbf{N} \leqslant A$ if A is infinite.

The axiom was used frequently and without mention in various branches of mathematics until 1904, when Zermelo gave an explicit statement of the axiom. Once it surfaced, many investigators turned their attention to it. Some tried to prove it from simpler, more readily accepted principles; others deduced consequences from it that seemed to some paradoxical.

Today, the axiom has virtually universal acceptance among mathematicians. There are several reasons for this. First of all, Gödel proved in 1938 that no contradiction can be derived from the axiom of choice and the other basic assumptions about sets unless the other assumptions already lead to a contradiction. We shall discuss this more thoroughly in a later section.

Another reason is that the vast majority of mathematicians are convinced on intuitive grounds that the axiom is true. To these mathematicians the axiom is a valid assertion about sets.

Expediency provides another reason for acceptance. In several branches of mathematics there are theorems that have long, complicated, and non-intuitive proofs not using the axiom of choice, and short, easily comprehended proofs using the axiom. There are cases where the first proof discovered used the axiom of choice, and only later was a proof found not using the axiom. In fact, the first proof of the Cantor-Bernstein theorem used the axiom, although the proof given here does not.

Aesthetic considerations motivate acceptance also. Frequently, the axiom of choice yields an organization of a theory that is easy to grasp and sits well with the intuitions. This is particularly true in set theory. For example, we shall use it in proving that given any sets A and B either $A \leqslant B$ or $B \leqslant A$.

In the remainder of this section, and in the next, we shall present some of the consequences of the axiom of choice.

Theorem 9.1. *Every set can be well ordered, i.e., for every set A there is a binary relation $<$ such that $<$ is a well ordering of A.*

The intuitive idea behind the proof that follows is this. A choice function f on $P(A) - \{\emptyset\}$ can be used to well-order parts of A as follows. Take $f(A)$ to be the least element a_0. The next element is $f(A - \{a_0\})$; call it a_1. The immediate successor of a_1 is $f(A - \{a_0, a_1\})$; call it a_2, and so on. We collect all orderings obtained in this way in a set $\mathcal{K}$ and show that any two members of $\mathcal{K}$ fit together in the sense that one is an initial segment of the other. From this it follows that $\bigcup \mathcal{K}$ is a well ordering of A.

PROOF: Let f be a choice function on $P(A) - \{\emptyset\}$. Let $\mathcal{K}$ be the set of all well orderings $<_B$ where $<_B$ has field B for some $B \subseteq A$ and $b = f(A - \{x : x <_B b\})$ for all $b \in B$. (So given an initial $<_B$ segment Y, f picks the next element in the $<_B$ ordering from $A - Y$.) We show that $\mathcal{K}$ satisfies the hypotheses of Theorem 8.12 and then that $\bigcup \mathcal{K}$ well-orders A.

First notice that $\mathcal{K} \neq \emptyset$, since the empty relation well-orders $\{f(A)\}$.

Now suppose that $<_B$ and $<_C$ belong to $\mathcal{K}$. By Theorem 8.11 we can assume that we have an order preserving function $S: B \to C$ such that $\operatorname{Ran} S$ is an initial segment of C. We claim that $S(x) = x$ for all $x \in B$. If not, there is a least b (in the sense of $<_B$) such that $S(b) \neq b$, say $S(b) = d$. But $d = f(A - \{x : x <_C d\})$, and $x <_C d$ implies $S^{-1}(x) <_B b$, so $S^{-1}(x) = x$. Hence $\{x : x <_C d\} = \{x : x <_B b\}$, and $d = b$—a contradiction. Therefore $S(x) = x$ for all $x \in B$, and $<_C$ is an initial segment of $<_B$. We now know that $\mathcal{K}$ satisfies the hypotheses of Theorem 8.12, and hence $\bigcup \mathcal{K}$ is a well ordering; call it $<$.

Now let b belong to the field of $<$. Then for some $<_B \in \mathcal{K}$, $b \in B$. Hence $b = f(A - \{x : x <_B b\})$, and so $b = f(A - \{x : x < b\})$. Therefore $< \in \mathcal{K}$.

Let A^* be the field of $<$. We claim that $A^* = A$. For if not, we can add $d = f(A - A^*)$ at the top to extend $<$ to $<^{\#}$, i.e., we define $x <^{\#} y$ if $x < y$ when both $x, y \in A^*$, and $x <^{\#} d$ when $x \in A^*$. Clearly $<^{\#} \in \mathcal{K}$, and so $d \in A^*$—a contradiction. Therefore $<$ well-orders A. □

Recall that we postponed the proof of Theorem 5.4v, which states that for any sets A and B, either $A \leqslant B$ or $B \leqslant A$. Assuming the axiom of choice, we can now supply the proof. By Theorem 9.1, there is a well ordering $<_A$ on A and a well ordering $<_B$ on B. Hence by Theorem 8.11 there is an order preserving function f on A into B or on B into A. Since f is 1-1, we have $A \leqslant B$ or $B \leqslant S$ as required. □

1.9 The Axiom of Choice

The next theorem is also a consequence of the Axiom of Choice and is used widely throughout algebra and analysis.

Definition 9.2. Let $(A, <)$ be a partially ordered structure. A subset X of A is a *chain* if $(X, < \restriction X)$ is a linear ordering. A chain X is *maximal* if no chain properly contains X.

Theorem 9.3 (Maximal Principle). *Every partially ordered structure has a maximal chain.*

PROOF: Let $(A, <)$ be a partially ordered structure. By Theorem 9.1 there is a well ordering (call it $<^*$) on A. Since we need to distinguish between the two orderings we now have on A, we shall write '$<^*$-least element' when we mean 'least element with respect to the ordering $<^*$', and so on.

Let a^* be the $<^*$-least element of A. Take K to be the set of all functions f such that $\mathrm{Dom} f$ is a $<^*$-initial segment of A and $\mathrm{Ran} f$ is a $<$-chain, and for each $a \in \mathrm{Dom} f$

i. $f(a) = a$ if $\{a\} \cup \{f(b): b <^* a\}$ is a $<$-chain, and
ii. $f(a) = a^*$ if $\{a\} \cup \{f(b): b <^* a\}$ is not a $<$-chain.

First we claim that if $f, g \in K$, then $f \subseteq g$ or $g \subseteq f$. For if not, then there is a $<^*$-least $c \in \mathrm{Dom} f \cap \mathrm{Dom} g$ such that $f(c) \neq g(c)$, which is impossible, since $\{f(b): b <^* c\} = \{g(b): b <^* c\}$.

Hence $\bigcup K$ is a function; call it F. $\mathrm{Dom} F$ is a $<^*$-initial segment by Lemma 8.10. It is also easy to see that F satisfies the other conditions for membership in K, so $F \in K$.

We claim that $\mathrm{Ran} F$ is a maximal chain. For if not, let d be the $<^*$-least element of $A - \mathrm{Ran} F$ such that $\{d\} \cup \mathrm{Ran} F$ is a chain. Define

$F^*(b) = b$ for all $b \in \mathrm{Dom} F$,
$F^*(b) = a^*$ for all $b \in \{x : x <^* d$ and $x \notin \mathrm{Dom} F\}$,
$F^*(d) = d$.

It is easy to see that $F^* \in K$, so that $d \in \mathrm{Ran} F^* \subseteq \mathrm{Ran} F$, contradicting the choice of d. Hence $\mathrm{Ran} F$ is a maximal chain in $(A, <)$. □

Definition 9.4. A partially ordered structure $(A, <)$ is a *tree* if for all $a \in A$, $\{b : b < a\}$ is $<$-well ordered.

Trees are frequently encountered in many branches of mathematics. An amusing application to game theory will be found in Exercise 10. The following theorem is particularly useful and has led to generalizations that are still being investigated.

Theorem 9.5 (König's Infinity Lemma). *Suppose $(A, <)$ is a tree such that A is infinite but each $a \in A$ has only finitely many immediate successors. Then $(A, <)$ has an infinite chain.*

PROOF: Well-order A by $<^*$. Let $a^> = \{b : b > a\}$. Now by recursion on $\mathbf{N}$ define a_n to be the $<^*$-least $a \in A$ such that a is an immediate successor of a_{n-1} and $a^>$ is infinite. It is then easy to see that $\{a_n : n \in \mathbf{N}\}$ is an infinite chain. □

EXERCISES FOR §1.9

1. Prove: If $f : A \underset{\text{onto}}{\to} B$, then $B \leqslant A$.

2. Let $(A, <)$ be a linearly ordered structure. An infinite descending chain in $(A, <)$ is a countable subset $\{a_n : n \in \mathbf{N}\}$ of A such that $a_{n+1} < a_n$ for each $n \in \mathbf{N}$. Show that $<$ is a well ordering of A iff there is no infinite descending chain in $(A, <)$.

3. Show that the axiom of choice is equivalent to the following statement: If for each $X, Y \in A$ either $X = Y$ or $X \cap Y = \emptyset$, then there is a set B such that for each $X \in A$, $X \cap B$ has exactly one element.

4. Deduce the axiom of choice from the well ordering principle.

5. Prove that the following statements are equivalent:
   (a) A is infinite.
   (b) A is equinumerous to a proper subset of itself.
   (c) $\mathbf{N} \leqslant A$.

6. The maximal principle implies that every set can be well ordered. [*Hint*: Let F be the set of well orderings that have a domain which is a subset of X. If $<_1$ and $<_2$ belong to F and $<_1$ is an initial segment of $<_2$, write $<_1 \triangle <_2$. F is partially ordered by $\triangle$. Let B be a maximal branch in F. Then $\bigcup B$ well-orders X.]

7. The maximal principle implies the axiom of choice. Of course this is immediate from Exercises 4 and 6, but give a direct proof along the following lines. Let X be a set of non-empty sets, and let F be the set of all functions f such that $\text{Dom} f \subseteq X$ and f is a choice function on its domain. Now partially order F by proper set inclusion and consider a maximal branch.

8. Let G be a group, and let $x \in G$, x not the identity. Then G contains a subgroup H such that $x \notin H$ and no subgroup K of G is such that $x \notin K$ and $H \subset K$.

9. Prove that every vector space has a basis.

10. Let G be a two player game in which the players move alternately, each player has only finitely many alternatives on each move, and each play of the game ends in a win for one or the other after finitely many steps. Show that either the first player has a strategy which guarantees a win for every play of the game, or the second player has such a strategy. [*Hint:* Consider a tree in which the first level consists of the finitely many moves that player 1 can make. For each point at level 1 let the immediate successors (at level 2) of that point be the possible moves of player 2 in response to this particular move by player 1. Construct the third level analogously and continue. Now apply the infinity lemma.]

1.10 Transfinite Numbers

The size of a finite set is the natural number that is equinumerous to it. What is the size of an infinite set? What we want is a generalization of the notion of number so that each set will be equinumerous to some number, finite or transfinite. In order to obtain a hint on how to proceed, we first turn our attention to a definition of the set of finite numbers and some of its set theoretic consequences. The definition we adopt is due to von Neumann.

The idea is to define 0 to be $\emptyset$, 1 to be $\{0\}$, 2 to be $1 \cup \{1\}$ (in other words $\{\emptyset, \{\emptyset\}\}$), 3 to be $2 \cup \{2\}$ (i.e., $\{\emptyset, \{\emptyset\}\}, \{\{\emptyset, \{\emptyset\}\}\}$, and so on. In general $n+1$ is $n \cup \{n\}$.

More precisely, we take **N** to be the smallest set X (in the sense of $\subseteq$) such that

i. $\emptyset \in X$,

ii. whenever $x \in X$, then $x \cup \{x\} \in X$.

This definition justifies "proofs by induction." To show that every element of **N** has property P, one need only show that 0 has the property, and whenever n has the property, so does $n \cup \{n\}$.

This definition has several pleasant features. The $<$ relation among the natural numbers can be defined to be the $\in$ relation (see Theorem 10.3 below). The successor function is then $f(n) = n \cup \{n\}$. With this definition n has exactly n elements, namely $n = \{0, 1, \ldots, n-1\}$.

Elementary number theory is often given an axiomatic foundation based on the Peano axioms. These state that there is a set **N** and a unary function f (call it the successor function) with domain N such that:

i. f is 1-1.
ii. There is a unique element $0 \in \mathbf{N}$ such that $0 \notin \mathrm{Ran} f$.
iii. If $X \subseteq \mathbf{N}$ and $0 \in X$ and $f[X] \subseteq X$, then $X = \mathbf{N}$.

Another attractive feature of von Neumann's definition is the ease with which Peano's axioms are shown to be satisfied, taking $f(x)$ to be $x \cup \{x\}$ (see Exercise 1). One can then proceed in the usual way (see Exercises 2, 3, and 4) to define addition and multiplication of integers, and then to develop number theory.

Our next few theorems will tell us more about the set theoretic properties of **N**.

Definition 10.1. x is $\in$-*transitive* if whenever $z \in x$ and $y \in z$, then $y \in x$.

Theorem 10.2. *Each* $n \in \mathbf{N}$ *is* $\in$-*transitive, and so is* **N**.

PROOF: Let X be the set of $\in$-transitive members of **N**. Surely $0 \in X$. By clause ii of the definition of **N**, if we show that $x \cup \{x\} \in X$ whenever $x \in X$, then $X = \mathbf{N}$ and we are done. So let $x \in X$. Suppose that $y \in x \cup \{x\}$ and $z \in y$. We must show $z \in x \cup \{x\}$. But $y \in x \cup \{x\}$ implies $y \in x$ or $y = x$. Since x is $\in$-transitive, $y \in x$ and $z \in y$ gives $z \in x$. $y = \{x\}$ yields $z \in x$ immediately. Hence $x \cup \{x\}$ is $\in$-transitive. Hence each $n \in \mathbf{N}$ is $\in$-transitive.

Now consider the set K of all $x \in \mathbf{N}$ such that $y \in x$ implies $y \in \mathbf{N}$. Clearly $0 \in K$. Suppose $x \in K$, and let $y \in x \cup \{x\}$. Either $y \in x$ or $y = x$; in either case $y \in K$. Hence by the definition of $\mathbf{N}$, $K = \mathbf{N}$, and so $\mathbf{N}$ is $\in$-transitive. □

Theorem 10.3. *Each $n \in \mathbf{N}$ is well ordered by $\in$, and so in $\mathbf{N}$.*

PROOF: The set X consisting of those members of $\mathbf{N}$ that are well ordered by $\in$ certainly contains 0. Suppose $x \in X$. We need to see that $x \cup \{x\} \in \mathbf{N}$. $x \notin x$, for otherwise $\{x\}$ would be a subset of x having no $\in$-minimal member, contrary to the assumption that $\in$ well-orders x. A similar argument shows that there is no $y \in x$ such that $x \in y$. Also, if $z \in y$ and $y \in x$, then $z \in x$ by the preceding lemma. These observations together with our assumption that $\in$ well-orders x shows that $\in$ linearly orders $x \cup \{x\}$. Now let y be a non-empty subset of $x \cup \{x\}$. If $y \cap x \neq \varnothing$, then y has an $\in$ least member, since $\in$ well-orders x. If $y \cap x = \varnothing$, then $y = \{x\}$, and x is the $\in$-minimal member (we already noticed that $x \notin x$). Thus every subset of $x \cup \{x\}$ has an $\in$-minimal member, and so $x \cup \{x\}$ is well ordered by $\in$. Using this, similar considerations show that $\in$ well-orders $\mathbf{N}$ also. □

What about extending the sequence $1, 2, 3, \ldots$ to include "infinite numbers"? Since $n = \{0, 1, \ldots, n-1\}$ for each $n \in N$, a reasonable candidate for the first number greater than each "finite number" is $\mathbf{N}$ itself, $\{0, 1, 2, \ldots\}$. Then why not go on and consider $\mathbf{N} \cup \{\mathbf{N}\}$ a number, the immediate successor of $\mathbf{N}$? Calling this number $\mathbf{N} + 1$, we go on to the successor of $\mathbf{N} + 1$, namely $\mathbf{N} + 1 \cup \{\mathbf{N} + 1\}$, which we call $\mathbf{N} + 2$. Continuing, we get $\mathbf{N} + 3, \mathbf{N} + 4$, and so on. Continuing, let the number immediately following $\mathbf{N}, \mathbf{N} + 1, \mathbf{N} + 2, \ldots$ be $\{0, 1, 2, \ldots, \mathbf{N}, \mathbf{N} + 1, \mathbf{N} + 2, \ldots\}$ which we can call $\mathbf{N} \cdot 2$. Then continue from there, letting $\mathbf{N} \cdot 2 + 1 = \mathbf{N} \cdot 2 \cup \{\mathbf{N} \cdot 2\}$, and so on. This begins the sequence of ordinal numbers, but what we need is an explicit way of defining them all.

It is tempting to define the ordinals as being the members of the least set X such that

i. $0 \in X$,
ii. $x \in X$ implies $x \cup \{x\} \in X$,
iii. $x \subseteq X$ implies $\cup x \in X$.

Unfortunately, such a set X is an impossibility, as we shall see in §1.11.

Although this approach fails, there is a satisfactory alternative. Rather than attempting to define the collection of all ordinals, we define the property of being an ordinal, taking as our guideline Theorems 10.2 and 10.3.

1.10 Transfinite Numbers

Definition 10.4. α is an *ordinal* if

i. α is $\in$-transitive, and
ii. α is well ordered by $\in$.

We shall use $\alpha, \beta, \gamma, \delta$ to denote ordinals, and we write Ord α as an abbreviation of 'α is an ordinal'. We shall frequently write $\alpha < \beta$ instead of $\alpha \in \beta$.

By Theorems 10.2 and 10.3, each $n \in \mathbf{N}$ is an ordinal. Other examples are given in the following.

Theorem 10.5.

i. Ord $\mathbf{N}$.
ii. Ord α *implies* Ord$(\alpha \cup \{\alpha\})$.
iii. *If* $\alpha \in \beta$ *and* Ord β, *then* Ord α.

PROOF OF i. This is immediate by Theorems 10.2 and 10.3. □

PROOF OF ii. See the proof of Theorem 10.3. □

PROOF OF iii. Suppose $\alpha \in \beta$ and Ord β. Let $x \in y$ and $y \in \alpha$. Since β is $\in$-transitive, $y \in \beta$, and hence so is x. Since β is linearly ordered by $\in$, $x \in \alpha$. Hence α is $\in$-transitive. Since $\alpha \subseteq \beta$ by the $\in$-transitivity of β, it follows that α is well ordered by $\in$. Hence Ord α. □

From Theorem 10.5iii we have immediately

Corollary 10.6. *If* α *is an ordinal, then* $\alpha = \{\beta : \beta \in \alpha$ *and* Ord $\beta\}$.

As we mentioned before, one cannot consistently talk about the set of all ordinals. (We shall consider this problem again in §1.11.) Hence, the $\in$-relation restricted to the ordinals cannot be considered a set either, since then the domain of $\in$ restricted to the ordinals would be a set, but this is the collection of all ordinals. Nevertheless, $\in$ is essentially a well ordering of the ordinals, as we now show.

Theorem 10.7. *Let* α, β, γ *be ordinals. Then*:

i. $\alpha \notin \alpha$.
ii. $\alpha \in \beta$ *implies* $\beta \notin \alpha$.
iii. $\alpha \in \beta$ *and* $\beta \in \gamma$ *implies* $\alpha \in \gamma$.
iv. *Either* $\alpha \in \beta$ *or* $\beta \in \alpha$ *or* $\alpha = \beta$.
v. *If X is a set of ordinals and* $X \neq \varnothing$, *then X has an $\in$-least element*.

PROOF OF i AND ii. If $\alpha \in \alpha$ or if $\alpha \in \beta$ and $\beta \in \alpha$, then $\{\alpha\}$ is a non-empty subset of α with no $\in$ minimal member, contrary to the assumption that α is an ordinal. □

PROOF OF iii. Clearly, since Ord γ, γ is $\in$-transitive. □

PROOF OF iv. By Theorem 8.11 we can assume that there is an order preserving function $f:\alpha\to\Gamma$ where Γ is an initial segment of β (or f takes β onto an initial segment of α). We show that f is the identity function. If not, then there is an $\in$-minimal element δ of α such that $f(\delta)\neq\delta$. But $f(\delta)=\{f(\xi):\xi\in\delta\}$ by our assumption on f and Corollary 10.6, and $\{f(\xi):\xi\in\delta\}=\{\xi:\xi\in\delta\}=\delta$. Hence f is the identity function. If $\operatorname{Ran} f=\beta$, then $\alpha=\beta$. If not, let γ be the least element in $\beta-\operatorname{Ran} f$. Clearly $\Gamma=\gamma$, and since f is the identity on α, $\gamma=\alpha$. So $\alpha\in\beta$. □

PROOF OF v. Suppose X is a set of ordinals and $\alpha\in X$. If $\alpha\cap X\neq\varnothing$, then it has an $\in$-least member β, since α is well ordered by $\in$. Clearly β is the $\in$-least member of X. If $\alpha\cap X=\varnothing$, then α is the $\in$-least member of X. □

We may write $\alpha+1$ instead of $\alpha\cup\{\alpha\}$. The notation is suggested by the fact that $n+1=n\cup\{n\}$ for $n\in\mathbf{N}$ with '+' the usual addition, and the following.

Corollary 10.8. i. $\alpha+1$ *is the immediate successor of* α, *i.e.*, $\alpha+1$ *is the least ordinal* β *such that* $\alpha\in\beta$.

ii. *If X is a set of ordinals then* $\operatorname{Ord}\bigcup X$.

PROOF OF i. Certainly $\alpha\in\alpha+1$, and by Theorem 10.5ii, $\alpha+1$ is an ordinal. Suppose $\gamma\in\alpha+1$. Then $\gamma\in\alpha$ or $\gamma=\alpha$ by the definition of $\alpha+1$. □

PROOF OF ii. Let X be a set of ordinals, and let $z=\bigcup X$. Suppose that $x\in y$ and $y\in z$. For some ordinal $\alpha\in X$, we have $y\in\alpha$. Since α is $\in$-transitive, $x\in\alpha$. Hence $x\in z$. Thus z is $\in$ transitive. Suppose $s,t,u\in z$. Then for some $\alpha,\beta,\gamma\in X$ we have $s\in\alpha$, $t\in\beta$, $u\in\gamma$. Since the ordinals are $\in$-transitive and linearly ordered by $\in$, we have that $s,t,u\in\delta$, where δ is the largest ordinal among α,β,γ. It then follows that $s\notin s$, $s\in t$ implies $t\notin s$; $s\in t$ and $t\in u$ implies $s\in u$; and either $s\in t$ or $t\in s$ or $s=t$. Hence z is linearly ordered by $\in$. Now let w be a non-empty subset of z. Then $w\cap\alpha\neq\varnothing$ for some $\alpha\in X$. It follows that the $\in$-minimal member of $w\cap\alpha$ is the $\in$-minimal member of w. Hence z is well ordered by $\in$. □

Given any well-ordered structure $(A,<)$, there is an ordinal α such that the well ordering $\in\restriction\alpha$ has the same mathematical properties as the ordering $<$. To make this more precise we need the following definition.

We say that two partial orderings $(A,<_A)$ and $(B,<_B)$ are *isomorphic* if there is a function $f:A\xrightarrow[\text{onto}]{1\text{-}1} B$ such that for each $a,a'\in A$ we have $a<_A a'$ iff $f(a)<_B f(a')$. We call such a function an *isomorphism*. We write $(A,<_A)\cong(B,<_B)$ if $(A,<_A)$ is isomorphic to $(B,<_B)$. A structure that is isomorphic to $(A,<_A)$ can be thought of as a sort of copy of $(A,<_A)$ that will have all of the mathematical properties of $(A,<_A)$ that depend only on the ordering $<_A$ and not on the nature of the elements of A (see Exercise 9). In this sense, every well-ordered structure is a copy of some ordinal, as we now show.

1.10 Transfinite Numbers

Theorem 10.9. *if $(A, <)$ is a well-ordered structure, then there is an ordinal α such that $(A, <) \cong (\alpha, \in \restriction \alpha)$.*

PROOF: Suppose that X and Y are initial segments of A with $X \subseteq Y$, and that f_X is an isomorphism on X onto $\alpha \in \text{Ord}$ and f_Y is an isomorphism on Y onto $\beta \in \text{Ord}$. It is easy to see that $f_Y \restriction X$ is an isomorphism on X onto some ordinal. By Theorem 8.11 we must have $f_Y \restriction X = f_X$. Hence $f_X \subseteq f_Y$.

Now let F be the set of all isomorphisms f such that $\text{Dom} f$ is an initial segment of A and $\text{Ran} f \in \text{Ord}$. By Theorems 8.12 and 10.8ii, $\bigcup F$ is an isomorphism f^* on an initial segment X^* of A onto some ordinal α^*. We claim that $X^* = A$. If not, let a^* be the least element in $A - X^*$. Now extend f^* to an isomorphism $f^\#$ on $X^* \cup \{a^*\}$ onto $\alpha^* + 1$ by defining $f^\# = f^* \cup \{(a^*, \alpha^*)\}$. Clearly $f^\# \in F$, so that $a^* \in \text{Dom} f^*$, contradicting the choice of a^*. Therefore $X^* = A$, and $(A, <)$ is isomorphic to $(\alpha^*, \in \restriction \alpha^*)$. □

Often an argument will be divided into cases, one corresponding to well orderings with a maximal element and one to well orderings without. An ordinal α with a maximal element β is called a *successor ordinal* because $\alpha = \beta + 1$; otherwise α is a *limit ordinal*. Notice that if α is a limit ordinal, then $\alpha = \bigcup \alpha$.

The inductive definitions given for addition and multiplication of natural numbers can be extended to all the ordinals as follows.

Addition of ordinals:

$$\alpha + 0 = \alpha,$$
$$\alpha + (\beta + 1) = (\alpha + \beta) + 1,$$
$$\alpha + \lambda = \bigcup_{\beta \in \lambda} (\alpha + \beta) \text{ if } \lambda = \bigcup \lambda.$$

Multiplication of ordinals:

$$\alpha \cdot 0 = 0,$$
$$\alpha \cdot (\beta + 1) = (\alpha \cdot \beta) + \alpha,$$
$$\alpha \cdot \lambda = \bigcup_{\beta \in \lambda} (\alpha \cdot \beta) \text{ if } \lambda = \bigcup \lambda.$$

Exponentiation of ordinals:

$$\alpha^0 = 1,$$
$$\alpha^{(\beta+1)} = \alpha^\beta \cdot \alpha,$$
$$\alpha^\lambda = \bigcup_{\beta \in \lambda} \alpha^\beta \text{ if } \lambda = \bigcup \lambda.$$

Many of the familiar properties that the arithmetic operations have when applied to finite numbers no longer hold in the more general setting. For example, $1 + \omega = \omega \neq \omega + 1$, and $\omega \cdot 2 = \omega + \omega \neq \omega - 2 \cdot \omega$, and $(1 + 1)\omega = 2 \cdot \omega = \omega \neq \omega + \omega$. Hence, the extended operations are not commutative or distributive. On the other hand, many of the familiar properties of plus and times do carry over (see Exercises 7 and 8). Although there are many interesting theorems about ordinal arithmetic, they are outside our main interest, and we go on to consider another notion of number.

The finite numbers have an attribute that some of the ordinals do not have. No two members of **N** are equinumerous, and so each finite set has a unique number associated with it, its magnitude. However, this is not a property that all ordinals enjoy. In fact, as we shall see, if α is infinite and β is no greater than α, then $\alpha \sim \alpha + \beta \sim \alpha \cdot \beta$. However, there is a more restrictive notion of number that does not have this defect.

Definition 10.10. A *cardinal* is an ordinal that is not equinumerous to any smaller ordinal. We write $\mathrm{Card}\, x$ if x is a cardinal.

We shall use the letters κ, λ, μ and ν to denote cardinals.

Clearly, since each cardinal is an ordinal, $\in$ well-orders the cardinals in the sense of Theorem 10.7.

Theorem 10.11.

i. *Each $n \in \mathbf{N}$ is a cardinal, and $\mathbf{N}$ is also.*
ii. *If κ is an infinite cardinal, then $\{\alpha : \mathrm{Ord}\, \alpha \text{ and } \alpha \leqslant \kappa\}$ is a cardinal, and in fact is the smallest cardinal larger than κ.*
iii. *If x is a set of cardinals, then $\bigcup x$ is a cardinal.*

PROOF OF i is clear. □

PROOF OF ii. Since $\{\alpha : \mathrm{Ord}\, \alpha \text{ and } \alpha \leqslant \kappa\}$ is $\in$-transitive and well ordered, it is an ordinal β. Since we cannot have $\beta \in \beta$, we must have $\kappa \prec \beta$ and $\beta \not\sim \alpha$ for any $\alpha \in \beta$. Hence β is a cardinal greater than κ. If $\lambda \in \beta$, then $\lambda \leqslant \kappa$, so β is the least cardinal greater than κ. □

PROOF OF iii. Let x be a set of cardinals. Then $\bigcup x$ is an ordinal by Corollary 10.8ii. Now suppose that $\alpha \in \bigcup x$ and $\alpha \sim \bigcup x$. But then $\alpha \in \kappa$ for some $\kappa \in x$. Also, by the $\in$-transitivity of κ, $\alpha \subseteq \kappa$. Hence, by the Cantor-Bernstein theorem (actually Theorem 6.1) we see that $\alpha \sim \kappa$—impossible, since $\mathrm{Card}\, \kappa$. Hence $\bigcup x$ is a cardinal. □

When viewed as a cardinal, **N** is usually denoted by ω or ω_0 or $\aleph_0$. With each ordinal α associate a cardinal ω_α such that ω_α is the immediate cardinal successor of ω_β if $\alpha = \beta + 1$ and $\omega_\alpha = \bigcup_{\beta \in \alpha} \omega_\beta$ if $\alpha = \bigcup \alpha$. $\aleph_\alpha$ is often used in place of ω_α.

We use κ^+ to denote the cardinal successor of κ. If $\kappa = \lambda^+$ for some λ, then κ is called a *successor cardinal*; otherwise κ is a *limit cardinal*.

As a consequence of the axiom of choice, every set has its magnitude.

Theorem 10.12. *For every x there is a unique cardinal κ such that $x \sim \kappa$.*

1.10 Transfinite Numbers

PROOF: Let $<$ be a well ordering of x. By Theorem 10.9, $(x, <)$ is isomorphic to $(\alpha, \in \restriction \alpha)$ for some ordinal α. Let κ be the least ordinal such that $\kappa \sim \alpha$. Then $x \sim \kappa$. Of course, there can be no other cardinal λ such that $x \sim \lambda$, since this implies $\kappa \sim \lambda$, which in turn implies that either κ or λ is not a cardinal. □

Definition 10.13. If $x \sim \kappa$ and Card κ, we write $c(x) = \kappa$ and say that the cardinality of x is κ.

Although cardinals are ordinals, the arithmetic of ordinals does not specialize to an arithmetic for cardinals. For example, the ordinal sum of two cardinals need not be a cardinal, as is the case for $\omega + \omega$. Similarly for ordinal multiplication of cardinals: $\omega \cdot 2$ is not a cardinal. This suggests that we should define the cardinal sum of κ and λ to be $c(\kappa + \lambda)$, where '+' here is ordinal addition; and define the cardinal product of κ and λ to be $c(\kappa \cdot \lambda)$ where '·' is ordinal multiplication. These definitions are unduly complicated. Equivalent (see Exercise 9) but simpler definitions that do not depend on those for ordinal addition and multiplication are the following.

Definition 10.14.
  i. The cardinal sum of κ and λ is $c(\kappa \cup \{(0,\alpha): \alpha \in \lambda\})$.
  ii. The cardinal product of κ and λ is $c(\kappa \times \lambda)$.

We denote the cardinal sum of κ and λ and the cardinal product of κ and λ by $\kappa + \lambda$ and $\kappa \cdot \lambda$.

Although we are using '+' and '·' to denote the ordinal addition and multiplication as well as cardinal addition and multiplication, it should be clear which is intended by the context and by our convention of using $\kappa, \lambda, \mu, \nu$ for cardinals and $\alpha, \beta, \gamma, \delta$ for ordinals.

Much of cardinal arithmetic is extremely simple because of the following fact.

Lemma 10.15. If κ is infinite, then $\kappa \cdot \kappa = \kappa$.

PROOF: Suppose that the theorem is true for all $\lambda \in \kappa$. Let $X = \kappa \times \kappa$, let $X_\alpha' = \{(\alpha, \beta): \beta \leq \alpha\}$, let $X_\alpha'' = \{(\beta, \alpha): \beta < \alpha\}$, and let $X_\alpha = X_\alpha' \cup X_\alpha''$. Clearly $X = \bigcup_{\alpha \in \kappa} X_\alpha$. Next we need a particular well ordering of X, which the diagram in Figure 10.1 will help explain.

If $x, y \in X$, say that $x <^* y$ iff one of the following holds:

  i. $x \in X_\alpha, y \in X_\beta$ and $\alpha \in \beta$.
  ii. $x \in X_\alpha', y \in X_\alpha''$.
  iii. $x, y \in X_\alpha'$ and $x = (\alpha, \beta)$ and $y = (\alpha, \gamma)$ and $\beta < \gamma$.
  iv. $x, y \in X_\alpha''$ and $x = (\beta, \alpha)$ and $y = (\gamma, \alpha)$ and $\beta < \gamma$.

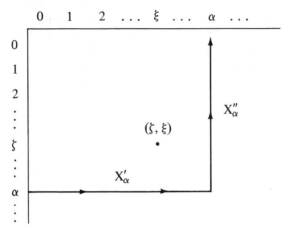

Figure 10.1

Since X'_α and X''_α are well ordered by $<^*$, so is X_α (by Theorem 8.6), and hence so is X (by Theorem 8.6 again). We show that $(X, <^*)$ is isomorphic to $(\kappa, \in \restriction \kappa)$.

First let S be a proper initial segment of X. Then $X-S$ has a least element x^*. Let $x^* \in X_{\alpha^*}$. Then $S \subseteq (\alpha^*+1) \times (\alpha^*+1) = \{(\gamma, \delta): \gamma, \delta \in \alpha^*+1\}$. Let $Y = (\alpha^*+1) \times (\alpha^*+1)$. Since $Y \sim c(\alpha^*+1) \times c(\alpha^*+1)$, and since $c(\alpha^*+1) \cdot c(\alpha^*+1) = c(\alpha^*+1)$ by the induction hypotheses, we have $cS \in \kappa$. Hence, $<^*$ is a well ordering in which every initial segment has cardinality less than κ.

Let β be that ordinal such that $(\beta, \in \restriction \beta) \sim (X, <^*)$ (by Theorem 10.9). Clearly $\beta \notin \kappa$ (otherwise Cardκ is false, since $\beta \sim X \geqslant \kappa$). From what we have just shown, $\kappa \notin \beta$. Hence $\kappa = \beta$, and so $\kappa \sim X$ as needed. □

Theorem 10.16. *Let κ and λ be cardinals with $\omega \leqslant \kappa$ and $\lambda \leqslant \kappa$. Then*

i. $\kappa + \lambda = \lambda + \kappa = \kappa$, *and*
ii. *if $\lambda \neq 0$, then $\kappa \cdot \lambda = \lambda \cdot \kappa = \kappa$.*

PROOF: Since $\kappa + \lambda \sim \kappa \cup \{(0, \alpha): \alpha \in \lambda\} \sim \lambda \cup \{(0, \beta): \beta \in \kappa\} \sim \lambda + \kappa$, we have $\kappa + \lambda = \lambda + \kappa$. Also $\kappa \times \lambda \sim \lambda \times \kappa$, so that $\kappa \cdot \lambda = \lambda \cdot \kappa$. Since $\kappa \leqslant \kappa + \lambda \leqslant \kappa + \kappa \leqslant \kappa \cdot \kappa$ and $\kappa \leqslant \kappa \cdot \lambda \leqslant \kappa \cdot \kappa$, we have by Lemma 10.15 that $\kappa + \lambda = \kappa \cdot \lambda = \kappa$. □

Definition 10.17. Let F be a function with domain α. By $\prod_{\beta \in \alpha} F(\beta)$ we mean the set of all functions f such that Dom$f = \alpha$ and $f(\beta) \in F(\beta)$ for each $\beta \in \alpha$.

One can think of this definition as a generalization of the finite Cartesian product.

The next theorem is extremely useful in making computations with cardinals.

1.10 Transfinite Numbers

Theorem 10.18 (König's Lemma). *Suppose $\kappa_\beta < \lambda_\beta$ for each $\beta \in \alpha$. Then $\bigcup_{\beta \in \alpha} \kappa_\beta < c\prod_{\beta \in \alpha} \lambda_\beta$.*

PROOF: For each $\delta \in \bigcup_{\beta \in \alpha} \kappa_\beta$ let $H(\delta)$ be that function h in $\prod_{\beta \in \alpha} \lambda_\beta$ such that for all $\beta \in \alpha$, $h(\beta) = \delta$ if $\delta \in \kappa_\beta$ and $h(\beta) = 0$ otherwise. Clearly $H: \bigcup_{\beta \in \alpha} \kappa_\beta \xrightarrow{1-1} \prod_{\beta \in \alpha} \lambda_\beta$. We need only show that $\bigcup_{\beta \in \alpha} \kappa_\beta$ is not equinumerous to $\prod_{\beta \in \alpha} \lambda_\beta$. For suppose $G: \bigcup_{\beta \in \alpha} \kappa_\beta \xrightarrow{1-1} \prod_{\beta \in \alpha} \lambda_\beta$. Let $X_\beta = \{(G(\delta))(\beta) : \delta \in \kappa_\beta\}$. Then $X_\beta \subseteq \lambda_\beta$ and $\lambda_\beta - X_\beta \neq \emptyset$. Let $f(\beta)$ be the least element in $\lambda_\beta - X_\beta$. Then $f \in \prod_{\beta \in \alpha} \lambda_\beta$, but clearly $f \in \operatorname{Ran} G$. Hence G is not onto. Therefore $\bigcup_{\beta \in \alpha} \kappa_\beta < c\prod_{\beta \in \alpha} \lambda_\beta$. □

By analogy with finite arithmetic, one might conjecture that if $1 < \kappa_\beta \leq \lambda_\beta$ for all $\beta \in \alpha$, then $\bigcup \kappa_\beta < c\prod \lambda_\beta$, but this is false (see Exercise 13).

The definition of cardinal exponentiation does *not* follow the pattern used for addition and multiplication; κ^λ in the cardinal sense is not defined as the cardinality of κ^λ in the ordinal sense. Instead, the definition is motivated by the fact that for finite cardinals $m^n \sim {}^n m$.

Definition 10.19. If κ and λ are cardinals, then $\kappa^\lambda = c({}^\lambda \kappa)$.

Again our notation involves some ambiguity, since κ^λ now has two different meanings depending on whether ordinal exponentiation or cardinal exponentiation is used. For example, 2^ω in the ordinal sense is countable, but 2^ω in the cardinal sense is not. However, in the remainder of the text, *exponentiation will always mean cardinal exponentiation*.

The proof of the next lemma is trivial and so is omitted.

Lemma 10.20.

i. If $\lambda \geq \nu$, then $\kappa^\lambda \geq \kappa^\nu$.
ii. If $\kappa \geq \lambda$, then $\kappa^\nu \geq \lambda^\nu$.

Theorem 10.21.

i. $\kappa^\lambda \kappa^\nu = \kappa^{\lambda+\nu}$.
ii. $(\kappa^\lambda)^\nu = \kappa^{\lambda \cdot \nu}$.
iii. $\kappa^\lambda \nu^\lambda = (\kappa \cdot \nu)^\lambda$.

PROOF OF i. We consider only the case where $\lambda \geq \nu$ and $\lambda \geq \omega$ and $\kappa \geq 2$, since the other cases are analogous or trivial. By the lemma, $\kappa^\lambda \geq \kappa^\nu$ and $\kappa^\lambda \geq \omega$. By Theorem 10.16, $\kappa^\lambda \kappa^\nu = \kappa^\lambda$. Also $\lambda + \nu = \lambda$, so $\kappa^{\lambda+\nu} = \kappa^\lambda$. □

PROOF OF ii. Let $f \in {}^\nu({}^\lambda \kappa)$. Then for every $\alpha \in \nu$, $\beta \in \lambda$ we have $(f(\alpha))(\beta) \in \kappa$. Let H map f to the function g_f defined $g_f(\alpha, \beta) = (f(\alpha))(\beta)$. Then $H: {}^\nu({}^\lambda \kappa) \xrightarrow{1-1} {}^{\lambda \times \nu} \kappa$. It is also clear that each $g \in {}^{\lambda \times \nu} \kappa$ is g_f for some $f \in {}^\nu({}^\lambda \kappa)$, and so H is onto ${}^{\lambda \times \nu} \kappa$. Hence $(\kappa^\lambda)^\nu = \kappa^{\lambda \cdot \nu}$. □

PROOF OF iii. We consider the case where $\lambda \geq \omega$, since the other case is trivial. We may assume that $\kappa \geq \nu$. Then $\kappa^\lambda \nu^\lambda = \kappa^\lambda$ by the lemma and

Theorem 10.16. If $\kappa \geqslant \omega$, then $\kappa \cdot \nu = \kappa$ by Theorem 10.16, and so $(\kappa \cdot \nu)^\lambda = \kappa^\lambda$ also. If $\kappa < \omega$, then $(\kappa \cdot \nu)^\lambda \leqslant (\kappa^2)^\lambda = \kappa^\lambda$ by the lemma and part ii. □

Here are two calculations that make use of Lemma 10.20 and Theorem 10.21 and show that Lemma 10.20 does not hold if we replace '$\leqslant$' by '$<$' throughout.

i. $(2^\omega)^\omega = 2^{\omega \cdot \omega} = 2^\omega = (2^\omega)^1$ (compare with Lemma 10.20i).
ii. $2^\omega = 2^{\omega \cdot \omega} = (2^\omega)^\omega \geqslant \omega^\omega$, which along with $2^\omega \leqslant \omega^\omega$ gives $2^\omega = \omega^\omega$ (compare with Lemma 10.20ii).

As we shall see in the next sections, the value of κ^λ in relation to κ and λ is very mysterious. For example, while it is consistent to assume that $2^\omega = \omega_1$, it is also consistent to assume that $2^\omega = \omega_2$. In fact, it is known that if $\kappa > \omega$ and there is no countable subset X of κ such that $\bigcup X = \kappa$, then one can consistently assume that $2^\omega = \kappa$. No principle we have used so far determines the value of 2^ω.

We end this section by considering extensions of two of the most commonly used principles in mathematics, definition by recursion and proof by induction on ω. What makes these principles work is the fact that ω is well ordered. Therefore it is not surprising that these principles have useful extensions to transfinite ordinals. We shall consider several extensions here and in the problems. Others, which are more complicated to state, will not be mentioned. For the most part, they are easy enough to devise when the need arises, following the form of the versions given here.

Theorem 10.22 (Transfinite Induction). *Suppose that for each $\beta \in \alpha$, $\beta \subseteq X$ implies $\beta \in X$. Then $\alpha \subseteq X$.*

PROOF: If the hypotheses is true but the conclusion false for X and α, let γ be the least element in $\alpha - X$. But then $\gamma \subseteq X$, and so $\gamma \in X$—a contradiction. Hence no such γ exists, i.e., $\alpha \subseteq X$. □

This theorem is frequently stated in the following form.

Theorem 10.22'. *Suppose that $X \subseteq \alpha$ and the following are true*:

i. $0 \in X$.
ii. $\beta \in X$ and $\beta + 1 \in \alpha$ implies $\beta + 1 \in X$.
iii. $\beta \subseteq X$ and $\beta \in \alpha$ implies $\bigcup \beta \in X$.

Then $X = \alpha$.

PROOF: Suppose γ is the least ordinal in $\alpha - X$. Then $\gamma \neq 0$, γ is not a successor ordinal, nor is γ a limit ordinal; hence no such γ exists. □

The set X above is usually specified by a property ψ, i.e., X is given as $\{\beta : \psi(\beta)\}$.

Another useful form of transfinite induction is the following.

1.10 Transfinite Numbers

Theorem 10.22″. *Suppose that ψ is a property such that $\psi(\alpha)$ whenever $\psi(\beta)$ for all $\beta \in \alpha$. Then $\psi(\alpha)$ for every ordinal α.*

PROOF: Like that of Theorem 10.22. □

The difference between Theorems 10.22 and 10.22″ is that α has been replaced by Ord (which, as we will see in the next section, is not a set), and X has been replaced by the property ψ, which does not correspond to a set either. Theorem 10.22′ can be stated with analogous modifications.

Next we consider various transfinite generalizations of definition by recursion. Again, we shall not try to give the most comprehensive versions of the theorem, since these are more complicated to state, but follow the same general outline as the simpler versions.

Theorem 10.23 (*Definition by Transfinite Recursion*). *Let G be a function such that for each $\beta \in \alpha$ and each $s \in {}^\beta(\mathrm{Ran}\, G)$ we have $(\beta, s) \in \mathrm{Dom}\, G$. Then there is a unique function F such that $\mathrm{Dom}\, F = \alpha$ and for each $\beta \in \alpha$*

$$F(\beta) = G(\beta, s_\beta),$$

where $\mathrm{Dom}\, s_\beta = \beta$ and for each $\gamma \in \beta$, $s_\beta(\gamma) = F(\gamma)$.

PROOF: Let **F** be the set of all functions f such that $\mathrm{Dom}\, f = \alpha_f$ for some $\alpha_f \in \alpha$, and for each $\beta \in \alpha_f$, $f(\beta) = G(\beta, s_\beta^f)$, where $\mathrm{Dom}\, s_\beta^f = \beta$ and $s_\beta^f(\gamma) = f(\gamma)$ for each $\gamma \in \beta$. **F** is non-empty, since the empty function is a member and **F** is partially ordered by $\subseteq$. By Theorem 9.3, $(\mathbf{F}, \subseteq)$ has a maximal chain; say B is such. Let $F = \bigcup B$. Clearly F is a function and $F(\beta) = G(\beta, s_\beta)$ for each $\beta \in \mathrm{Dom}\, F$. We need only see that $\alpha = \mathrm{Dom}\, F$. If not, let δ be the least element in $\alpha - \mathrm{Dom}\, F$. It is obvious that $F \cup \{(\delta, G(\delta, s_\delta))\} \supset F$, where $s_\delta(\gamma) = F(\gamma)$ for each $\gamma \in \delta$. This contradicts the maximality of B, and so $\mathrm{Dom}\, F = \alpha$.

We now prove the uniqueness of F. Suppose F and F' are two different functions with the properties stated in the theorem. Let δ^* be the least δ such that $F(\delta^*) \neq F'(\delta^*)$. But then $s_{\delta^*} = s'_{\delta^*}$, where $s_{\delta^*}(\gamma) = F(\gamma)$ and $s'_{\delta^*}(\gamma) = F'(\gamma)$ for each $\gamma \in \delta^*$. Hence $F(\delta^*) = G(\delta^*, s_{\delta^*}) = F'(\delta^*)$, and so no such δ^* exists, i.e., F is unique. □

Another version of definition by transfinite recursion can be given, which is in the spirit of Theorem 10.22″, with Ord in place of α. To facilitate the statement of this version we need some definitions. A property ψ is said to be *functional* if for each x there is at most one y such that ψxy [i.e., such that (x,y) has the property ψ]. If ψ is functional and ψxy, we say that x is in the domain of ψ and y is in the range of ψ, and we write $\psi x = y$. Of course this notation is in complete agreement with that used for functions. The difference is that functions are sets, whereas some properties (such as $\mathrm{Ord}\, \alpha$, or $\beta = \alpha + 1$ and $\mathrm{Ord}\, \alpha$) do not correspond to sets, as we shall see in the next section.

Theorem 10.23'. *Suppose that ψ is a functional property such that (α, s) is in the domain of ψ whenever s is a function from α into the range of ψ. Then there is a unique functional property θ such that*
$$\theta(\alpha) = \psi(\alpha, s_\alpha)$$
whenever Ord α *and s_α is the function on α such that $s_\alpha(\gamma) = \theta(\alpha)$ for each $\gamma \in \alpha$.*

PROOF: Like that of Theorem 10.23. □

The content of this theorem for the time being is as vague as the notion of property which will not be defined precisely until §3.4.

EXERCISES FOR §1.10

1. Let $f(x) = x \cup \{x\}$. With von Neumann's definition of **N** show that Peano's axioms are satisfied.

2. Let $+$ be the least set X such that:
   i. $(n, 0, n) \in X$ for each $n \in \mathbf{N}$.
   ii. $(n, m, k) \in X$ implies $(n, m+1, k+1) \in X$ for each $n, m, k \in \mathbf{N}$.
   Write $n + m = l$ if $(n, m, l) \in X$. Show that X is a function with Dom $X = \mathbf{N} \times \mathbf{N}$. Show that $n + m = m + n$ for each $n, m \in \mathbf{N}$.

3. Let $\cdot$ be the least set X such that:
   i. $(n, 0, 0) \in X$ for each $n \in \mathbf{N}$.
   ii. $(n, m, k) \in X$ implies $(n, m+1, k+n) \in X$ for each $n, m, k \in \mathbf{N}$.
   Write $n \cdot m = l$ if $(n, m, l) \in X$. Show that X is a function with Dom $X = \mathbf{N} \times \mathbf{N}$. Show that $n \cdot m = m \cdot n$ for each $n, m \in \mathbf{N}$.

4. Prove that $n(m+k) = n \cdot m + k \cdot m$ for each $n, m, k \in \mathbf{N}$.

5. Prove that '$\cong$' is an equivalence relation, i.e., prove that:
   (a) $(A, <_A) \cong (A, <_A)$.
   (b) $(A, <_A) \cong (B, <_B)$ implies $(B, <_B) \cong (A, <_A)$.
   (c) $(A, <_A) \cong (B, <_B)$ and $(B, <_B) \cong (C, <_C)$ implies $(A, <_A) \cong (C, <_C)$.

6. Show that if $(A, <_A) \cong (B, <_B)$ and $<_A$ is a linear ordering, then so is $<_B$. Similarly for $<_A$ a dense ordering and a well ordering.

7. Show that $\alpha \cdot (\beta + \gamma) = \alpha \cdot \beta + \alpha \cdot \gamma$, where α, β, γ are ordinals. However, it is not always true that $(\alpha + \beta) \gamma = \alpha \cdot \gamma + \beta \cdot \gamma$.

8. Prove the following laws of exponents for ordinal exponentiation:
   $$(\alpha^\beta)^\gamma = \alpha^{\beta \cdot \gamma},$$
   $$\alpha^\beta \cdot \alpha^\gamma = \alpha^{\beta + \gamma}.$$
   However, it is not true that $\alpha^\gamma \cdot \beta^\gamma = (\alpha \cdot \beta)^\gamma$ for all ordinals α, β, γ. (Consider $2^\omega \cdot 2^\omega \stackrel{?}{=} 4^\omega$; we already know that $2^\omega \cdot 2^\omega = 2^{\omega + \omega}$.)

9. Show Definition 10.14 is equivalent to the definition for cardinal sum and product suggested in the preceding paragraph.

10. Suppose $cX = cX'$, $cY = cY'$, and X and Y are infinite. Show that $c(X \cup Y) = c(X' \cup Y')$, $c(X \times Y) = c(X' \times Y')$, and $c^Y X = c^{Y'} X'$.

11. Suppose κ, λ, and μ are cardinals. Show that $(\kappa + \lambda)\mu = \kappa \cdot \mu + \lambda \cdot \mu$.

12. Suppose that for each $n \in \omega$, $\kappa_n < 2^\omega$. Show that $\bigcup_{n \in \omega} \kappa_n < 2^\omega$.

13. A relation R is well founded if for each x either $\{y : yRx\} = \emptyset$ or there is a y^* such that $y^* R x$ and there is no z such that xRy^* and zRy^*. (Note that every well ordering is well founded.) Prove the following extension of Theorem 10.22: Suppose R is well rounded and that X has the property that whenever $x \in \text{Dom } R$ and $\{y : yRx\} \subseteq X$, then $x \in X$. Then $\text{Dom } R \subseteq X$.

14. Extend Theorem 10.22″ to well-founded properties along the lines of Exercise 13.

1.11 Paradise Lost, Paradox Found (Axioms for Set Theory)

One of the most appealing features of set theory is its generality. Its broad applicability stems from the fact that so many notions of classical mathematics can be formulated within it, so that set theory can be used as a foundation of all of classical mathematics. However the unbridled use of 'set' to refer any collection of objects quickly leads to mathematical nonsense.

For example, one cannot say that there is a set whose elements are exactly the ordinal numbers. Such a set would be well ordered by $\in$ (Theorem 10.7) and transitive (Theorem 10.5iii), and so would itself be an ordinal α. But then $\text{Ord } \alpha$ and $\alpha \in \alpha$, an impossibility. Exactly the same argument shows that no set contains all the cardinals.

A more direct example was given by Bertrand Russell. Say that a set x is normal if $x \notin x$; otherwise say that x is abnormal. Is there a set X whose members are precisely the normal sets? If so, then X is either normal or abnormal and not both. X is not normal, for if it were, then $X \in X$ (by the definition of X) and so would be abnormal. On the other hand, X is not abnormal, for if it were, then $X \in X$, and by the definition of X, X would be normal. Hence X is neither normal nor abnormal—a contradiction.

Of course, the set of all normal sets, if it existed, would contain all ordinals, and so this example can be reduced to the one preceding it. However, Russell's example does not use the notion of ordinal number, but merely set membership.

Exactly what is it about the assumptions 'Ord is a set' or 'the set of all normal sets exists' that leads to nonsense? This problem has been much discussed among philosophers and mathematicians for many decades, but no completely satisfactory answer has been given.

What should one do with set theory in view of these paradoxical 'sets'? Should we abandon its use as a notational and conceptual framework for classical mathematics? And what about Cantor's proof that most numbers are transcendental? Should that be abandoned because of the paradoxes even though none of the paradoxical "sets" are mentioned in the proof? In fact, the paradoxical "sets" of the examples never arise in any branch of classical mathematics, and the sets that do arise seem to be completely innocuous as far as giving rise to contradictions is concerned.

In the early part of this century, Russell and Whitehead in their *Prinicpia*, and Zermelo in a series of papers, attempted to axiomatize a significant portion of set theory in a way that would avoid the paradoxes. The axiomatization given by Zermelo and later modified by Fraenkel is the one most frequently encountered today. This axiomatization appears to be highly successful. The axioms have strong intuitive appeal, apparently asserting simple truths about sets. The axiomatization seems to be free of contradiction, and moreover, is strong enough to provide a base for all of classical mathematics. One axiom, the axiom of extensionality, says that a set is determined by its members. The remaining axioms either state that a certain set exists or that a set is obtained from a given set by a specified operation. In developing set theory within such an axiomatic framework, the only sets that can be asserted to exist are those that can be proven to exist by a valid argument whose only premises about sets are those given by the axioms.

The remainder of this section is devoted to the *Zermelo-Fraenkel* axiomatization (abbreviated ZF).

Axiom of Extensionality. If x and y have the same elements then $x = y$.

So this axiom says that a set is completely determined by its members.

Axiom of Regularity. Every non-empty set has an $\in$-least member, i.e., if there is some $y \in x$, then there is some $z \in x$ for which there is no $w \in z \cap x$.

We could have appealed to this axiom in Definition 10.4, the definition of ordinal. In view of the axiom of regularity, clause ii can be replaced by

ii. α is linearly ordered by $\in$.

This shortens the proofs of Theorem 10.5 and Corollary 10.6 somewhat, but we did not want to mention this axiom at that time.

Axiom of the Null Set. There is a set with no members.

This is what we have been calling $\emptyset$ or 0. Of course, the axiom of extensionality implies that there is only one such set.

Axiom of Pairing. If x and y are sets, then there is a set z such that for all w, $w \in z$ iff $w = x$ or $w = y$.

1.11 Paradise Lost, Paradox Found (Axioms for Set Theory)

This axiom says that for every x and y, $\{x,y\}$ exists.

Axiom of Union. For every x there is a y such that $z \in y$ iff there is a $w \in x$ with $z \in w$.

So this axiom says that if x exists, then so does $\bigcup x$.

Axiom of the Power Set. For every x there is a y such that for all z, $z \in y$ iff $z \subseteq x$.

In other words if x is a set, so is $P(x)$.

Axiom of Infinity. There is a set x such that $0 \in x$ and whenever $y \in x$, then $y \cup \{y\} \in x$.

This axiom assures us that there is a set containing ω. The next axiom will allow us to extract ω from such a set.

Axiom of Replacement. If P is a functional property and x is a set, then $\text{Ran}(P \restriction x)$ is a set, i.e., there is a set y such that for every z, $z \in y$ iff there is a $w \in x$ such that $P(w) = z$.

Recall from §1.10 that a functional property P is one such that for every x there is at most one y such that Pxy. In our examples of paradoxes the pathological sets are inordinately large. If y is a set obtained by replacement as the range of P restricted to x, then the magnitude of y is no greater than that of x, which provides some intuitive justification of replacement as an axiom.

Our statement of the axiom of replacement is somewhat sloppy in that the notion of a property is undefined. At this point we shall take it to mean any statement about sets mentioning only the $\in$-relation. A precise version of this axiom (and more elegant statements of the others) will be given in §3.4.

We used the axiom of replacement in proving Theorem 10.9, although an alternative proof can be given that avoids its use. In our proof we need to know that there is a set F whose elements are the f_a's for $a \in A$. Let $P(u,v)$ hold iff $u \in A$ and $v = f_u$, or $u \notin A$ and $v = \emptyset$. Then P is functional, and the range of P restricted to A is the needed set F. The axiom of replacement was used again in Theorem 10.11ii.

A very useful consequence of the axiom of replacement is the *axiom of separation*, which says that a definable subset of a set is a set. In other words, if x is a set and S a property then there is a set y whose elements are exactly those of x which satisfy S. To deduce the axiom of separation from replacement, first choose some $a \in x$ such that a has property S (if no such a exists, then the axiom of the null set gives $\{y : y \in x \text{ and } Sy\}$). Now let Puv be the property

Either $u \in x$ and Su and $v = u$, or $u \notin x$ and $v = a$, or not Su and $v = a$.

Clearly, for every u there is a unique v such that Puv. Now the axiom of replacement applied to x and this P gives the set $\{z : z \in x$ and $Sz\}$ immediately.

On the other hand, we will see in Exercise 7 of §1.12 that the axiom of separation does not imply replacement and hence is a weaker axiom.

Since the axiom of separation is a theorem of ZF as we have just seen, there is no need to include it as an additional axiom.

Zermelo's original axiomatization included separation but not regularity. Fraenkel later modified the axiomatization by deleting separation and adding replacement.

Separation has been used implicitly several times in previous sections. As another example of its use, we can prove that if x is a set, then $\bigcap x$ is a set. If $x = \emptyset$, then we are done, and if there is some $y \in x$, then separation gives us the existence of the set $\{z : z \in y$ and $z \in w$ for all $w \in x\}$, which is $\bigcap x$. Letting s be the set specified in the axiom of infinity, and letting x be the set of all $z \subseteq s$ such that $\emptyset \in z$ and whenever $u \in z$ then $u \cup \{u\} \in z$ (x exists by the axioms of power set and separation), we see that $\bigcap x$ is a set, so the axioms give the existence of ω.

The axiom of separation is strong enough to develop most of classical mathematics within our set theoretic framework. However, quite recently, Martin found an assertion in analysis that can be proved from replacement but not separation. The assertion is 'Every Borel set of reals is determined'. (The set of Borel sets is the smallest set containing the open intervals that is closed under complementation and countable unions. A set X is determined if one of the two players in the following game has a winning strategy: The players a and b move alternately beginning with the a player. Each chooses an integer between 0 and 9 inclusive; say a chooses a_i on the ith move and b chooses b_j on the jth. Then player a wins just in case $0.a_0 b_1 a_2 b_3 a \ldots \in X$.)

The axioms mentioned so far, including replacement, constitute the Zermelo-Fraenkel axiomatization, and as we have said are sufficient to form a foundation for classical mathematics. Moreover, the axiomatization is extremely elegant in that it can be stated in terms of $\in$ alone (see §3.4). On the other hand, there are statements of relatively recent vintage that have considerable mathematical interest but cannot be proved or disproved in ZF. Some of these have been considered as additional axioms. The one that has gained the broadest acceptance is the axiom of choice. The axiomatization obtained by adjoining the axiom of choice to the Zermelo-Fraenkel axioms will be denoted by ZFC.

At this point we want to stress that everything done in the preceding sections can be justified within ZFC, and most of it within ZF. For example, consider the notion of the ordered pair (x, y). The only property of the ordered pair that has mathematical significance is the following: If $(x, y) = (u, v)$, then $x = u$ and $y = v$. If we are to develop a suitable notion of ordered pair within our axiomatic framework, then (x, y) has to be given a

1.11 Paradise Lost, Paradox Found (Axioms for Set Theory)

definition in terms of x, y, and $\in$, and then using ZFC the ordered pair must be shown to exist and to have the requisite property. This we now do.

The axiom of pairing asserts that for every x and y there is a z such that $w \in z$ iff $w = x$ or $w = y$. By extensionality, there is only one z such that $w \in z$ iff $w = x$ or $w = y$. This z is, of course, the unordered pair of x and y, which we denote by $\{x,y\}$. For $x = y$ we write $\{x\}$ instead of $\{x,x\}$. Now define (x,y) to be $\{\{x\},\{x,y\}\}$. Another application of pairing gives

Theorem 11.1. *For every x and y, (x,y) exists.*

Theorem 11.2. *If $(x,y) = (r,s)$, then $x = r$ and $y = s$.*

PROOF: Suppose $(x,y) = (r,s)$. By extensionality, (x,y) and (r,s) have the same members. Since $\{x\} \in (x,y)$ [by definition of (x,y)], $\{x\} \in (r,s)$. Similarly $\{x,y\} \in (r,s)$. But $z \in (r,s)$ iff $z = \{r\}$ or $z = \{r,s\}$ [by definition of (r,s)]. The argument now breaks up into two cases.

Case 1: $\{x\} = \{r\}$. Then $x = r$ by extensionality. $\{x,y\} = \{r\}$ or $\{x,y\} = \{r,s\}$. The first implies that $y = r = x$ (extensionality again), and so $\{x,y\} = \{x\}$ and $(x,y) = \{\{x\}\}$. Since $\{r,s\} \in (x,y)$, extensionality implies that $\{r,s\} = \{x\}$ and so $s = x = y$ (extensionality), and we are done. If $\{x,y\} = \{r,s\}$, then $y = r$ or $y = s$ by extensionality. If $y = s$, we are done. If $y = r$, then $y = x$ and the argument just given shows $r = s = y = x$.

Case 2: $\{x\} = \{r,s\}$. Then by extensionality, $r = s = x$, so $(r,s) = \{\{x\}\}$. Since $\{x,y\} \in (r,s)$, extensionality implies $y = x = s$ and we are done.

Thus, arguing within ZF, we have shown that for every x and y the ordered pair (x,y) exists and has the requisite properties. □

Given x, let $\bigcup x$ be the unique y whose existence is assured by the union axiom (the unicity of y is a consequence of extensionality). The unique y of the power set axiom is denoted by $P(x)$.

Theorem 11.3. *For every X and Y, there is a Z such that for all z, $z \in Z$ iff there is an $x \in X$ and a $y \in Y$ such that $z = (x,y)$.*

PROOF: If X and Y are sets, so is $\{X, Y\}$ by the axiom of pairing. So $X \cup Y$ (i.e., $\bigcup \{X, Y\}$) exists by the axiom of union. Two applications of the power set axiom give us the existence of $P(P(X \cup Y))$. For every $x \in X$ and $y \in Y$ we have $\{x\} \in P(X \cup Y)$ and $\{x,y\} \in P(X \cup Y)$; hence $(x,y) \in P(P(X \cup Y))$. The axiom of replacement (in fact substitution is enough) gives us the existence of $\{(x,y) : x \in X, y \in Y,$ and $(x,y) \in P(P(X \cup Y))\}$.

Denoting the Z of the theorem by $X \times Y$, we have just proved in ZF that the Cartesian product of any two sets exists (i.e., is a set). □

We could now go on and repeat all the definitions and theorems of the preceding sections, justifying each step by an argument from the axioms of

ZFC. However, what we have just done should help convince the reader that such a task is a straightforward, albeit tedious, exercise. At the end of this section there are several problems that continue in this direction.

There are other statements that have been considered as axioms but none of them has gained the wide acceptance of the axiom of choice. Moreover, it has been shown by Gödel that if ZF is free of contradiction, then so is ZFC (we shall say more about this in the next section). With some of these other axioms, either no such proof of relative consistency can be given or else the axiom does not appeal to the intuitions as strongly as the axiom of choice (at least at this time) or are not as useful.

Perhaps the most famous of these axioms is the *generalized continuum hypothesis*, abbreviated GCH. This states that the cardinal successor of the cardinal κ is 2^κ. The special case for $\kappa = \omega$, namely the assertion $\omega_1 = 2^\omega$, is called the *continuum hypothesis* (CH). Since $2^\omega = cP(\omega)$ and $P(\omega) \sim \mathbf{R}$ (see §1.5), the CH says that every subset of $\mathbf{R}$ either is equinumerous to $\mathbf{R}$ or is countable. As concrete a statement as this seems, Gödel's work in 1938 and Cohen's work in 1963 show that CH and GCH cannot be proved or refuted within ZFC.

A very active area of investigation in the past few years has centered around axioms of infinity. Roughly speaking, these axioms assert the existence of extremely large cardinals. The axiom of infinity of ZF is an axiom of infinity and gives us the existence of ω, a cardinal which has the following two properties:

i. $n \in \omega$ implies $2^n \in \omega$, and
ii. if $n \in \omega$ and $m_0, m_1, \ldots, m_{n-1} \in \omega$, then $\bigcup_{i \in n} m_i \in \omega$.

It is natural to ask if there are any other cardinals κ that enjoy these properties, i.e., cardinals κ such that

i. $\lambda \in \kappa$ implies $2^\lambda \in \kappa$, and
ii. $\lambda \in \kappa$ and $\nu_\alpha \in \kappa$ for every $\alpha \in \lambda$ implies $\bigcup_{\alpha \in \kappa} \nu_\alpha \in \kappa$.

Such a κ is called a *strong inaccessible*, and the statement that there is a strong inaccessible other than ω we denote by IC. IC is an example of an axiom of infinity. We shall see in the next section that IC cannot be proved in ZFC even if we adjoin the GCH. However, there are many mathematicians who find this axiom appealing. Their point of view is that any axiom consistent with ZFC that enlarges the domain of set theory by introducing new sets should be accepted. The consistency of ZFC with IC seems highly likely, but it has been shown that a proof of relative consistency like that given for AC or GCH is impossible.

In recent years there has been a proliferation of axioms of infinity, asserting the existence of larger and larger cardinals, so that in comparison a strongly inaccessible cardinal is quite small. One such axiom asserts the existence of a weakly compact cardinal larger than ω. Let $[X]^n$ denote the set of n-element subsets of X, $[X]^n = \{\{x_1, \ldots, x_n\} : x_i \in X$ and $x_i \neq x_j$ if

$i \neq j$}. Say that κ is weakly compact if whenever $Y \subseteq [\kappa]^2$, then there is some $Z \subseteq \kappa$ such that $cZ = \kappa$ and either $[Z]^2 \subseteq Y$ or $[Z]^2 \subseteq [\kappa]^2 - Y$. ω is weakly compact, by a famous theorem of Ramsey (see Exercise 3). However, the axiom asserts the existence of a larger weakly compact cardinal. Moreover, it can be shown that if $\omega \in \kappa$ and κ is weakly compact, then κ is strongly inaccessible, but much larger than the first strongly inaccessible cardinal. No proof of relative consistency of this axiom with ZF can be given, since no such proof can be given for IC.

Far larger than the first uncountable weakly compact cardinal is the first measurable cardinal, a notion introduced by Ulam. Say that a function $f: P(\kappa) \to \{0,1\}$ is a two valued measure on κ if $f(\kappa) = 1$, and whenever $\lambda < \kappa$ and $\{X_\alpha : \alpha \in \lambda\}$ is a set of pairwise disjoint subsets of κ, then $f \bigcup_{\alpha \in \lambda} X_\alpha = \sum_{\alpha \in \lambda} f(X_\alpha)$. A cardinal κ is measurable if there is a measure f whose domain is κ.

Other cardinals have been considered which dwarf the first measurable cardinal (huge, super-huge, Vopěnka, etc.), and the game of finding ever larger axioms of infinity continues.

Before we close this section we want to mention several famous statements that, like the axiom of choice, are independent of ZF, i.e., can be neither proven nor refuted within ZF. The first concerns the reals with the usual ordering $(\mathbf{R}, <)$. This ordering is dense, is complete (every set having an upper bound has a least upper bound), and has the property that no uncountable set of intervals can be pairwise disjoint. Souslin's hypothesis asserts that any other ordering $(A, <)$ having these three properties is isomorphic to the reals. This statement, earthy as it seems, is independent not only of ZF but also of ZFC plus the GCH.

Another question, even more relevant to analysis, is the following. Is every set of reals measurable in the sense of Lebesgue? Lebesgue measurability is a standard notion that is treated in most texts dealing with real analysis, and we shall not discuss it here except to say that the Lebesgue measure assigns to certain subsets of the reals a length, and this notion of length generalizes the usual one as applied to, say, a union of disjoint intervals. In courses of real analysis, it is proved that sets exist that are not Lebesgue measurable, and these proofs invariably use the axiom of choice. In fact, choice must be used in a strong form, for Solovay in 1964 proved that the assertion 'all sets are Lebesgue measurable' is consistent with ZF and a restricted version of choice. This restricted version of choice, called the countable axiom of choice, asserts that every countable set of non-empty sets has a choice function. Countable choice and the assumption that all sets of reals have length in the sense of Lebesgue yields a very slick development of the early portion of real analysis.

Finally, we want to mention an axiom that is not known to be consistent with ZF. This is Mycielski's axiom of determinacy. Recall the game mentioned in our discussion of Martin's theorem: A set X of reals is fixed and two players a and b alternately choose integers between 0 and 9 inclusive; say a chooses a_1, then b chooses b_2, then a chooses a_3, etc. If the number $0.a_1b_2a_3b_4\ldots$ belongs to X, then player a is declared the winner; otherwise b wins. The axiom of determinacy states that the game is determined regardless of X, i.e., one of the players has a winning strategy (of course, which player has a winning strategy will depend on X). ZFC implies the negation of the axiom of determinacy, although ZF plus determinacy implies that every countable set of non-empty subsets of **R** has a choice function. Moreover, it is conjectured that ZF plus determinacy is consistent with the countable axiom of choice. Much of the interest in determinacy stems from its implications for real analysis. For example, determinacy implies that every set of reals is Lebesgue measurable. In addition determinacy implies that every set of reals has the property of Baire and every uncountable set of reals contains a perfect set (we leave the definition of these notions to a course in real analysis). Determinacy is also known to imply the consistency of 'there is a measurable cardinal', which, as we shall see in the next section, shows that determinacy is not a consequence of ZF.

EXERCISES FOR §1.11

1. Prove in ZF that if x and y are sets, then so is ^yx, i.e., there is a set z such that for all w, $w \in z$ iff $w : y \to x$.

2. Show that if κ is strongly inaccessible and if λ and ν_α ($\alpha \in \lambda$) are less than κ, then $c\prod_{\alpha \in \lambda} \nu_\alpha \in \kappa$. (Recall that $\prod_{\alpha \in \lambda} \nu_\alpha$ is the set of all functions f on λ such that $f(\alpha) \in \nu_\alpha$ for each $\alpha \in \lambda$.)

3. Ramsey's theorem states that whenever $X_1 \cup X_2 \cup \cdots \cup X_m = [\omega]^n$, then there is an infinite $Z \subseteq \omega$ such that $[Z]^n \subseteq X_j$ for some j. This theorem has been remarkably useful in algebra, number theory, and logic. Moreover, its generalizations initiated an entire subject, combinatorial set theory, and motivated several large cardinal axioms. We sketch the proof, leaving the details as exercises for the reader. We consider only the case for $m=2$, since the theorem says nothing when $m=1$, and the case for $m>2$ reduces to the case $m-1$. The case for $n=1$ is easy. Suppose $m=2$, $n=1$. Let $x_0=0$. Let $Z_{1i}=\{n:n \neq 0$ and $\{0,n\} \in X_i\}$ for $i=1,2$. Either Z_{11} or Z_{12} is infinite. Let Z_1 be Z_{11} if $cZ_{11}=\omega$, and let $p_1=1$; otherwise let $Z_1=Z_{12}$ and $p_1=2$. Let $x_1 \in Z_1$. Let $Z_{2j}=\{n:n \neq x_1, n \in Z_1,$ and $\{x_1,n\} \in X_j\}$. If $cZ_{21}=\omega$, take $Z_2=Z_{21}$ and $p_2=1$; otherwise take $Z_2=Z_{22}$ and $p_2=2$. Now choose $x_2 \in Z_2$. Continuing, we get $x_0,x_1,x_2,\ldots;Z_1,Z_2,Z_3,\ldots;p_1,p_2,p_3,\ldots$. Either for infinitely many i's, $p_i=1$, or for infinitely many i's, $p_i=2$. Suppose the first (the second case yields a completely analogous argument). Let $Z=\{x_j:p_j=1\}$. Take $x_i,x_j \in Z$ where $i<j$. Then $x_j \in Z_i$, and since $p_i=1$, we have $\{x_i,x_j\} \in X_1$. Thus $[Z]^2 \subseteq X_1$. An analogous argument reduces the case $m+1$ to m when $m \geq 2$.

4. Prove that any set of pairwise disjoint open intervals of reals is countable.

5. Consider the game described on pp. 49 and 50. For which of the following choices of X does the first player have a winning strategy? $\emptyset$, **R**, $\{\frac{1}{4},\frac{1}{2}\}$, **Q**.

1.12 Declarations of Independence

This section is a meager introduction to what is now a vast and highly technical branch of set theory, the study of independence and relative strength of axioms. We prove here, among other things, that IC is not implied by ZFC. We also discuss the famous results of Gödel and Cohen.

We need several notions from model theory, which we shall meet again in much greater detail and generality in §3. A *structure* is an ordered pair (A,e) where $A \neq 0$ and e is a binary relation on A. Let σ be a statement in the language of set theory. Say that (A,e) is a *model* of σ if σ is true in (A,e) when $\in$ is interpreted as e, and 'for all x' is interpreted as 'for all $x \in A$', and 'there is an x' is interpreted as 'there is an x in A'.

For example, $(\mathbf{R}, <)$ is a model of the axiom of extensionality, since if x and y are real numbers such that $z < x$ iff $z < y$ for all real numbers z, then $x = y$. However, $(\mathbf{R}, <)$ is not a model of the axiom of the null set. It is also easy to see that $(\mathbf{R}, <)$ is a model of union but not of power set (see Exercise 1).

If Σ is a set of assertions in the language of set theory, then we say that $(A, <)$ is a model of Σ if each sentence in Σ is true in $(A, <)$. We write $(A,e) \models \sigma$ or $(A,e) \models \Sigma$ if (A,e) is a model of σ or (A,e) is a model of Σ respectively.

Once e is defined in a given context, we may refer to (A,e) simply as A.

Our main interest in this section is in consistency results. We say that an assertion σ is *consistent* with a set of assertions Σ just in case $\Sigma \cup \{\sigma\}$ has a model. If both σ is consistent with Σ and the negation of σ is consistent with σ, then σ is said to be *independent* of Σ. So to prove that σ is independent of Σ it is enough to display two models of Σ, one in which σ is true and one in which σ is false. We say that Σ *implies* σ if every model of Σ is a model of σ. Hence Σ implies σ iff the negation of σ is not consistent with Σ.

As an example, let Σ be the axioms defining a group, and let σ be the commutative or Abelian axiom. Since there are groups (i.e., models of Σ) that are Abelian and others that are not, σ is independent of Σ.

It is easy to find examples of groups, but the situation is quite different regarding models of ZF, and even fragments of ZF. The consistency of $\text{ZF} \cup \{\sigma\}$ implies the existence of a model for ZF, but Gödel has shown that the existence of a model of ZF cannot be proved from ZF. Since ZF is a sufficient framework for all of classical mathematics, this means that the existence of a model of ZF is a new assumption, to be taken on faith, that is not provable within classical mathematics. Nevertheless, the intuitive appeal of the axioms and the fact that no contradiction has been derived from them in the forty plus years following their discovery gives us confidence that such a model exists, but this (or the existence of a model for some fragments of ZF) has to be taken as an additional assumption, which we do without further mention.

We begin with a string of definitions and lemmas that will be fundamental to all our consistency proofs.

Definition 12.1. Fix x. For each ordinal α define $P_\alpha x$ as follows:

$$P_0(x) = x,$$
$$P_{\alpha+1}(x) = P(P_\alpha(x)),$$
$$P_\gamma(x) = P\left(\bigcup_{\beta \in \gamma} P_\beta\right) \text{ for } \gamma = \bigcup \gamma.$$

(This is an example of a recursive definition on Ord as discussed in §1.10.)

The next lemma lists some of the elementary properties of the P_α's that we need.

Lemma 12.2. *Suppose that x is ε-transitive. Then*

i. $P_\alpha(x)$ *is ε-transitive.*
ii. *If $\beta \in \alpha$, then $P_\beta(x) \in P_\alpha(x)$ and so $P_\beta(x) \subset P_\alpha(x)$.*
iii. $\beta \in P_\alpha(x)$ *iff $\beta \in \alpha$.*

PROOF: The proof of each clause is by induction on α. Clearly, parts i through iii are true when $\alpha = 0$. Suppose i through iii hold for all $\delta \in \alpha$. There are two cases to consider: $\alpha = \gamma + 1$ for some γ, and $\alpha = \bigcup \alpha$. We prove the first and leave the second case, which is similar, as an exercise.

Let $z \in w$ and $w \in P_\alpha(x)$. By the definition of $P_\alpha(x)$, we have $w \subseteq P_\gamma(x)$ and so $z \in P_\gamma(x)$. By the induction hypothesis $P_\gamma(x)$ is $\in$-transitive, and so $z \subseteq P_\gamma(x)$. Hence $z \in P_\alpha(x)$, giving part i. □

The definition of $P_\alpha(x)$ immediately gives $P_\gamma(x) \in P_\alpha(x)$, so by part i, $P_\gamma(x) \subset P_\alpha(x)$. □

If $\beta < \alpha$, then $\beta \leq \alpha$, and so by the induction hypothesis $\beta \subseteq P_\gamma(x)$. So $\beta \in P_\alpha(x)$. Conversely, if $\beta \in P_\alpha(x)$, then $\beta \subseteq P_\gamma(x)$, and so $\beta \leq \gamma < \alpha$. This proves part iii. □

Lemma 12.3. *Suppose that $\alpha = \bigcup \alpha$ (i.e., that α is a limit ordinal). Let $M = \bigcup_{\beta \in \alpha} P_\beta(s)$, where s is $\in$-transitive. Then $(M, \in \restriction M)$ is a model of the following axioms:*

i. *null set,*
ii. *extensionality,*
iii. *regularity,*
iv. *pairing,*
v. *union,*
vi. *power set,*
vii. *choice.*

PROOF OF i follows from Lemma 12.2iii. □

PROOF OF ii. Suppose $x, y \in M$ and $x \neq y$. Then there is a $z \in (x-y) \cup (y-x)$. By Lemma 12.2i, $z \in M$. Hence there is a $z \in M$ such that $z \in (x-y) \cup (y-x)$, and so extensionality holds in M. □

PROOF OF iii. Let $x \in M$, and let y be $\in$-minimal in x. Then $y \in M$ by Lemma 12.2i and is an $\in$-minimal member of x. □

PROOF OF iv. Suppose $x, y \in M$. Then $x \in P_\beta(s)$ and $y \in P_\gamma(s)$ for some $\beta, \gamma \in \alpha$. Let $\delta = \beta \cup \gamma$. Then $\{x, y\} \subseteq P_\delta(s)$, and so $\{x, y\} \in P_{\delta+1}(s) \subseteq M$. □

PROOF OF v. Let $x \in M$; say $x \in P_\beta(s)$, where $\beta \in \alpha$. If $z \in \bigcup x$, then $z \in P_\beta(s)$ by two applications of Lemma 12.2i. Hence $\bigcup x \subseteq P_\beta(s)$ and so $\bigcup x \in P_{\beta+1}(s) \subseteq M$. □

PROOF OF vi. Again let $x \in P_\beta(s)$, where $\beta \in \alpha$. If $y \subseteq x$, then $y \subseteq P_\beta(s)$ by Lemma 12.2i. Hence $y \in P_{\beta+1}(s)$, and so $P(x) \in P_{\beta+2}(s) \subseteq M$. □

PROOF OF vii. Suppose x is a set of non-empty sets and $x \in P_\beta(s)$, where $\beta \in \alpha$. Let f be a choice function on x. Then f is a set of ordered pairs $(y, f(y))$ where $y \in x$ and $f(y) \in y$. Since $(y, f(y)) = \{\{y\}, \{y, f(y)\}\}$ and since $y, f(y) \in P_\beta(s)$ by Lemma 12.2i, we have $(y, f(y)) \in P_{\beta+2}(s)$. Hence $f \subseteq P_{\beta+2}(s)$, and so $f \in P_{\beta+3}(s) \subseteq M$. So M satisfies the axiom of choice. This concludes the proof of Lemma 12.3. □

Our next theorem shows that the axiom of infinity is not redundant. In fact, none of the axioms of ZFC are redundant. There are several problems at the end of this section that provide a partial proof of this claim.

Theorem 12.4. *Let $M = \bigcup_{i \in \omega} P_i(0)$. Then $(M, \in \upharpoonright M)$ is a model of ZFC except for the axiom of infinity. Hence, the axiom of infinity is not implied by the other axioms of ZFC.*

PROOF: In view of the preceding lemma, we need only see that M satisfies the axiom of replacement but not the axiom of infinity.

To see that replacement holds, let ψ be a property such that for every $y \in M$ there is a unique $z \in M$ such that ψyz. Let $x \in M$. An easy induction on i shows that $P_i(0)$ is finite for each $i \in \omega$. Hence x is finite (by Lemma 12.2i. For each $y \in x$ let $f(y)$ be the first $i \in \omega$ such that ψyz for some $z \in P_{f(y)}(0)$. Then the range of f is finite and has an upper bound $m \in \omega$. So $\{z : z \in M$ and ψyz for some $y \in x\} \subseteq P_m(0)$, and hence $\{z : z \in M$ and ψyz for some $y \in x\} \in P_{m+1}(0) \subseteq M$. Therefore, replacement holds in M.

Now suppose $x \in M$ and x satisfies the axiom of infinity, i.e., $0 \in x$ and $y \cup \{y\} \in x$ whenever $y \in x$. Hence $x \supseteq \omega$. Let $x \in P_i(0)$ where $i \in \omega$. Since $i+1 \in x$ and $x \subseteq P_i(0)$ (by Lemma 12.2i), we have $i+1 \in P_i(0)$, contradicting Lemma 12.2iii. Hence no such x exists, and M is not a model of the axiom of infinity. □

It may happen that an element of M has a property ψ when viewed within the structure (M, e) but does not have the property when considered in the context of all sets. For example, $\omega+1$ is a cardinal in $P_{\omega+2}(0)$ but not in the context of all sets (Exercise 9). In the work of Gödel and Cohen, x may be the power set of y in the sense of one model but not in the sense of another or in the context of all sets. However, there are many important properties that behave more uniformly, at least with respect to structures of the following kind.

Definition 12.5. The structure (M,e) is *standard* if M is $\in$-transitive and $e = \in \restriction M$.

Definition 12.6. The n-ary property ψ is *absolute* if for every standard structure M and every $x_1,\ldots,x_n \in M$, we have $\psi x_1,\ldots,x_n$ is a true statement about sets iff $M \vDash \psi x_1,\ldots,x_n$.

Lemma 12.7. *The following are absolute:*

i. $x = y$,
ii. $x \in y$,
iii. $x \subseteq y$,
iv. $x = \{y,z\}$,
v. $x = (y,z)$,
vi. x *is an ordered pair*,
vii. R *is a relation*,
viii. f *is a function*,
ix. f *is a 1-1 function with* $\mathrm{Dom} f = x$ *and* $\mathrm{Ran} f = y$.
x. x *is* $\in$-*transitive*,
xi. $\mathrm{Ord}\, x$.

PROOF: Let (M,e) be a standard structure so that M is $\in$-transitive and e is $\in \restriction M$. Then parts i and ii are immediate, and iii follows directly from ii. □

PROOF OF iv. Let $x,y,z \in M$. By part ii and the $\in$-transitivity of M, for any $w \in M$, $w \in x$ iff $w \in M$ and $M \vDash w \in x$. Hence the following statements are equivalent:

$M \vDash x = \{y,z\}$.
For every $w \in M$, $w \in x$ iff $w = y$ or $w = z$.
For every w, $w \in x$ iff $w = y$ or $w = z$.
$x = \{y,z\}$. □

PROOF OF v follows from iv and the definition of (y,z) as $\{\{y\},\{y,z\}\}$. □

PROOF OF vi. Let $x \in M$. The following statements are equivalent:

$M \vDash x$ is an ordered pair.
There exist $y,z \in M$ such that $x = (y,z)$.
There exist y,z such that $x = (y,z)$.
x is an ordered pair.

1.12 Declarations of Independence   55

To get from the third statement to the second note that if $x=(y,z)$ and $x \in M$, then $\{y\}, \{y,z\} \in M$ and so $x,y \in M$. □

PROOF OF vii follows from ii and vi. □

PROOF OF viii follows from vii. □

PROOF OF ix. Let $f \in M$. By part viii we may suppose that f is a function and that $M \vDash f$ is a function. Notice that $x \neq y$ and $(x,z),(y,z) \in f$ iff $(x,z),(y,z) \in M$ and $(x,z),(y,z) \in f$. Hence f is 1-1 just in case $M \vDash f$ is 1-1. Also $x \in \text{Dom} f$ iff for some y $(x,y) \in f$, iff $(x,y) \in M$ and $(x,y) \in f$, iff $M \vDash x \in \text{Dom} f$. Hence $z = \text{Dom} f$ iff $M \vDash z = \text{Dom} f$. Similarly for Ran f. □

PROOF OF x. Let $x \in M$. Then $z \in y$ and $y \in x$ iff $z,y \in M$ and $z \in y$ and $y \in x$. □

PROOF OF xi follows from x and ii. □

Now let's look at the other axioms of infinity discussed in §1.11. All are known to imply IC, the axiom that asserts the existence of a strongly inaccessible cardinal. Since our next theorem states that IC is not implied by ZFC, it follows that the other axioms of infinity are not implied by ZFC either.

Theorem 12.8. *ZFC does not imply IC.*

PROOF: Suppose ZFC implies IC, and let κ be the first strongly inaccessible cardinal. Let $M = \bigcup_{\beta \in \kappa} P_\beta(0)$. We claim that $(M, \in \restriction M)$ is a model of ZFC but not of IC. In view of Lemma 12.3 we need only show that $(M, \in \restriction M)$ is an $\in$-model of infinity and replacement but not of IC.

Since $\omega \in M$ and M is $\in$-transitive, it follows that M is a model of the axiom of infinity.

Now suppose ψ is a property such that for each $y \in M$ there is a unique $z \in M$ such that ψyz. Let $x \in M$. Our argument now is quite similar to that given for Theorem 12.4. For each $y \in x$ we let $f(y)$ be the least $\alpha \in \kappa$ such that ψyz for some $z \in P_\alpha(0)$. Let Z be the range of f. An easy induction on β, using the fact that κ is strongly inaccessible, shows that $cP_\beta(0) \in \kappa$ for all $\beta \in \kappa$. Hence $cx \in \kappa$, and so $cZ \in \kappa$. Again using the strong inaccessibility of κ we get that $\bigcup Z \in \kappa$. Let $\gamma = \bigcup Z$. Then $\{z : \psi yz \text{ for some } y \in x\} \subseteq P_\gamma(0)$ and so is a member of M. Hence replacement is true in M.

We now show that IC is false in M. Let $\alpha \in \kappa$ (so that α is an ordinal in M). By Lemma 12.7xi, α is an ordinal and $\alpha \in \kappa$. Hence either α is not a cardinal or $\alpha \leq 2^\lambda$ for some cardinal $\lambda \in \alpha$, or $\alpha = \bigcup \{\lambda_\beta : \beta \in \lambda\}$ where $\lambda \in \alpha$ and $\{\lambda_\beta : \beta \in \lambda\} \subseteq \alpha$. In the first two cases there is a function $f: \alpha \xrightarrow{1\text{-}1} P(\lambda)$ for some $\lambda \in \alpha$. Since $\lambda, \alpha \in M$ and since $f \subseteq P_{\alpha+3}(0) \subseteq M$, we have by Lemma 12.7iii, ix that $f: \alpha \xrightarrow{1\text{-}1} P(\lambda)$ in the sense of M. Hence α is not strongly inaccessible. In the remaining case, λ and each λ_β for $\beta \in \alpha$

belong to $P_\alpha(0)$, and so $\{\lambda_\beta : \beta \in \lambda\} \in P_{\alpha+1}(0)$. Hence $\bigcup \{\lambda_\beta : \beta \in \lambda\} \in M$, so α is not strongly inaccessible in this case either. □

We shall now use the terms structure and model in a more general context that involves an abuse of notation. We now think of a structure as an ordered pair (M,e) where M is a unary property and e is a binary property. The abuse of notation arises because M and e need not be sets, in which case (M,e) no longer denote an ordered pair of sets. However, we think of M as the collection of all sets x having the property M. In general, this collection, like the collection of all sets, or the collection of all cardinals, will be "too large" to be a set.

The notion of truth in (M,e) and the notion of model are then extended in the obvious way. For example, to say that pairing is true in (M,e), or that (M,e) is a model of pairing, or that $(M,e) \vDash$ pairing, means that for every x and y having the property M, there is a z having the property M such that for all w having the property M, $w \in z$ iff $w = x$ or $w = y$. Clearly, there is a need to further abuse notation, and we write '$x \in M$' instead of 'x has the property M'; also, we will write '$x \in y$' instead of '(x,y) has the property e' (except for one exercise, e will be $\in$, so this takes care of itself).

The definitions of '$\in$-transitive' and 'standard model' are generalized in the obvious way.

Now let Σ be the axioms of ZFC other than regularity. Let x have the property M just in case $x \in P_\alpha(0)$ for some α. Now consider $(M, e \restriction M)$. Trivial modifications of Lemma 12.3 and the proof of Theorem 12.4 show that M is a model of ZFC. Hence if the axioms of ZF other than regularity have a model, then ZF has a model. On the other hand, there is a model of the axioms of ZF other than regularity in which regularity fails. Thus the regularity axiom is not redundant, and in fact we have

Theorem 12.9. *The axiom of regularity is independent of the other axioms of ZF.*

For many years, the status of the axiom of choice relative to ZF remained a mystery. Is choice independent of ZF, or provable from ZF, or perhaps even refutable from ZF? The same question arises regarding the generalized continuum hypothesis and ZF. As far as the axioms of ZF itself are concerned, each can be shown to be independent of the remaining ones without too much effort (see Exercises 2 and 3 for a couple of instances). For choice and especially the GCH, the problem is much more difficult and more crucial—more crucial because these statements are far less intuitive to most mathematicians than the statements of ZF; heuristic arguments for or against these axioms as valid assertions about the universe of all sets are not very compelling. A proof of AC from ZF would cause AC to be accepted as a valid statement about sets; a disproof from ZF would cause its rejection; and similarly for GCH. But such proofs are impossible, for in 1938, Gödel assuming the existence of a model for ZF, described a model of ZFC plus GCH; then, in 1963, Cohen, assuming the existence of a model for ZF, produced a model of ZF in which choice fails, and a model of ZFC in which GCH fails. (One can show that choice is a consequence of ZF plus GCH, and so no model of ZF plus GCH exists in which choice is false.) Together these results prove the following.

Theorem 12.10.

i. *The axiom of choice is independent of ZF.*
ii. *GCH is independent of ZFC.*

The constructions of Gödel and Cohen are much too involved to give here, although it might appear at first glance that part of Gödel's contribution has already been dealt with in Theorem 12.8 and Theorem 12.9, where we constructed a model of ZFC. However, our proofs that choice holds in these models depended on choice being used in the universe of all sets. Thus assuming the truth of ZFC in the universe of all sets, we produced models of ZFC having additional properties. But suppose that some doubter believes that ZF is true in the universe of all sets but that choice is not—even more, he suspects that choice can be disproved from ZF. Gödel produced a model of ZFC assuming only that ZF has a model and hence showed that ZFC is as consistent as ZF, and so choice cannot be disproved from ZF.

Not only are these theorems of Gödel and Cohen milestones in the foundations of mathematics, but the techniques used to prove them have been extremely fruitful in the last decade, yielding consistency results that answered longstanding problems in logic, topology, analysis, algebra, and other branches of mathematics. Work in this direction is still continuing at an enormous rate.

Exercises for §1.12

1. Show that $(\mathbf{R}, <)$ is a model of union but not power set.

2. The power set axiom cannot be proved from the remaining axioms of ZF. Let M be the set of all x such that x, $\bigcup x$, $\bigcup \bigcup x$, $\bigcup \bigcup \bigcup x,\ldots$ are all countable. Show that all the axioms of ZF are true in M except for the power set axiom.

3. Prove the remaining half of Lemma 12.2, for $\alpha = \bigcup \alpha$.

4. (a) Show that there is no finite sequence of sets $x_0, x_1, \ldots, x_n$ such that $x_0 \in x_1 \in \cdots \in x_n \in x_0$.
   (b) Show that there is no sequence of sets $x_0, x_1, \ldots$ such that $\ldots x_2 \in x_1 \in x_0$. (*Hint:* Use regularity.)

5. Suppose that ψ is a property such that for every x, ψx whenever ψy for all $y \in x$. Show that ψx for all x. An argument based on this principle is called an induction on sets. (*Hint:* Use regularity.)

6. Let $x'=x$ for all x other than $x=0$ or $x=1$. Let $0'=1$ and $1'=0$. Let $x\in'y$ iff $x\in y'$. With M the collection of all sets, prove that $(M, \in')$ is a model of ZF in which regularity fails. This along with the argument preceding Theorem 12.9 gives a proof of 12.9 and shows that regularity is independent of ZF.

7. The axiom of separation is weaker than the axiom of replacement. To show this let $M = P_{\omega+\omega}(0)$ and consider $(M, \in \restriction M)$. Verify that this is a model of separation and all the axioms of ZF except replacement. To see that replacement fails, consider the property ψ_{xy} which holds iff $x \in \omega$ and $y = P_{\omega+x}(0)$. As we have seen in §1.11, separation is a consequence of ZF, and so this example shows that separation is weaker than replacement.

8. Show that the following are absolute:
$$x = y \times z,$$
$$x = \bigcup y,$$
$$x = \bigcap y,$$
$$x = 2,$$
$$x = \omega.$$

9. Card x is not absolute. [*Hint:* Consider the structure $(P_{\omega+2}(0), \in \restriction P_{\omega+2}(0))$ and Card $\omega + 1$.]

PART II
An Introduction to Computability Theory

2.1 Introduction

What are the capabilities and limitations of computers? Are they glorified adding machines capable of superfast arithmetic computations and nothing else? Can they outdo man in the variety of problems they can handle? Let's narrow the question a bit. Consider the class of number theoretic functions that a computer can be programmed to compute or that a man can be instructed to compute. Are any of these functions computable by a computer but not by a man, or by a man but not by a computer? Is there a number theoretic function that is not computable by any computer, and if so, can such functions be described? Is there a computer that can be programmed to compute any function that any other computer can compute? Is man such a computer?

In order to make these questions amenable to mathematical analysis, Alan Turing in 1942, introduced a purely abstract mathematical notion of computer and presented heuristic arguments in support of the view that his "machines" have exactly the same computational powers as a "real computer" or a man, at least if speed of computation is ignored.

The reason that his arguments are necessarily heuristic is that neither the notion of "man computable" nor "real computer" is mathematically defined. His mathematical machines are intended to be an abstraction of "real computers" that allows precise mathematical analysis, just as the integral is an attempt to make rigorous our intuitive notion of area.

We begin this section with a description of Turing machines, and later take up some of the various intuitive arguments that support the thesis that man, computers, and Turing machines are computationally equivalent. Most of the work will be aimed at delimiting the capabilities of Turing machines, culminating with Gödel's famous incompleteness theorem of 1931, one of the milestones of mathematics. Some time will be spent on the consequences of the incompleteness theorem and in particular its bearing on the problem of placing arithmetic, or any other significant branch of mathematics, on a sound axiomatic foundation.

2.2 Turing Machines

The kind of machine we have in mind performs computations by printing and erasing checks on a tape. The tape is partitioned into cells, and each cell either is blank or has a single check. At any given step in the computation, the machine scans a single cell. This scanned cell is then checked or left blank, and the scanner may move one cell to the left or to the right or remain stationary. Exactly which of these six alternatives occurs at a given step of the computation is completely determined by two things: the internal state of the machine (we assume the machine has only finitely many states), and whether the scanned cell is checked or blank. The situation can be pictured as follows:

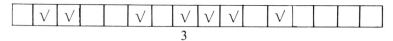

Here, the machine is in state 3, and cell 8 of this tape is checked and is the scanned cell. This notation is a bit clumsy and some simplification is called for.

Recall that $\mathbf{N}$ is the set of natural numbers $\{0, 1, 2, \ldots\}$ and $\mathbf{N}^+$ is the set of strictly positive natural numbers. If f is a function, then $\mathrm{Dom} f$ is its domain and $\mathrm{Ran} f$ is its range. Most of the other symbols used here are defined in §1.2.

Definition 2.1. A *tape* is a sequence $a_1 a_2 a_3 \ldots$ where each a_i is 0 or 1. A *marker* is a pair (j, k) where $y \in \mathbf{N}^+$ and $k \in \mathbf{N}^+$. A *tape position* is a marker (j, k) and a tape $a_1 a_2 a_3 \ldots$; the jth term is the *scanned term*, and k is the *state* for this tape position.

Here we have substituted 0's and 1's for the blanks and checks of our initial description. We now need three functions: one to tell us how to alter the scanned term, one to tell us which term to scan next, and one to tell us the next state of the machine.

Definition 2.2. A *Turing machine* is an ordered triple of functions (d, p, s), each with the same domain D, where D is a finite set of tuples of the form (i, k) with $i \in \{0, 1\}$ and $k \in \mathbf{N}^+$, and where $\mathrm{Ran} d \subseteq \{0, 1\}$, $p(i, k) \in$

2.2 Turing Machines

$\{-1, 0, 1\}$, and $\operatorname{Ran} s \subseteq \mathbf{N}^+$. If $M = (d, p, s)$, then the domain of M, $\operatorname{Dom} M$, is D.

We may refer to a Turing machine simply as a machine.

The functions d, p, s are the "print-erase" function, the "next position" function, and the "next state" function respectively. For example, if the tape position t is given by

$$(3, 2): 1 1 1 1 0 1 0 \ldots$$

and the machine $M = (d, p, s)$ is such that

$$d(1, 2) = 0,$$
$$p(1, 2) = -1,$$
$$s(1, 2) = 4,$$

then one application of M to t yields a new tape position

$$(2, 4): 1 1 0 1 0 \ldots .$$

Originally the third cell is scanned and 2 is the state of the machine. This is denoted by the marker $(3, 2)$. d tells us to place a 0 in the scanned cell; p tells us to move the scanner to the left, so that now the second cell is the scanned cell; and s tells us that the new state is 4. More generally, we have

Definition 2.3. Let t be the tape position with marker (j, k) and tape a, and let M be the machine (d, p, s). Then $M(t)$, the *successor tape position*, has marker $(j + p(a_j, k), s(a_j, k))$ and tape $a_1 a_2 \ldots a_{j-1} \, d(a_j, k) \, a_{j+1} a_{j+2} \ldots$ provided that $(a_j, k) \in \operatorname{Dom} M$ and $j + p(a_j, k) > 0$. A *partial computation* of M is a sequence $t_1, t_2, \ldots, t_m$ of tape positions such that $t_{i+1} = M(t_i)$ for each $i < m$. The sequence is a computation if $M(t_m)$ is not defined. If $t_1, t_2, \ldots, t_m$ is a computation, then t_1 is the *input* and t_m the *output*. If $t_1, \ldots, t_r$ is a partial computation, we may write $M'(t_1)$ for t_r.

There is a convenient way of writing a machine as a table. For example

		0	1
M:	1	1L2	0R1
	2	1O2	

denotes the machine (d, p, s) with domain $\{(0, 1), (1, 1), (0, 2)\}$ where

$$d(0, 1) = 1, \quad p(0, 1) = -1, \quad s(0, 1) = 2,$$
$$d(1, 1) = 0, \quad p(1, 1) = 1, \quad s(1, 1) = 1,$$
$$d(0, 2) = 1, \quad p(0, 2) = 0, \quad s(0, 2) = 2.$$

Since a p-value of 1 indicates a scanner shift to the right, a p-value of 0 indicates no movement of the scanner, and a p-value of -1 indicates a scanner shift to the left, we use R, O, and L in the table instead of 1, 0, and -1.

As an example, take M to be the machine above and t the tape position

$$(2,1): 0100000\ldots.$$

Then $M(t)$ is

$$(3,1): 0000000\ldots.$$

[Since the second term is the scanned term and its value is 1, and the state is 1, $d(1,1)=0$ is the value of the second term in $M(t)$. $p(1,1)=1$, which says that the term to the right of the second term, i.e., the third term, is the new scanned term. Finally $s(1,1)=1$, which says that the new state is 1.] It should be clear that the following is a computation:

$$(2,1): 0100000\ldots$$
$$(3,1): 0000000\ldots$$
$$(2,2): 0010000\ldots$$
$$(2,2): 0110000\ldots.$$

An easy induction on n shows that with M as above there is a computation that begins with

$$(2,1): 0\underbrace{1\ldots1}_{n \text{ consecutive 1's}}000\ldots$$

and ends with

$$(n+1,2): \underbrace{00\ldots0}_{n-1 \text{ consecutive 0's}}11000\ldots.$$

If we think of n as being represented by n consecutive 1's, then this machine computes the constant function $f(n)=2$.

In abbreviating tapes we may use 1^n to denote an n-term subsequence of 1's and 0^n an n-term subsequence of 0's. For example

$$0111001111000\ldots$$

may be written as

$$0\,1^3\,0^2\,1^4,$$

suppressing mention of the tail of 0's.

We may write $0^m(n_1,\ldots,n_k)$ for $0^m01^{n_1}01^{n_2}0\ldots01^{n_k}$, where $m \in \mathbb{N}$. For $m=0$ this becomes $(n_1,\ldots,n_k)$, and if $k=1$, then we write n_1 instead of (n_1). The sequence or tape $(n_1,\ldots,n_k)$ may be written as $\bar{n}$.

The *input* $(n_1,\ldots,n_k)$ is the marker $(2,1)$ with the tape $(n_1,\ldots,n_k)$. The *output* n has the leftmost 1 of the tape $0^k n$ as the scanned term, where $k \geq 1$.

We have introduced several obvious ambiguities in our notation, confusing the n-term sequence of 1's with the number 1 $(=1^n)$, the tape $(n_1,\ldots,n_k)$ with the sequence $(n_1,\ldots,n_k)$, and so on. This may however increase the readability of what follows, and the particular meaning intended will be apparent from the context.

2.2 Turing Machines

Definition 2.4. Suppose that M is a machine and g is a k-ary function such that for each $(n_1,\ldots,n_k) \in {}^k(\mathbf{N}^+)$ there is a computation with respect to M with the input $(n_1,\ldots,n_k)$ and the output $g(n_1,\ldots,n_k)$. We then say that g is *computable* and that M computes g.

We have already seen that the constant function $f(n) = 2$ is computable and is computed by the machine

	0	1
1	1L2	0R1
2	1O2	

The following theorem establishes the computability of several basic functions.

Theorem 2.5. *The following functions are computable*:
  i. *the addition function* $\mathrm{Sum}(m,n) = m+n$,
  ii. *for each k and d in* $\mathbf{N}^+$, *the constant k-function* $C_{k,d}(n_1,\ldots,n_k) = d$,
  iii. *for each k and t in* $\mathbf{N}^+$ *with* $t \leq k$, *the projection k-function* $P_{k,t}(n_1,\ldots,n_k) = n_t$,
  iv. *the 1-function*

$$\mathrm{Pred}(m) = \begin{cases} m-1 & \text{if } m \geq 2, \\ 1 & \text{if } m = 1. \end{cases}$$

[*Read* 'Pred(m)' *as* 'Predecessor m'.]

PROOF OF i. Consider the machine

	0	1
1		0R2
2	1L3	1R2
3	0R4	1L3

It is easy to see that the following is a subsequence of the computation beginning with (m,n):

$$(2,1): \quad 0\,1^m\,0\,1^n$$
$$(3,2): \quad 0\,0\,1^{m-1}\,0\,1^n$$
$$(m+2,2): \quad 0\,0\,1^{m-1}\,0\,1^n$$
$$(m+1,3): \quad 0\,0\,1^{m-1}\,1\,1^n$$
$$(2,3): \quad 0\,0\,1^{m+n}$$
$$(3,4): \quad m+n$$

Hence $m+n$ is computable. $\square$

PROOF OF ii. We give the machine for $d=2$ only, but this can be easily modified to handle any given d:

	0	1
1	0R2	0R1
2	1L3	0R1
3	1O3	

□

PROOF OF iii. The general idea will be quite clear from a discussion of the special case $k=4$, $t=3$. For this case we use the following machine:

	0	1
1	0R2	0R1
2	0R3	0R2
3	0R4	1R3
4	0L5	0R4
5	0L5	1L6
6	0R7	1L6

If (n_1, n_2, n_3, n_4) is the input, then a computation results in which the first two blocks of 1's are erased, the third is passed over, the fourth is erased, and then the scanner returns to the first 1 of the third block. □

PROOF OF iv. It is easy to verify that $\text{Pred}(m)$ is computed by the following machine:

	0	1
1		0R2
2	1O3	

This completes the proof of Theorem 2.5. □

Two different machines may compute the same function. For example, the machine given above for the sum $x+y$ computes the same unary function as the following:

	0	1
1		1O2

namely, the identity function $f(n)=n$. In fact it is easy to see that any computable function is computed by infinitely many different Turing machines (Exercise 5).

Our machine for addition computes both a unary function and a binary function but not a ternary function. Some machines do not compute any function; others compute k-functions for all k.

2.2 Turing Machines

A machine M may fail to compute a k-ary function for one of two reasons: there is some k-ary input t such that $M^r(t)$ exists for all $r \in \mathbf{N}^+$ (so the partial computation does not end), or there is a computation beginning with t that does not end in an output. An example of the first phenomenon is given by

	0	1
1	0R1	1R1

An example of the second is

	0	1
1		1L1

If a function f is computable, then there is a Turing machine M that, given the input $\bar{n}$, prints out $f(\bar{n})$. In this sense we can consider M as answering the question: '*What is $f(n)$?*' It is natural to ask whether Turing machines can answer other types of mathematical questions. For example, can these machines answer questions such as 'Is n a prime?' or 'Is m the greatest common divisor of m_1 and m_2?'? Alternatively, we can view these questions as queries about *set membership* in the following way: Let X be the set of all primes, and let Y be the set of all triples (m, m_1, m_2), where m is the greatest common divisor of m_1 and m_2. The first question is equivalent to the question 'Is $n \in X$?', while the second is equivalent to 'Is $(m, m_1, m_2) \in Y$?'. More generally, these considerations lead us to the following notion of a *computable set*.

Definition 2.6. Let X be a set of k-tuples of natural numbers. The *representing function of X*, R_X, is the k-function defined by $R_X(\bar{n}) = 1$ if $\bar{n} \in X$, $R_X(\bar{n}) = 2$ if $\bar{n} \notin X$. We say that X is *computable* if R_X is a computable k-function.

If one interprets an output of 1 as being the answer 'yes' and the output 2 as being the answer 'no', then a Turing machine which computes R_X answers the question 'Is $\bar{n} \in X$?'.

Notice that the above definition is made only for sets of k-tuples for fixed k, i.e., we do not consider sets containing both k-tuples and t-tuples when $k \neq t$.

A set P of k-tuples is called a k-relation (recall §1.3). For example, the set of all ordered triples (m, m_1, m_2) such that m is the greatest common divisor of m_1 and m_2 is a 3-relation. If P is a k-relation, it is often more convenient to write $P(n_1, \ldots, n_k)$ when we mean $(n_1, \ldots, n_k) \in P$. Thus if P is the set of all ordered triples (m, m_1, m_2) such that m is the greatest common divisor of m_1 and m_2, we have $P(3, 15, 12)$ but not $P(2, 8, 23)$. For binary relations P, more specialized notation is frequently used, and we may write xPy instead of $(x,y) \in P$ or $P(x,y)$. So, for example, if P is the set of all ordered doubles (x,y) such that $x \leq y$, then $3P8$ but not $5P2$. Actually, this is the familiar notation, for if we write '$\leq$' instead of 'P' this becomes $3 \leq 8$ but not $5 \leq 2$.

Thus to say that P is a *computable k-relation* means that there is a Turing machine which correctly answers all questions of the form 'Is $(n_1,\ldots,n_k)$ in the relation P?'.

As usual, the 1-tuple (n) is identified with n, and the 1-relation P is identified with $\{n:(n)\in P\}$.

We next consider some examples of computable sets and relations.

EXAMPLE 2.7. Let X denote the set of all odd numbers. Then $R_X(n)$ equals 1 if n is odd and equals 2 if n is even. It is easily seen that the following machine computes R_X:

	0	1
1	1L2	0R2
2	1O3	0R1

Hence, X is computable.

EXAMPLE 2.8. Let X denote the set of all positive integers $\geqslant 2$, i.e., $X = \{m : m \geqslant 2\}$. $R_X(m) = 1$ if $m \geqslant 2$ and $R_X(1) = 2$. It is easy to see that the following machine computes R_X:

	0	1
1		1R2
2	1L5	0R3
3	0L4	0R3
4	0L4	1O1

Hence X is computable.

EXAMPLE 2.9. Let P denote the relation $\leqslant$. That is, $P = \{(m,n) : m \leqslant n\}$. The representing function R_P for this relation is defined by $R_P(m,n) = 1$ if $m \leqslant n$, and $R_P(m,n) = 2$ otherwise.

The Turing machine below computes R_P in the following way: Given the input (m,n), the machine erases the first check on the tape and then the last check on the tape, and then repeats the procedure until either the block of 1's on the left has been erased but not the entire block on the right (in which case all 1's are then erased and then 1 is printed), or the right block of 1's is erased first (in which case all 1's are then erased and two 1's are printed). The dotted line divides the table into halves: the top half dictates the erasing of the leftmost 1 and then movement to the right after checking that all of the 1's in the left hand block have not been erased; the lower half of the table dictates a dual operation but from right to left.

2.2 Turing Machines

	0	1
1		0R2
2	0R5	1R3
3	0R4	1R3
4	0L6	1R4
5	1O11	0R5
6		0L7
7	0L10	1L8
8	0L9	1L8
9	0R1	1L9
10	1L5	0L10

Much more interesting examples of computable functions and relations will be found in the sections that follow.

The notion of a Turing machine can be formulated in many equivalent ways—equivalent in the sense that the resulting set of computable functions will be the same as the set of functions that are computable according to our present definitions. For example, two way infinite tapes of the form $...a_{-2}a_{-1}a_0a_1a_2...$ can be used in place of one way infinite tapes. The terms of the tape might be restricted to $\{0,1,...,n\}$ instead of $\{0,1\}$. There are many more alternative formulations, each being advantageous in certain circumstances and disadvantageous in others. Our choice is motivated by personal preference and expediency in the present development.

EXERCISES FOR §2.2

1. Give a Turing machine that computes $C_{2,3}$ (see Theorem 2.5). Write down the computation with respect to this machine, beginning with the input (2,2).

2. Give a Turing machine that computes $P_{3,2}$ (see Theorem 2.5). Write down the computation beginning with (2,3,4).

3. Give the computation arising from the inputs 1 and 4 with respect to the machine in the text that computed Pred(m).

4. Give the complete sequence of tape positions arising from the input (2,1) with respect to the machine that computed Pred(m). Is there a tape output?

5. Let f be a computable k-function. Show that there are infinitely many different machines that compute f.

6. Determine the 1-function and 2-function computed by the following machine:

	0	1
1	1L2	1R1
2	0R3	1L2

7. Consider the following machine:

	0	1
1	0R1	1R2
2	1L3	1R2

For which inputs is there an output?

8. Find a machine that computes a k-function for all $k \in \mathbf{N}^+$.
9. Find a machine that computes the function $f(n) = 2n$.
10. Find a machine that computes the function $f(n) = 3n$.
11. Show that the set of all even numbers is computable.
12. Show that the set $\{1, 4, 7, 10, 13, \ldots\}$ is computable.
13. Find a machine that computes the 2-function $f(n, m) = nm$. Briefly describe the behavior of your machine, indicating why it works.
14. Find a machine that computes the 2-function

$$\text{Diff}'(n, m) = \begin{cases} m - n & \text{if } m - n \in \mathbf{N}^+, \\ 1 & \text{otherwise.} \end{cases}$$

2.3 Demonstrating Computability Without an Explicit Description of a Turing Machine

To prove that a particular function, set, or relation is computable by exhibiting a Turing machine that computes it is usually a long and tedious task even for simple examples. In this section we shall describe procedures for proving the computability of many functions, sets, and relations without explicitly exhibiting machines that compute them. The proofs of these theorems are constructive in the sense that if a function is shown to be computable by one of these theorems, then, following the proof of the theorem, an explicit machine for computing the function can be exhibited. Most of the proofs will be deferred until §2.4.

Suppose $g_1, \ldots, g_r$ are k-functions and f is an r-function. The *composition* of f with $g_1, \ldots, g_r$ is the k-function h defined by $h(\bar{m}) = f(g_1(\bar{m}), g_2(\bar{m}), \ldots, g_r(\bar{m}))$ for any k-tuple $\bar{m}$. For example, if $g_1(m, n) = m + n$, $g_2(m, n) = mn$, $g_3(m, n) = m^2 + n$, $f(u, v, w) = u^3 + v^2 + w$, then

$$h(m, n) = f(g_1(m, n), g_2(m, n), g_3(m, n)) = (m + n)^3 + (mn)^2 + m^2 + n.$$

Theorem 3.1. *If f is a computable r-function and $g_1, \ldots, g_r$ are computable k-functions, then the composition of f with $g_1, \ldots, g_r$ is a computable k-function.*

2.3 Demonstrating Computability without an Explicit Description of a Turing Machine

This result states, in other words, that the composition of computable functions is computable. The proof of this result is given in §2.4. In this section we shall be primarily concerned with showing how results of this type are used to demonstrate computability without recourse to machines.

EXAMPLE 3.2. Using Theorem 3.1, we see that the function $f(n)=2n$ is computable, since $f(n) = \text{Sum}(P_{1,1}(n), P_{1,1}(n)) = \text{Sum}(n,n) = 2n$ and both the functions Sum and $P_{1,1}$ are computable by Theorem 2.5.

EXAMPLE 3.3. Suppose g is a computable 1-function. Define g' by $g'(m) = g(m+1)$. Then $g'(m) = g(\text{Sum}(P_{1,1}(m), C_{1,1}(m)))$. Let $h(m) = \text{Sum}(P_{1,1}(m), C_{1,1}(m))$. h is computable by Theorems 2.5 and 3.1. Hence $g(h(m))$ is computable by Theorem 3.1. But $g(h(m)) = g'(m)$, so g' is computable.

It is tempting to consider the following argument as a proof of the computability of $f(n)=2n$: By Theorem 2.5, there is a machine that computes the sum function $m+n$. To compute $2n$ we need only input (n,n) into this machine. The flaw in this argument is that we are inputting a 2-tuple (n,n), and not a 1-tuple as the definition of a computable 1-function requires.

EXAMPLE 3.4. We show that, for every n, the 1-function Prod_n, defined by $\text{Prod}_n(m) = nm$, is computable. (In Example 3.2 above we showed that Prod_2 is computable.) The proof is by induction on n. For $n=1$, we have $\text{Prod}_1 = P_{1,1}$, which is computable by Theorem 2.5. Now suppose that Prod_k is computable. Then $\text{Prod}_{k+1}(m) = \text{Sum}(\text{Prod}_k(m), P_{1,1}(m)) = km + m = (k+1)m$. Since Sum and $P_{1,1}$ are computable by Theorem 2.5 and Prod_k is computable by assumption, we see by Theorem 3.1 that Prod_{k+1} is computable. This completes the induction.

As a corollary to Theorem 3.1 we have the following useful result which, loosely speaking, allows us to change variables in a computable function:

Corollary 3.5. *Let f be a computable r-function, and let $k \in \mathbf{N}^+, i_1 \leq k, i_2 \leq k, \ldots, i_r \leq k$. Then the k-function h defined by $h(\bar{n}) = f(P_{k,i_1}(\bar{n}), P_{k,i_2}(\bar{n}), \ldots, P_{k,i_r}(\bar{n}))$ is computable.*

PROOF: The result follows immediately from Theorems 2.5 and 3.1. □

EXAMPLE 3.6. Suppose f is a computable 2-function. Then the 3-function h_1 defined by $h_1(a,b,c) = f(a,b)$ is computable, since $h_1(a,b,c) = f(P_{3,1}(a,b,c), P_{3,2}(a,b,c))$. Similarly the following 2 and 3-functions are computable:

i. the 3-function h_2 defined by $h_2(a,b,c) = f(a,c)$,
ii. the 3-function h_3 defined by $h_3(a,b,c) = f(a,a)$,
iii. the 2-function h_4 defined by $h_4(a,b) = f(b,a)$.

In Example 3.4, we showed that for every n, Prod_n is computable. The reader might be inclined to think that this also shows that the multiplication function, Mult, defined by $\text{Mult}(m,n) = mn$, is also computable. For to compute $\text{Mult}(m,n)$ we need only go to the machine that computes Prod_m and let n be the input. The fallacy in this argument is that Mult is a 2-function: to show that Mult is computable we need to find one machine such that if the input is (m,n), the output is mn. The above argument would involve going to infinitely many machines, one for each value of n. While it is not particularly difficult to write down a machine that computes Mult directly, we instead consider another method of defining a function in terms of given functions, a method which we will use to define Mult from Sum. As with composition, this method, called definition by recursion, defines a computable function when the given functions are computable.

We can define Mult (by recursion) as the unique function satisfying the following pair of equations:

$$\text{Mult}(1,n) = n, \tag{1}$$

$$\text{Mult}(m+1,n) = \text{Mult}(m,n) + n. \tag{2}$$

For example, using (2) repeatedly and then using (1), we see that $\text{Mult}(4,7) = \text{Mult}(3,7) + 7 = \text{Mult}(2,7) + 7 + 7 = \text{Mult}(1,7) + 7 + 7 + 7 = 7 + 7 + 7 + 7$. We can rewrite (1) and (2) as follows:

$$\text{Mult}(1,n) = P_{1,1}(n), \tag{1'}$$

$$\text{Mult}(m+1,n) = \text{Sum}(\text{Mult}(m,n),n). \tag{2'}$$

We have already shown that $P_{1,1}$ and Sum are computable (Theorem 2.5). An easy application of the next theorem gives us the computability of Mult.

Theorem 3.7 (Definition by Recursion). *Let f be a $(k+2)$-function and let g be a k-function. There is a unique $(k+1)$-function h satisfying the following two equations:*

$$h(1,\bar{n}) = g(\bar{n}) \quad (\bar{n} \text{ denotes an arbitrary } k\text{-tuple}), \tag{1}$$

$$h(m+1,\bar{n}) = f(h(m,\bar{n}),m,\bar{n}). \tag{2}$$

Furthermore, if f and g are computable, then so is h.

As with our definition of Mult, we note that these equations may be viewed as a set of directions that when followed will produce the sequence $h(1,\bar{n}), h(2,\bar{n}), \ldots$. The first equation tells us what $h(1,\bar{n})$ is; the second equation tells us how to get from a given term of the sequence $h(1,\bar{n}), h(2,\bar{n}), \ldots$ to the next term. The existence and uniqueness of the function h is a special case of Theorem 10.23 in Part I, but a self-contained proof is sketched in Exercise 4. The proof that h is computable if f and g are is deferred until Section 2.4.

2.3 Demonstrating Computability without an Explicit Description of a Turing Machine 71

We can now use Theorem 3.7 to get the computability of Mult. We first observe that the function f defined by $f(a,b,c) = \text{Sum}(a,c)$ is computable by Corollary 3.5. Now apply Theorem 3.7, taking $k=1$, $g=P_{1,1}$, and this f.

EXAMPLE 3.8. Consider the two equations

$$\text{Pow}(1,n) = n = P_{1,1}(n), \tag{1}$$

$$\text{Pow}(m+1,n) = \text{Mult}(\text{Pow}(m,n),n). \tag{2}$$

These equations have the form specified in Theorem 3.7 if we take $g = P_{1,1}$ and define a 3-function f by $f(a,b,c) = \text{Mult}(a,c)$. We know that g is computable by Theorem 2.5, and that f is computable by Corollary 3.5 and the computability of Mult (obtained above). Hence by Theorem 3.7 the unique 2-function Pow satisfying these equations is computable. Notice that from (1) and (2) we have

$$\text{Pow}(1,n) = n, \quad \text{Pow}(2,n) = \text{Mult}(n,n) = n^2, \quad \text{Pow}(3,n) = \text{Mult}(n^2,n) = n^3,$$

and so on. An easy induction on m shows that $\text{Pow}(m,n) = n^m$. The reader who does not yet appreciate the power of these methods should attempt to find a machine that computes Pow directly.

EXAMPLE 3.9. Let $f(a,b,c) = \text{Pred } a$. f is computable by Theorem 2.5 and Corollary 3.5. Consider these two equations:

$$h(1,n) = \text{Pred } n, \tag{1}$$

$$h(m+1,n) = f(h(m,n),m,n). \tag{2}$$

From Theorem 3.7 we see that these equations define a unique computable 2-function h. We next show, by induction on m, that $h(m,n) = \text{Diff}'(m,n)$, where Diff' is defined as in Exercise 19, §2, that is,

$$\text{Diff}'(m,n) = \begin{cases} n-m & \text{if } n-m \geq 1, \\ 1 & \text{otherwise.} \end{cases}$$

For $m=1$, n arbitrary, we have $h(1,n) = \text{Pred } n = \text{Diff}'(1,n)$. Suppose for some k and all n we have $h(k,n) = \text{Diff}'(k,n)$. Then $h(k+1,n) = \text{Pred}(h(k,n)) = \text{Pred}(\text{Diff}'(k,n))$. To complete the induction we need only show that $\text{Diff}'(k+1,n) = \text{Pred}(\text{Diff}'(k,n))$. If $\text{Diff}'(k+1,n) \geq 2$, then $\text{Diff}'(k+1,n) = n-(k+1) = (n-k)-1 = \text{Pred}(\text{Diff}'(k,n))$. If $\text{Diff}'(k+1,n) = 1$, then $n-(k+1) \leq 1$, so $n-k \leq 2$. Thus $\text{Pred}(\text{Diff}'(k,n)) = 1$ and the induction is completed.

We may write $m \dotdiv n$ instead of $\text{Diff}'(n,m)$.

Theorem 3.7 yields computable $(k+1)$-functions for $k > 1$ only. A useful analog of Theorem 3.3 that yields computable 1-functions is the following:

Corollary 3.10. *Let f be a computable 2-function and d a fixed positive integer. Then the unique 1-function h satisfying the following equations is*

computable:
$$h(1) = d, \quad (1)$$
$$h(n+1) = f(h(n), n). \quad (2)$$

PROOF: Define a 3-function f' by $f'(a,b,c) = f(a,c)$. The two equations
$$h'(1,p) = C_{1,d}(p), \quad (1')$$
$$h'(n+1,p) = f'(h'(n,p), n, p) \quad (2')$$
define a unique computable 2-function h'. We shall prove, by induction on n, that $h(n) = h'(n,1)$ for all n. For $n = 1$, we have $h(1) = d = C_{1,d}(1) = h'(1,1)$. Suppose $h(k) = h'(k,1)$. Then $h(k+1) = f(h(k), k) = f'(h(k), 1, k) = f'(h'(k,1), 1, k) = h'(k+1, 1)$. This completes the induction and shows that $h(n) = h'(n,1)$ for all n. The computability of h now follows from the computability of h', $C_{1,1}$, $P_{1,1}$ and the fact that $h(n) = h'(n,1) = h'(P_{1,1}(n), C_{1,1}(n))$. □

EXAMPLE 3.11. The factorial function $n! = n \cdot (n-1) \cdot \ldots \cdot 1$ can be defined by recursion as follows:
$$1! = 1,$$
$$(n+1)! = (n+1) \cdot n!.$$
The second equation may be written as
$$(n+1)! = \text{Mult}(n!, \text{Sum}(n, C_{1,1}(n))).$$
Since $\text{Mult}(y, \text{Sum}(x, C_{1,1}(x)))$ is computable by Theorem 2.5 and Theorem 3.1, it follows by Corollary 3.10 that $n!$ is computable.

EXAMPLE 3.12. Let g be a computable 1-function, and define $S_g(n) = g(1) + g(2) + \cdots + g(n) = \sum_{i=1}^{n} g(i)$. Giving an alternative description of S_g and applying Corollary 3.10 gives the computability of S_g:
$$S_g(1) = g(1),$$
$$S_g(n+1) = S_g(n) + g(n+1)$$
$$= \text{Sum}(S_g(n), g(\text{Sum}(n, C_{1,1}(n))))$$
[here, the f of 3.10 is $\text{Sum}(y, g(\text{Sum}(x, C_{1,1}(x))))$].

Starting with g a computable $(k+1)$-function and defining
$$S_g(m, \bar{n}) = g(1, \bar{n}) + g(2, \bar{n}) + \cdots + g(m, \bar{n}) = \sum_{i=1}^{m} g(i, \bar{n}),$$
we get a computable function S_g by arguing as above but using Theorem 3.7 instead of Corollary 3.10.

A similar argument gives the computability of H_g defined by
$$H_g(m, \bar{n}) = g(1, \bar{n}) \cdot g(2, \bar{n}) \cdot \ldots \cdot g(m, \bar{n}) = \prod_{i=1}^{m} g(i, \bar{n})$$
(see also Exercise 8).

EXAMPLE 3.13. We prove that every polynomial with positive integral coefficients is computable. We first show by induction on m that each polynomial of the form $f(n_1,\ldots,n_m) = an_1^{k_1}n_2^{k_2}\cdots n_m^{k_m}$ is computable. For $m=1$ we have $f(n_1) = \text{Mult}(a, n_1^{k_1}) = \text{Mult}(C_{1,a}(n_1), \text{Pow}(C_{1,k_1}(n_1), P_{1,1}(n_1)))$. Now suppose $an_1^{k_1}n_2^{k_2}\cdots n_{m-1}^{k_{m-1}}$ is computable. Then so is $an_1^{k_1}n_2^{k_2}\cdots n_m^{k_m}$, since this is

$$\text{Prod}(an_1^{k_1}n_2^{k_2}\cdots n_{m-1}^{k_{m-1}}, n_m^{k_m})$$
$$= \text{Prod}(an_1^{k_1}n_2^{k_2}\cdots n_{m-1}^{k_{m-1}}, \text{Pow}(C_{1,k_m}(n_m), P_{1,1}(n_m))).$$

Thus any polynomial with positive integral coefficients having only one term is computable. Now we use induction on the number of terms. If h has r terms, $r>1$, then $h(n_1,\ldots,n_m) = g(n_1,\ldots,n_m) + f(n_1,\ldots,n_m)$, where g is a polynomial with $r-1$ terms and f is a polynomial with 1 term. So f is computable, as we have just shown, and g is computable by the induction hypotheses. Hence $h(n_1,\ldots,n_m) = \text{Sum}(g(n_1,\ldots,n_m), f(n_1,\ldots,n_m))$ and is computable by Theorems 2.5 and 3.1.

We next consider the computability of some relations. Recall that if P is a relation, then R_P denotes the representing function for P, and P is computable just in case R_P is.

EXAMPLE 3.14. The relations $>$ and $<$ are computable. We saw in Example 2.8 that the set $\{m: m \geq 2\}$ is computable. Let f be the representing function for this set, i.e., $f(m)=1$ if $m \geq 2$ and $f(1)=2$. Then f is computable. It follows that the 2-function g defined by

$$g(m,n) = f((m+1) \dotdiv n)$$

is computable, since g is obtained by composition of computable functions [$\dotdiv$ was shown computable in Example 3.9, and $n+1 = \text{Sum}(n, C_{1,1}(m))$]. Note that $g(m,n)=1$ if and only if $(m+1) \dotdiv n \geq 2$, i.e., if and only if $m - n \geq 1$, or $m > n$. This proves that $g(m,n)=1$ if $m > n$ and $g(m,n)=2$ otherwise. Thus g is the representing function of the relation $>$ (which is the set $\{(m,n): m>n\}$). Since g is computable, $>$ is computable. If we define $h(m,n) = g(n,m)$, then h is the representing function of $<$, i.e., $h(m,n)=1$ if $m<n$ and $h(m,n)=2$ otherwise. We see from Example 3.6 that h is computable and so the relation $<$ is also computable.

EXAMPLE 3.15. The relations $\leq$ and $\geq$ are computable: Let f be the representing function for the relation $>$, i.e., $f(m,n)=1$ if $m>n$, and $f(m,n)=2$ otherwise. Let g be defined by $g(m,n) = 3 \dotdiv f(m,n) = C_{2,3}(m,n) \dotdiv f(m,n)$. Then g is computable (by Example 3.9, Example 3.14, Theorem 2.5, and Theorem 3.1). Also, g is the representing function for $\leq$, for $g(m,n)=1$ if and only if $3 \dotdiv f(m,n) = 1$, and this occurs if and only if $m \not> n$, i.e., if and only if $m \leq n$. Setting $h(m,n) = g(n,m)$ shows that the relation $\geq$ is computable.

EXAMPLE 3.16. The equality relation is the set $\{(x,x): x \in \mathbf{N}^+\}$, denoted as usual by $=$. The relation $=$ is computable: Let f be the representing function of $\geq$, g the representing function of $\leq$, and h the representing function of $>$. Define the 2-function w by $w(m,n) = h(2, \text{Mult}(f(m,n), g(m,n)))$. Then $w(m,n) = 1$ if and only if $2 > f(m,n)g(m,n)$, which happens if and only if $f(m,n) = 1$, $g(m,n) = 1$, i.e., $m \geq n$ and $n \leq m$. Thus w is the representing function of the equality relation.

We are now in a position to prove that the collection of computable sets is closed under the Boolean operations of union, intersection, and complementation. Recall that $^k(\mathbf{N}^+)$ is the set of all k-tuples whose terms belong to $\mathbf{N}^+$, and $X - Y = \{z : z \in X \text{ and } z \notin Y\}$.

Theorem 3.17. *Let X and Y be computable k-relations. Then $^k(\mathbf{N}^+) - X$, $X \cup Y$, and $X \cap Y$ are computable k-relations.*

PROOF: The representing function $R_{[^k(\mathbf{N}^+) - X]}$ for $^k(\mathbf{N}^+) - X$ is $3 \dotdiv R_X(\bar{n})$, since $\bar{n} \in {}^k(\mathbf{N}^+) - X$ implies $\bar{n} \notin X$, so $R_X(\bar{n}) = 2$ and $3 \dotdiv R_X(\bar{n}) = 1$; while if $\bar{n} \notin {}^k(\mathbf{N}^+) - X$, then $\bar{n} \in X$, so $R_X(\bar{n}) = 1$ and $3 \dotdiv R_X(\bar{n}) = 2$. The computability of $3 \dotdiv R_X(n)$ follows from Theorem 3.1.

$R_{X \cup Y} = R_>(3, R_X(\bar{n}) \cdot R_Y(\bar{n}))$, since $\bar{n} \in X \cup Y$ implies $\bar{n} \in X$ or $\bar{n} \in Y$, so $R_X(\bar{n}) = 1$ or $R_Y(\bar{n}) = 1$, and $R_X(\bar{n}) \cdot R_Y(\bar{n})$ is at most 2; on the other hand, $\bar{n} \notin X \cup Y$ implies $R_X(\bar{n}) = 2$ and $R_Y(\bar{n}) = 2$, so $R_X(\bar{n}) \cdot R_Y(\bar{n})$ is greater than 3. Computability follows from Theorem 3.1.

It is also easy to see that $R_{X \cap Y} = R_=(1, R_X(\bar{n}) \cdot R_Y(\bar{n}))$, and computability again follows from Theorem 3.1. □

As we have mentioned before, given a k-relation P and a k-tuple $\bar{n}$, we may write $P\bar{n}$ instead of $\bar{n} \in P$. This notational device is extended to Boolean combinations of relations as follows: Given k-relations P and Q, we may write $P\bar{n} \vee Q\bar{n}$ when $\bar{n} \in P \cup Q$, $P\bar{n} \wedge Q\bar{n}$ when $\bar{n} \in P \cap Q$, and $\neg P\bar{n}$ when $\bar{n} \notin P$.

The following theorem and corollary are analogs of Theorem 3.1 and Corollary 3.5 for relations.

Theorem 3.18. *Let P be a computable r-relation, and let $g_1, \ldots, g_r$ be computable k-functions; then the k-relation Q, defined by '$Q(\bar{n})$ if and only if $P(g_1(\bar{n}), \ldots, g_r(\bar{n}))$', is computable.*

PROOF: Clearly, $R_Q(\bar{n}) = R_P(g_1(\bar{n}), \ldots, g_r(\bar{n}))$. Now apply Theorem 3.1. □

Corollary 3.19. *Let P be a computable r-relation, and let $i_1 \leq k, i_2 \leq k, \ldots, i_r \leq k$. Then the k-relation Q defined by '$Q\bar{n}$ if and only if $P(P_{k,i_1}(\bar{n}), P_{k,i_2}(\bar{n}), \ldots, P_{k,i_r}(\bar{n}))$' is computable.*

PROOF: Immediate from Theorems 2.5 and 3.18. □

2.3 Demonstrating Computability without an Explicit Description of a Turing Machine

EXAMPLE 3.20. Every finite relation is computable: We first consider the special case of a k-relation P with only one member, say $(a_1,\ldots,a_k)$. Clearly, $\bar{n} \in P$ if and only if $(P_{k,1}(\bar{n}) = a_1) \wedge (P_{k,2}(\bar{n}) = a_2) \wedge \cdots \wedge (P_{k,k}(\bar{n}) = a_k)$. The computability of P follows from Theorem 2.5, Example 3.16, Theorem 3.18 [to get the computability of $P_{k,i}(\bar{n}) = a_i$] and Theorem 3.17. Now let Q be a k-relation with t members, say $Q = \{(a_1^1,\ldots,a_k^1), (a_1^2,\ldots,a_k^2),\ldots,(a_1^t,\ldots,a_k^t)\}$. Then $Q = P^1 \cup P^2 \cup \cdots \cup P^t$, where P^i is the relation whose only member is $(a_1^i,\ldots,a_k^i)$ for each $i \leq t$. The computability of Q is now immediate from Theorem 3.17.

If P is a $(k+1)$-relation, then $(\exists x \leq n_1) P(x, n_1, \ldots, n_k)$ is the relation consisting of those k-tuples $(n_1, n_2, \ldots, n_k)$ for which there is at least one m equal to or less than n_1 such that $P(m, n_1, n_2, \ldots, n_k)$. For example, the 2-relation 'z divides y', written $z|y$, can be expressed as $(\exists x \leq y)(x \cdot z = y)$. This relation is of the form $(\exists x \leq y) P(x, y, z)$, where $P(x, y, z)$ is $x \cdot z = y$. We say that $(\exists x \leq n_1) P(x, n_1, n_2, \ldots, n_k)$ is the result of applying the existential bounded quantifier $(\exists x \leq n_1)$ to $P(x, n_1, n_2, \ldots, n_k)$.

Applying the universal bounded quantifier $(\forall x \leq n_1)$ to $P(x, n_1, n_2, \ldots, n_k)$ gives the relation $(\forall x \leq n_1) P(x, n_1, n_2, \ldots, n_k)$ consisting of those k-tuples $(n_1, n_2, \ldots, n_k)$ such that for all m equal to or less than n_1 we have $P(m, n_1, n_2, \ldots, n_k)$. For example, the 1-relation 'y is a prime', written as Prime y, can be expressed as $(\forall x \leq y)(x = y \vee x = 1 \vee (\neg x|y))$.

Starting with a computable P and applying bounded quantification gives another computable relation; this is the content of the next theorem.

Theorem 3.21. *If P is a computable $(k+1)$-relation, then both $(\exists x \leq n_1) P(x, n_1, n_2, \ldots, n_k)$ and $(\forall x \leq n_1) P(x, n_1, n_2, \ldots, n_k)$ are computable k-relations.*

PROOF. The relation $(\exists x \leq n_1) P(x, n_1, n_2, \ldots, n_k)$ can be written as $\sum_{i=1}^{n_1} R_P(i, n_1, n_2, \ldots, n_k) < 2n_1$, and computability follows from the assumption that R_P is computable, the computability of the function $f(n_1, n_2, \ldots, n_k) = 2n_1$, Examples 3.12 and 3.15, and Theorem 3.18. The relation $(\forall x \leq n_1) P(x, n_1, n_2, \ldots, n_k)$ is the same as $\neg((\exists x \leq n_1)(\neg P(x, n_1, n_2, \ldots, n_k)))$, and so is computable by the computability of P, the line above, and Theorem 3.17. □

EXAMPLE 3.22. The relations $x|y$ and Prime x are computable, as is easily seen from the form in which these relations are expressed above and Theorems 3.17, 3.18 and 3.21.

Let P be a $(k+1)$-relation, and let $\bar{n} \in {}^k(\mathbf{N}^+)$. Suppose that there is an x such that $P(\bar{n}, x)$. Then there must be a smallest such x which we denote by $\mu x P(\bar{n}, x)$ [read "the least x such that $P(\bar{n}, x)$"]. If there is no x for which $P(\bar{n}, x)$, then $\mu x P(\bar{n}, x)$ has no meaning. Thus in general, for a given P, $\mu x P(\bar{n}, x)$ will be defined for certain $\bar{n}$ and undefined for others. However, should P have the property that for all $\bar{n} \in {}^k(\mathbf{N}^+)$ there is an x

such that $P(\bar{n},x)$, then $\mu x P(\bar{n},x)$ is defined for all $\bar{n} \in {}^k(\mathbf{N}^+)$. In this case the following equation defines a k-function [with domain ${}^k(\mathbf{N}^+)$]:

$$f(\bar{n}) = \mu x P(\bar{n},x).$$

As we shall see in the next section, if P is computable, there is a Turing machine that will search for the least x such that $P(\bar{n},x)$ when the input is $\bar{n}$. This is stated in the next theorem.

Theorem 3.23. *Let P be a computable $(k+1)$-relation. Then there is a Turing machine M such that for every input $\bar{n}$,*

i. *the output is $\mu x P(\bar{n},x)$ if there is an x such that $P(\bar{n},x)$, or*
ii. *there is no output and no x such that $P(\bar{n},x)$.*

Hence, if for every $\bar{n} \in {}^k(\mathbf{N}^+)$ there is an x such that $P(\bar{n},x)$, then the function f defined by $f(\bar{n}) = \mu x P(\bar{n},x)$ is computable, and in fact is computed by M.

The proof is deferred to the next section.

EXAMPLE 3.24. Let $\mathrm{Prm}(n)$ be the nth prime number in order of magnitude. So we have $\mathrm{Prm}(1)=2$, $\mathrm{Prm}(2)=3$, $\mathrm{Prm}(3)=5$, $\mathrm{Prm}(4)=7$, $\mathrm{Prm}(5)=11$, etc. [Do not confuse the 1-function $\mathrm{Prm}(n)$, which enumerates the primes, with the 1-relation $\mathrm{Prime}(n)$, 'n is a prime'.] $\mathrm{Prm}(n)$ can be defined as follows:

$$\mathrm{Prm}(1) = 2,$$
$$\mathrm{Prm}(n+1) = \mu x (\mathrm{Prime}(x) \wedge (x > \mathrm{Prm}(n))).$$

$(\mathrm{Prime}(x) \wedge (x>y))$ is computable by Examples 3.15 and 3.22 and Theorem 3.17. Moreover, for every n there is an m such that $P(n,m)$. Hence the function f defined by $f(n) = \mu x P(n,x)$ is computable by Theorem 3.23.

EXAMPLE 3.25. Let $\mathrm{Exp}'(m,n)$ be defined by

$$\mathrm{Exp}'(m,n) = \mu x ((\neg m^x | n) \vee m = 1).$$

Then Exp' is computable by examples 3.8, 3.16, and 3.22, and Theorems 2.5, 3.17, 3.18, and 3.23.

A set $\{P_1, \ldots, P_r\}$ of k-relations is said to be a *partition* of ${}^k(\mathbf{N}^+)$ if $P_1 \cup P_2 \cup \cdots \cup P_r = {}^k(\mathbf{N}^+)$ and $P_i \cap P_j$ is empty whenever $i \neq j$. As another application of Theorem 3.23 we have the following theorem, which gives conditions sufficient to conclude that a function defined by cases is computable.

Theorem 3.26. *Let $g_1, \ldots, g_r$ be computable k-functions, and let $P_1, \ldots, P_r$ be computable k-relations such that $\{P_1, \ldots, P_r\}$ is a partition on ${}^k(\mathbf{N}^+)$. Let f be* the k-function defined (by cases) as follows:

$$f(\bar{n}) = g_1(\bar{n}) \text{ if } P_1(\bar{n})$$
$$= g_2(\bar{n}) \text{ if } P_2(\bar{n})$$
$$\cdots$$
$$= g_r(\bar{n}) \text{ if } P_r(\bar{n}).$$

Then f is computable.

PROOF: $f(\bar{n})$ is the least y such that

$$(g_1(\bar{n}) = y \land P_1(\bar{n})) \lor (g_2(\bar{n}) = y \land P_2(\bar{n}))$$
$$\lor \cdots \lor (g_r(\bar{n}) = y \land P_r(\bar{n})).$$

The computability of f now follows from the assumption that the g_i and P_i are computable, Example 3.16, and Theorems 3.17, 3.18, and 3.23. □

EXAMPLE 3.27. Let $\text{Max}(n_1, n_2, n_3)$ be the largest member of the set $\{n_1, n_2, n_3\}$. Max can be defined by cases as follows:

$$\text{Max}(n_1, n_2, n_3) = \begin{cases} n_1 & \text{if } n_1 \geq n_2 \text{ and } n_1 \geq n_3, \\ n_2 & \text{if } n_2 \geq n_3 \text{ and } n_2 > n_1, \\ n_3 & \text{if } n_3 > n_1 \text{ and } n_3 > n_2. \end{cases}$$

The computability of Max follows easily from Example 3.15 and Theorems 3.17 and 3.26. (See also Exercise 16.)

For easier reference, we collect several of the examples in this section in the following theorem.

Theorem 3.28. *The functions and relations described below are computable:*

i. $\text{Mult}(m, n) = m \cdot n$.
ii. $\text{Pow}(m, n) = n^m$.
iii. *All polynomials with positive integral coefficients.*
iv. $m \dotminus n = \begin{cases} m - n & \text{if } m > n, \\ 1 & \text{otherwise.} \end{cases}$
v. $n! = 1 \cdot 2 \cdot \ldots \cdot n$.
vi. *Provided that g is computable, so is*

$$\sum_{i=1}^{m} g(i, \bar{n}) = g(1, \bar{n}) + g(2, \bar{n}) + \cdots + g(m, \bar{n}).$$

vii. *Provided that g is computable, so is*

$$\prod_{i=1}^{m} g(i, \bar{n}) = g(1, \bar{n}) \cdot g(2, \bar{n}) \cdot \ldots \cdot g(m, \bar{n}).$$

viii. $=, \leq, \geq, <, >$.
ix. *All finite relations.*
x. $m | n$ (m *divides* n).

xi. Prime(n) (n is a prime).
xii. Prm(n) = *the nth prime*.
xiii. Exp$'(m,n) = \mu x((\neg m^x | n) \lor m = 1)$.
xiv. Max($n_1, n_2, \ldots, n_k$) = *the largest member of* $\{n_1, n_2, \ldots, n_k\}$.

EXERCISES FOR §2.3

1. Let g be a computable 1-function and k a fixed positive integer. Prove that the 1-functions g', g'' defined below are computable:
   (a) $g'(m) = g(m+k)$.
   (b) $g''(m) = g(km)$.

2. Suppose f is a computable 3-function. Prove that the following functions are computable:
   (a) the 3-function h_1 defined by $h_1(a,b,c) = f(a,c,b)$,
   (b) the 4-function h_2 defined by $h_2(a,b,c,d) = f(b,a,c)$,
   (c) the 4-function h_3 defined by $h_3(a,b,c,d) = f(a+b,c,d)$,
   (d) the 2-function h_4 defined by $h_4(a,b) = f(a,b,a)$,
   (e) the 1-function h_5 defined by $h_5(a) = f(a,a,2a)$.

3. Prove that the 2-function f defined by $f(a,b) = 2a + b$ is computable.

4. Let f be a $(k+2)$-function, and let g be a k-function. Then there is a unique function h that satisfies the following two equations:

$$h(1, \bar{n}) = g(\bar{n}) \tag{1}$$

$$h(m+1, \bar{n}) = f(m, \bar{n}, h(m, \bar{n})) \quad \text{for all } m \in \mathbf{N}^+, \bar{n} \in {}^k(\mathbf{N}^+). \tag{2}$$

[*Hint:* Let A be the set of all functions l such that for some $m^* \in \mathbf{N}^+$

$$l(1, \bar{n}) = g(\bar{n}),$$
$$l(m+1, \bar{n}) = f(m, \bar{n}, h(m, \bar{n})),$$

whenever $m \leq m^*$ and $\bar{n} \in {}^k(\mathbf{N}^+)$. Then $A \neq \emptyset$, and whenever $l_1 \in A$ and $l_2 \in A$, then $l_1 \subseteq l_2$ or $l_2 \subseteq l_1$. Now let $h = \cup A$. This gives the existence of h; the proof of unicity is easier.]

5. Consider the function h defined by the two equations

$$h(1) = 1, \tag{1}$$

$$h(n+1) = \text{Sum}(2n, h(n)). \tag{2}$$

   (a) Find $h(6)$ by direct use of the equations.
   (b) Prove that h is computable.

6. (a) Prove that $h(n) = \sum_{i=1}^{n} i$ is computable.
   (b) Show $h(n,m) = \sum_{i=1}^{n} i^m$ is computable.

7. Define g by the two equations

$$g(1,n) = n \tag{1}$$

$$g(m+1,n) = n^{g(m,n)}. \tag{2}$$

   Show that g is compatible.

8. Let g be a computable 1-function. Show that the 1-function
$$Q_g(n) = \prod_{i=1}^{n} g(i) = g(1)g(2)\ldots g(n)$$
is computable.

9. Show that if f and g are computable 1-functions, then so is
$$Q_{f,g}(n) = \prod_{i=1}^{n} f(i)^{g(i)}.$$

10. Prove that all polynomials in any number of variables are computable.

11. Let $P = \{(a,b) : a^2 > b\}$. Prove that P is computable.

12. Let X be a subset of $\mathbf{N}^+$ such that $\mathbf{N}^+ - X$ is finite. Prove that X is computable.

13. Let $X = \{(a,b,c) : a^2 + b^2 = c^2\}$. Prove that X is computable.

14. Let $Y = \{(a,b,c) : a^2 \dotdiv b^2 > c^2\}$. Prove that Y is computable.

15. Let g.c.d.(m,n) be the greatest common divisor of m and n. Show that this function is computable. Do the same for the least common multiple of m and n, l.c.m.(m,n).

16. Let $f(n) = 1$ for all n if Fermat's last theorem is true. Otherwise let $f(n) = 2$ for all n. Prove that f is computable.

17. Find a relation $R(x,y)$ such that for some but not all n, there is an m such that $R(n,m)$. Let $f(n) = \mu x R(n,x)$. What is the domain of f? In what sense does the Turing machine mentioned in Theorem 3.23 compute this f?

18. Let
$$f(n) = \begin{cases} n & \text{if } n \text{ and } n+2 \text{ are both prime,} \\ 1 & \text{otherwise.} \end{cases}$$
Show that f is computable. Note that f is unbounded if and only if the following unsolved conjecture is true: There are infinitely many twin prime pairs, i.e., there are infinitely many primes p such that $p+2$ is also prime.

19. Let $X = \{1\} \cup \{n : 2n \text{ is the sum of two primes}\}$. Show that X is computable. A famous unsolved conjecture of Goldbach states that every even number greater than 2 is the sum of 2 primes. In other words, the conjecture states that $X = \mathbf{N}^+$.

2.4 Machines for Composition, Recursion, and the Least Operator

In this section we prove Theorems 3.1, 3.7, and 3.23. Each proof consists of the description of a machine whose existence is asserted in the corresponding theorem. For example, Theorem 3.1 asserts the existence of a machine that computes $f(g_1(\bar{n}),\ldots,g_r(\bar{n}))$ when f is a computable r-function and the g's are computable k-functions. Our descriptions of this machine will be given in terms of machines that compute f and the g's. The description is detailed enough so that a machine for the composite function can be written down explicitly whenever machines that compute f and the g's are given explicitly. Careful proofs that our machines do what we claim they do can be given by induction on some parameter of the tape (on m in Lemma 4.2 for example), but this is messy business, and we shall not torture the reader with these details. However, following the machines explicitly through one or two computations should demonstrate the workings of the particular machine under consideration.

Similarly, the proof of computability of a function h defined recursively in terms of computable functions g and f by

$$h(1,\bar{n}) = g(\bar{n}),$$
$$h(m+1,\bar{n}) = f(h(m,\bar{n}), m, \bar{n})$$

consists of a description of a machine that computes h, a description given in terms of machines that compute g and f. As before, one obtains an explicit description of a machine that computes h when provided with machines that compute g and f.

To prove Theorem 3.23 we describe a machine that searches for the least x such that $P(\bar{n}, x)$, provided that P is computable. The description is given in terms of a machine that computes R_P.

Next we describe some "component" machines and several methods of linking machines together. Using these methods with our component machines will yield the machines for composition, recursion, and the μ-operator.

Definition 4.1. Write $M|t \leadsto s$ if there is a computation with respect to M with input t and output s. If $M|t \leadsto s$ and M is initial and t is $(k+j, 1)$: $a_1 \ldots a_k \, b_1 \ldots b_j \ldots b_r \ldots c_1 c_2 \ldots$ and s is $(k+l, n)$: $a_1 \ldots a_k \, b'_1 \ldots b'_l \ldots b'_r \ldots c_1 c_2 \ldots$, then we may write M:
$\ldots b_1 \ldots \overset{*}{b_j} \ldots b_r \ldots \leadsto b'_1 \ldots \overset{*}{b'_l} \ldots b'_r \ldots$, where the $*$ indicates the scanned term.

We also write $\overset{*}{n}$ to indicate that the leftmost cell of the block of n 1's is scanned. Similarly for $\overset{*}{0^n}$ and $\overset{*}{1^n}$.

Lemma 4.2. *There is a machine, which we call*

$$\boxed{\text{compress}},$$

such that for each m and $n \in \mathbf{N}^+$

$$\boxed{\text{compress}} \, | \ldots 1 \overset{*}{0^m} 1 1^{(n-1)} 0 \ldots \leadsto \ldots 1 \overset{*}{0} 1 1^{(n-1)} 0^m \ldots .$$

2.4 Machines for Composition, Recursion, and the Least Operator

PROOF: Let $\boxed{\text{compress}}$ be the following machine:

	0	1
1		1L2
2	1L3	0R7
3	0R4	1R2
4	0L5	1R4
5		0L6
6	0R1	1L6

This machine first checks to see if $m \geq 2$. If not, then the output has the same tape as the input. If $m \geq 2$, then a partial computation yields the tape position

$$(k+m,1): \ldots 1 0^{m-1} \overset{*}{1^n} 0 0 \ldots,$$

and this process of printing a 1 to the left of 1^n and erasing a 1 on the right of 1^n is iterated until we get $\ldots 1 0 1^n 0^m \ldots$. $\square$

Let $M = (d, p, s)$. We may write $M(i,k)$ for $(d(i,k), p(i,k), s(i,k))$. Let $S(M)$ be the set of all k such that either $(0,k) \in \text{Dom}\, M$ (the domain of M), or $(1,k) \in M$, or $k \in \text{Ran}\, s$.

Definition 4.3. Let $u > S(M)$. Define $M_{\underset{\downarrow}{u}}$ to be the machine with domain

$$\{(i,k) : i \in \{0,1\} \text{ and } k \in S(M)\}$$

such that

$$M_{\underset{\downarrow}{u}}(i,k) = \begin{cases} M(i,k) & \text{if } (i,k) \in \text{Dom}\, M, \\ (i, 0, u) & \text{otherwise.} \end{cases}$$

Clearly $(j,k)a$ is an output for M arising from the input t iff $(j,u)a$ is an output of $M_{\underset{\downarrow}{u}}$ arising from the input t.

Definition 4.4. Let $[M,l]$ be the machine with domain $\{(i, l+k) : (i,k) \in \text{Dom}\, M\}$ such that

$$[M,l](i, l+k) = (d(i,k), p(i,k), s(i,k) + l)$$

for each $(i,k) \in \text{Dom}\, M$.

So the table for $[M,l]$ is the result of replacing each state k in the table of M by $l+k$.

Definition 4.5. Let $u = S(M) + 1$.

i. By $\underset{M_1}{\underset{\swarrow}{M}}$ we mean $M_u \cup [M_1, u-1]$. This machine is the result of joining M_1 to M in tandem so that an output from M becomes an input to M_1.

ii. Let M_0 be the machine

| u | $0\,0\,u+1$ | |

By $\underset{M_1}{\underset{\swarrow}{M}}$ we mean $M_u \cup M_0 \cup [M_1, u]$.

This machine takes all outputs $(i,j)a$ of M in which $a_j = 0$ and uses these as inputs to M_1.

iii. Now let M_0 be

| u | | $1\,0\,u+1$ |

Define $\underset{M_1}{\underset{\searrow}{M}}$ to be $M_u \cup M_0 \cup [M_1, u]$. Here all outputs of M in which a 1 is scanned are converted to inputs for M_1.

iv. Let $\underset{\downarrow}{\nearrow M}$ be $M_u \cup M_0$, where

$$M_0 = u\,\boxed{0\,0\,1\quad\quad}.$$

Roughly speaking, an output $(j,k)a$ for M in which $a_j = 0$ is converted to an input by $\nearrow M$ and looped back.

v. Take

$$M_0 = u\,\boxed{\quad\quad 1\,0\,1},$$

and define $M\underset{\downarrow}{\nwarrow}$ to be $M_u \cup M_0$. What would be an output $(j,k)a$ for M, where $a_j = 1$, is looped back to $M\nwarrow$.

vi. Now define

$$\underset{M_1 \quad M_2}{\underset{\swarrow \quad \searrow}{M}}$$

to be $\underset{\downarrow}{M_u} \cup M_0 \cup [M_1, u] \cup [M_2, u+v]$, where M_0 is

| u | $0\,0\,u+1$ | $1\,0\,u+v+1$ |

and $v = S(M_1)$. An output from M becomes an input to M_1 if the scanned term is 0, and an input to M_2 if the scanned term is 1.

2.4 Machines for Composition, Recursion, and the Least Operator

Lemma 4.6. *For each $k \in \mathbf{N}^+$ there is a machine*

$$\boxed{\text{copy } k}$$

such that

$$\boxed{\text{copy } k} \,\big|\, (n_1, \ldots, n_k) \overset{*}{\rightsquigarrow} \left(n_1, \overset{*}{n_2}, \ldots, n_k, n_1\right).$$

PROOF: We consider only the case $k=2$, which is enough to illustrate the general argument. Let M_1 be the machine

	0	1
1		1R2
2	0L3	0R2
3	0L3	1R4

Clearly $M_1 | (n_1, n_2) \overset{*}{\rightsquigarrow} 0\,1\,0^{n_1}\,1^{n_2}$.

Next we need a machine M_2 such that

$$M_2 | 0\,1^{j+1}\,0^{n_1-j}\,1^{n_2}\,0\,1^j \overset{*}{\rightsquigarrow} 0\,1^{j+1}\,0^{n_1-j}\,1^{n_2}\,0\,1^{j+1}.$$

For M_2 we can take

	0	1
1	0R1	1R2
2	0R3	1R2
3	1L4	1R3
4	0L5	1L4
5	0L6	1L5
6	0L6	1R7

Rows 1–3: Go right and add a 1 to 1^j.
Rows 4–6: Go left.

We also need a machine M_3 to check if the cell to the right of the scanned cell in the output of M_2 has a 1:

M_3:

	0	1
1	0R2	
2	0L3	1O3

Now we take $\boxed{\text{copy } 2}$ to be

$$\begin{array}{c} M_1 \\ \downarrow \\ \curvearrowright \begin{bmatrix} M_2 \\ \downarrow \\ M_3 \end{bmatrix} \end{array}.$$

$\square$

Definition 4.7. Define M^k by recursion on k as follows:
$$M^0 = M,$$
$$M^{k+1} = \begin{bmatrix} M^k \\ \downarrow \\ M \end{bmatrix}.$$

For example

$$\boxed{\text{copy } k}^k \Big| \bar{n} \rightsquigarrow \left(n_1, \ldots, n_k \overset{*}{n}_1, \ldots, n_k \right).$$

Lemma 4.8.

i. *There is a machine* $\boxed{\text{shift right}}$ *such that*

$$\boxed{\text{shift right}} \Big| \ldots 1\, 1^{m-1} 0\, 1^n \ldots \overset{*}{\rightsquigarrow} \ldots 1^m 0\, 1\, 1^{n-1} \ldots .$$

ii. *There is a machine* $\boxed{\text{shift left}}$ *such that*

$$\boxed{\text{shift left}} \Big| \ldots 0\, 1^m 0\, 1\, 1^{n-1} \ldots \overset{*}{\rightsquigarrow} \ldots 0\, 1\, 1^{m-1} 0\, 1^n \ldots .$$

iii. *There is a machine* $\boxed{\text{erase}}$ *such that*

$$\boxed{\text{erase}} \Big| \ldots 1\, 1^{n-1} 0\, 1 \ldots \overset{*}{\rightsquigarrow} \ldots 0^n 0\, 1 \ldots .$$

PROOF: Trivial. □

Lemma 4.9. *Let* $g_1, g_2, \ldots, g_r$ *be computable k-functions. Then there is a machine* $Mg_1 g_2 \ldots g_r$ *such that for all* $\bar{n} = (n_1, \ldots, n_k)$ *we have*

$$Mg_1 g_2 \ldots g_r \Big| \bar{n} \overset{*}{\rightsquigarrow} \bar{n} g_1(\bar{n}) g_2(\bar{n}) \ldots g_r(\bar{n}).$$

PROOF: If g is a compatible function we let $\boxed{g}$ denote a machine that computes g.

For Mg_1 take

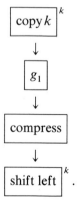

2.4 Machines for Composition, Recursion, and the Least Operator

For $Mg_1 \cdots g_{r-1} g_r$ take

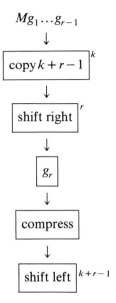

PROOF OF THEOREM 3.1. Let $g_1, g_2, \ldots, g_r$ be computable k-ary functions, and let f be a computable r-ary function. We want to show that the k-ary function h defined by

$$h(\bar{n}) = f(g_1(\bar{n}), g_2(\bar{n}), \ldots, g_r(\bar{n}))$$

is computable. Let $\boxed{f}$ compute f. Then h is computed by

$$\begin{array}{c} Mg_1 \cdots g_r \\ \downarrow \\ \boxed{\text{erase}}^k \\ \downarrow \\ M_f \end{array}$$ □

PROOF OF THEOREM 3.7. Let g be a k-function computed by $\boxed{g}$, and f a computable $(k+2)$-function. Define h by the equations

$$h(1, \bar{n}) = g(\bar{n})$$
$$h(m+1, \bar{n}) = f(h(m, \bar{n}), m, \bar{n}).$$

By Corollary 3.5, the function f' defined by

$$f'(m, \bar{n}, r) = f(r, m, \bar{n})$$

is computable, say by $\boxed{f'}$. This gives an alternate definition for h:

$$h(1, \bar{n}) = g(\bar{n}),$$
$$h(m+1, \bar{n}) = f'(m, \bar{n}, h(m, \bar{n})),$$

and this definition will guide our construction of a machine that computes M.

Let

$$M_1 | 0 \overset{*}{1} \bar{n} \rightsquigarrow \ldots 0 \overset{*}{1}{}^{g(\bar{n})},$$

$$M_1 | 0 \overset{*}{1}{}^m \bar{n} \rightsquigarrow 1 0 \overset{*}{1}{}^{m-1} \bar{n} \quad \text{for } m > 1.$$

For M_1 we can take

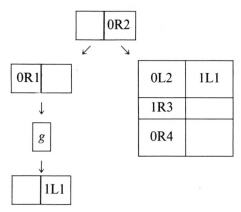

Next we need a machine M_2 such that

$$M_2 | 1 0 \overset{*}{1}{}^{m-1} \bar{n} \rightsquigarrow 1 0 1^{m-1} \bar{n} 0 0 1^{m-2} \bar{n} 0 0 \ldots 1 \bar{n} 0 \overset{*}{1}{}^{g(\bar{n})}.$$

M_2:

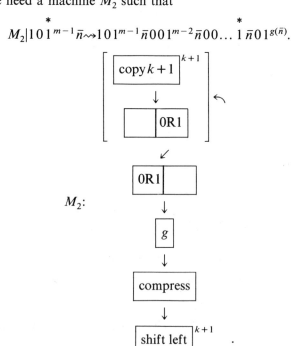

2.4 Machines for Composition, Recursion, and the Least Operator   87

Now we need a machine M_3 such that
$$M_3|0\,1^r\bar{n}00\,1^{r-1}\bar{n}01^p \overset{*}{\leadsto} 0\,1^r\bar{n}01^{f(r-1,\bar{n},p)},$$
and
$$M_3|1\,0\,1^r\bar{n}01^p \overset{*}{\leadsto} 0^l\,0\,1^{f(r-1,\bar{n},p)} \quad \text{for some } l.$$
For M_3 we can take

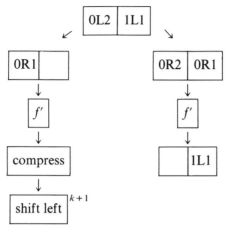

We are now ready to assemble a machine for h:

□

PROOF OF THEOREM 3.23. Let P be a computable $(k+1)$-relation where R_p is computed by $\boxed{R_p}$. We want a machine M such that, given $\bar{n}$, the machine will search for the least m for which $P(\bar{n},m)$.

Let $M_1|\bar{n} \overset{*}{\leadsto} \bar{n}01$.

Let $M_2|\bar{n}\,0\,1^m\,0^l\,1 \overset{*}{\leadsto} \bar{n}01^{m+1}$. For M_2 we can take

1L3	0L2
0L2	1R1
0R4	1L3

↓

$\boxed{\text{shift left}}^k$

Next we describe a machine M_3 such that given $\bar{n}01^{\overset{*}{m}}$, our machine first yields a copy and then uses the copy to compute R_p. If the answer is no, then the output is $\bar{n}01^{m+1}$; if the answer is yes, then the output is $\bar{n}01^m0^{r-1}0$ for some r.

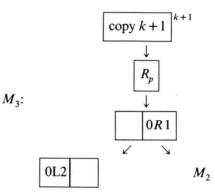

M_3:

We also need a machine $M_4 | \bar{n}01^{\overset{*}{m}} \rightsquigarrow 0^l 1^{\overset{*}{m}}$, where $l = \Sigma_{1 \leq j \leq k} n_j + k + 1$.

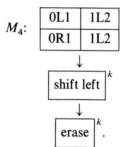

M_4:

We can now assemble our machine M:

$$M_1 \\ \downarrow \\ M_3 \leftarrow\!\!\!\frown \\ \downarrow \\ M_4 \ .$$

□

EXERCISES FOR §2.4

1. Let

$$M = \begin{array}{|c|c|} \hline 1L2 & 1R1 \\ \hline 0R3 & 1L2 \\ \hline \end{array}$$

What is the table for $M_4 \atop \downarrow M$? What is the table for $M \atop \downarrow M$? What 1-function does M^r compute?

2. Let

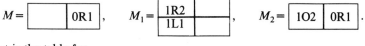

What is the table for

Show that this machine computes the representing function for $\{x : x \geq 2\}$.

3. Define multiplication by recursion as in §2.3. Following the proof of Theorem 3.7, describe a machine for Mult and use this machine to compute Mult(3,2).

4. Suppose that $R(\bar{n}, x)$ and $S(\bar{n}, x)$ are computable relations such that for each $\bar{n}$ there is either some x such that $R(\bar{n}, x)$ or some x such that $S(\bar{n}, x)$. Show that $\mu x(R(\bar{n}, x) \wedge S(\bar{n}, x))$ is a computable function.

2.5 Of Men and Machines

What do we mean when we say that a given set of directions is an algorithm for computing the n-ary function f? We mean that given any $(x_1, \ldots, x_n) \in {}^n(\mathbf{N}^+)$ we invariably obtain $f(x_1, \ldots, x_n)$ after a finite amount of time by following the directions of the algorithm. If there is an algorithm for f, we shall say that f is man computable. For example, the Euclidean algorithm tells us how to divide m by n, where $m \in \mathbf{N}$ and $n \in \mathbf{N}^+$, so as to obtain p and $q \in \mathbf{N}$ such that $pn + r = m$. Hence, the resulting functions $f(n, m) = p$ and $g(n, m) = r$ are man computable.

Given a Turing machine that computes a function f, we can obtain $f(x_1, \ldots, x_n)$ for arbitrary $(x_1, \ldots, x_n) \in {}^n(\mathbf{N}^+)$ by contructing the appropriate tape sequence in accordance with the machine's table. Thus the table provides an algorithm for the computation of f. In this sense any machine computable function is man computable. This, of course, is a philosophical argument, not a mathematical proof, since no definition of man computable function or algorithm has been previously given that is precise enough for mathematical analysis. Nevertheless, it seems that no matter what notion of man computable one might have in mind, a table for the computation of f is an algorithm for the computation of f.

What about the converse? Is there a man computable function that is not machine computable? Turing took the philosophical stand that there is no such function. Thus, in Turing's view a function is man computable iff it is machine computable. Machine computability is then, according to Turing, the correct mathematical definition of our vague intuitive notion of man computability, and a table is an appropriate abstraction of the notion of algorithm.

There are several heuristic arguments that bolster Turing's stand. Perhaps the most simple minded is that in over thirty years no function has been found that is man computable but not machine computable.

Another argument requires adequate faith in the genius of men like Post, Church and Kleene, Gödel, and Turing, along with a reluctance to accept coincidence as an explanation. Each of these men defined a set of number theoretic functions and proposed that their set be taken as the set of man computable functions. Although their definitions seemed to differ radically from each other, it later turned out that they all define the same set of functions, i.e., they all define the set of machine computable functions. Most of these alternative definitions were made before Turing proposed his thesis.

A third argument, and to me the most persuasive, attempts to analyze the intuitive notion of man computability into its simplest components. Let's say we have an algorithm for computing the unary function f. The algorithm is a set of directions given in some language, say English. By restating the directions if necessary, we can assume that the number n is written as n consecutive checks on a tape divided into cells. Thus the directions tell us how to pass from n to $f(n)$, each step of the computation depending on at most the preceding steps, which, as we have seen in previous sections, can be coded as the last step of the computation. It seems quite plausible that the directions can be written in such detail that the passage from one step in the computation to another is accomplished by erasing or writing checks on a tape and moving one cell at a time. This then gives us an algorithm that is very close to a Turing machine.

For these reasons, mathematicians regard the Turing machine as the mathematical definition of algorithm, and Turing machine computability as the precise analog of man computability.

With slight modifications, the above arguments offered in support of the computational equivalence of men and Turing machines can also be given in support of the computational equivalence of real computers and Turing machines. The speed of computation does not enter into these considerations; we require only that the computation end in a finite number of steps.

2.6. Non-computable Functions

Having suffered through the tedium of §2.3, the exhausted reader might be willing to concede that all k-functions are computable. If not, he might suspect that non-computable functions are in some sense rare, or difficult to describe, or even indescribable. However, this is not the case; in this section we show that most functions are not computable, and then we describe some specific examples of non-computable functions.

If one accepts Turing's theses, then these functions are not computable by any "real machine" in any technology, even allowing a computation to

2.6 Non-computable Functions

take an aribtrary (but finite) amount of time and tape. Nor are these functions computable by man; no finite set of directions will tell a man how to find $f(x)$ for arbitrary x if f is not computable by a Turing machine.

The following definition and theorem will be used to prove that among all the functions, relatively few are computable.

By $\#_k$ we mean the k-function with domain $^k\mathbf{N}$ defined by

$$\#_k(n_1, n_2, \ldots, n_k) = p_1^{n_1} \cdot p_2^{n_2} \cdot \ldots \cdot p_k^{n_k}$$

where p_i is the ith prime in order of magnitude. For example, $\#_3(4,1,3) = 2^4 \cdot 3^1 \cdot 5^3$.

Theorem 6.1. *If* $\#_k(\bar{n}) = \#_l(\bar{m})$, *then* $k = l$ *and* $\bar{n} = \bar{m}$.

PROOF: This is an immediate consequence of the unique factorization theorem of arithmetic which says that every $n > 1$ has one and only factorization of the form $n = q_1^{r_1} \cdot q_2^{r_2} \cdot \ldots \cdot q_l^{r_l}$ where $l, r_1, \ldots, r_l \in \mathbf{N}^+$, the q's are all prime, and for each $i < l$, $q_i < q_{i+1}$. (Our theorem uses only the fact that each n has at most one such representation.) Since the unique factorization theorem is usually proved in courses on algebra, we shall not take the time to give a proof of it here. □

Now we assign a number to each machine in such a way that different machines are assigned different numbers. First define

$$\#1, \quad \#0, \quad \#R, \quad \#L, \quad \#O, \quad \#j \text{ for } j \geq 1$$

to be

$$1, \quad 2, \quad 1, \quad 2, \quad 3, \quad j$$

respectively. If r is the row of a machine—say r is

| UVW | XYZ |

—we define $\#r$ to be $\#_6(\#U, \#V, \#W, \#X, \#Y, \#Z)$. If M is a k-row machine, the ith row being r_i, we define $\#M$ to be $\#_k(\#r_1, \#r_2, \ldots, \#r_k)$. For example, if M is the machine

0R2	1R1
1R3	0L2
1R4	

then

$$\#r_1 = 2^2 3^1 5^2 7^1 11^1 13^1,$$
$$\#r_2 = 2^1 3^1 5^3 7^2 11^2 13^2,$$
$$\#r_3 = 2^1 3^1 5^4,$$

and

$$\#M = 2^{\#r_1} 3^{\#r_2} 5^{\#r_3}.$$

The fact that different machines are assigned different numbers is immediate from Theorem 6.1.

From now on, we shall not mention Theorem 6.1 when making use of it.

We can now easily show that there are functions that are not computable. For each computable k-function f, define $G_k(f)$ to be the smallest number m such that $m = \#M$ for some machine M that computes f. Then G_k is a 1-1 function on the set of all computable k-functions into $\mathbf{N}^+$. As was seen in §1.5, there is no 1-1 function whose domain is the set of all k-functions on $\mathbf{N}^+$ and whose range is contained in $\mathbf{N}^+$. Therefore, some k-functions are not computable.

By Theorem 4.13, in Part I we see that there are $\aleph_0$ computable k-functions. (Hence, there are $\aleph_0$ computable functions, because a countable union of countable sets is countable, and the set of computable functions is $\bigcup_{k \in \mathbf{N}^+} C_k$, where C_k is the set of computable k-functions.) Since there are $2^{\aleph_0}$ k-functions, we see that there are $2^{\aleph_0}$ k-functions that are not computable; so most k-functions are not computable.

We shall now see that there are non-computable functions that are very easy to describe.

EXAMPLE 6.2. First we make a list (with repetitions) of all the computable 1-functions as follows. If m is the number of a machine that computes a 1-function, we let f_m be that 1-function; if m is not the number of a machine that computes a 1-function, we let $f_m(n) = 1$ for all n. Then each computable 1-function occurs at least once in the sequence $f_1, f_2, \ldots$. Now define a 1-function F as follows:

$$F(n) = f_n(n) + 1.$$

We claim that F is not computable. For suppose F is computable. Then F is f_l for some l. Hence $F(l) = f_l(l)$, but at the same time $F(l) = f_l(l) + 1$—a contradiction. Therefore F is not computable. (This is another example of a diagonal argument; see the second proof of Theorem 5.1 in Part I.)

EXAMPLE 6.3. As above, let $f_1, f_2, \ldots$ be an enumeration of the computable 1-functions. Let F be the 1-function defined by

$$F(n) = \sum_{i=1}^{n} f_i(n).$$

Then F is not computable. Indeed, if f is a computable 1-function, then there is an m such that for all $n \geq m$, $F(n) > f(n)$. For if f is a 1-function, then $f = f_l$ for some l, and for all $n > l$ we have $F(n) = \sum_{i=1}^{n} f_i(n) > f_l(n)$. In other words, F eventually majorizes every computable 1-function.

EXAMPLE 6.4 (The Self-Halting Problem). We say that M halts for the input A if the complete sequence of tape positions determined by M that begins with A is finite. Let K' be the set of all machines M such that M halts for the input $\#M$. Let $K = \{\#M : M \in K'\}$. The self-halting problem

2.6 Non-computable Functions

can be stated as follows: Is K computable, i.e., is there a machine that, when given the code number $\#M$ of any machine M, prints out 1 if M halts for the input $\#M$ and prints out 2 otherwise?

The answer is no: K is not computable. To see this, we argue by contradiction. Suppose K is computable. This means that R_K (the representing function for K) is computable, where

$$R_K(n) = \begin{cases} 1 & \text{if } n \in K, \\ 2 & \text{otherwise.} \end{cases}$$

Let M_1 be a machine that computes R_K. By making trivial modifications of M_1 if necessary, we can assume that the state of all outputs is the same. Let M_2 be

	1R2
0R2	0L3

Let

$$M = \begin{matrix} M_1 \\ \downarrow \\ M_2 \end{matrix}.$$

Clearly, if $M_1|n \leadsto 1$, then M does not halt when the input is n, and if $M_1|n \leadsto 2$, then $M|n \leadsto 1$. Hence M halts on $\#M$ iff $M_1|\#M \leadsto 1$ iff M does not halt on $\#M$—a contradiction. Therefore the assumption that R_K is computable is untenable, and we must conclude that K is not computable.

EXAMPLE 6.5. Let K_1 be the set of all numbers m such that for some M, $m = \#M$, and M yields the output 1 when the input is $\#M$. A slight modification of the argument used in Example 6.4 proves the non-computability of K_1.

EXAMPLE 6.6. Let K_2 be the set of all numbers of the form $\#M$, where M computes $C_{1,1}$, the constant 1-function whose value is always 1. We shall show that K_2 is not computable. Given a machine M, let $\hat{M}$ be the machine
$\begin{matrix} M_1 \\ \downarrow \\ M \end{matrix}$ where M_1 computes $C_{1,\#M}$; say M_1 is as given in the proof of
Theorem 2.5. It is easy to see that there is an effective procedure for getting $\hat{M}$ from M; hence assuming Turing's theses, the function g, defined as follows, is computable:

$$g(m) = \begin{cases} \#\hat{M} & \text{if } m = \#M, \\ 1 & \text{if for no } M \text{ is } m = \#M. \end{cases}$$

g can be shown computable by the techniques of §2.3 without recourse to Turing's thesis, but we shall not take the time to go through this tedious but straightforward bit of work. Notice that $\#\hat{M} \in K_2$ if and only if $\#M \in K_1$ (we defined K_1 in the preceding example). Hence, $R_{K_2}(g(m)) = R_{K_1}(m)$ for all m. Since g is computable and R_{K_1} is not by Example 6.5, we see by Theorem 3.1 that R_{K_2} is not computable either. Hence, K_2 is not computable.

EXAMPLE 6.7. Let K_3 be the 2-relation consisting of those 2-tuples (m,n) such that m and n are numbers of machines that compute the same 1-function. K_3 is not computable: for let a be the number of a machine which computes $C_{1,1}$. Then $K_3(m,a)$ if and only if $K_2(m)$, and since K_2 is not computable, neither is K_3 (by an application of Theorem 3.1).

EXAMPLE 6.8. If $m = \#M$ and M computes a 1-function, we let $F(m)$ be the smallest number n such that for some M', $n = \#M'$ and M' computes the same 1-function that M computes. If m is not the number of a machine, we let $F(m) = 1$. F is not computable, for otherwise the relation $(F(m_1) = F(m_2)) \wedge (F(m_1) \neq 1)$ would be computable; but this relation is K_3, shown to be non-computable above.

EXAMPLE 6.9 (The Halting Problem). Is there a machine that decides whether or not an arbitrarily given machine always halts? More precisely, let K_4 be the set of those numbers $\#M$ such that M halts for each input. The halting problem asks if K_4 is computable. Using the function g of Example 6.6, we see that

$$K_4(g(\#M)) \text{ if and only if } K(\#M),$$

where K is the non-computable set of Example 6.4. Hence K_4 is not computable.

Assuming Turing's thesis, Example 6.6 shows that there is no decision procedure for determining whether or not an arbitrarily given machine computes the constant function $f(x) = 1$. Example 6.7 shows that no algorithm exists for deciding whether two arbitrarily given machines compute the same 1-function. Example 6.8 may be interpreted as saying that there is no effective procedure for finding the smallest machine that computes the same 1-function as a given machine. The last example shows that no effective procedure exists for testing whether or not a machine halts for all inputs.

EXERCISES FOR §2.6

1. Find non-computable 1-functions f and g such that $f+g$ and $f \cdot g$ are both computable. [Recall that $f+g$ and $f \cdot g$ are the functions whose value at n are $f(n)+g(n)$, and $f(n) \cdot g(n)$ respectively.]

2. Is is true that whenever f and g are non-computable 1-functions, then so is the function h, defined by

$$h(n) = f(g(n))?$$

3. Prove: If $A \subseteq \mathbf{N}^+$ and A is infinite, then there are non-computable sets B and C such that $A = B \cup C$ and $B \cap C$ is empty. (*Hint:* Try a cardinality argument. How many ways are there of partitioning A into two subsets?)
4. Let K be the set of all numbers of machines M that yield some number k as an output for the input $\#M$. Prove that K is not computable.
5. Let E be the set of all numbers of machines that compute bounded 1-functions. Show that K is not computable.

2.7 Universal Machines

Man is a universal computer in the sense that he can compute any function that a machine can. In this section we show that there is a machine U that is just as versatile. Given any k and any k-function f and any machine M that computes f, U yields the output $f(\bar{n})$ when the input is $(\#M, \bar{n})$. Thus U can be "programmed" to compute f, and $\#M$ is such a program. So by Turing's thesis, this machine U can be programmed to compute any function that a man can compute.

We begin by assigning numbers to tape positions. A given tape position A is completely specified by indicating the marker, and which terms have value 1. If only finitely many terms have value 1, then all of this information can be retrieved from any number of the form $\#_t A$, defined to be $2^{c_1} \cdot 3^{c_2} \cdot 5^{d_1} \cdot 7^{d_2} \ldots \mathrm{Prm}^{d_t}(t+2)$, where:

i. c_1 is the number of the scanned term.
ii. c_2 is the state of A.
iii. d_i is 1 if the ith term is 1, and d_i is 2 otherwise.
iv. $t = \max\{c_1, k\}$, where k is the least number such that the $(k+l)$th term of the tape has value 0 for all $l \in \mathbf{N}^+$.

As an illustration, if A is $(5,8)$ $0 1 0 1 1 1 0 0 0 \ldots$, then one $\#_6 A$ is $2^5 \cdot 3^8 \cdot 5^2 \cdot 7 \cdot 11^2 \cdot 13 \cdot 17 \cdot 19$. As another example, $\#_8 A$ is $2^5 \cdot 3^8 \cdot 5^2 \cdot 7 \cdot 11^2 \cdot 13 \cdot 17 \cdot 19 \cdot 23^2 \cdot 29^2$. When writing $\#_t A$ we shall always assume that t satisfies condition iv above.

If l is $\#_t A$ and M is a machine, let $l_M = \#_t M(A)$, and otherwise take l_M to be 1.

We need a computable 2-function STP (read "successor tape position") such that $\mathrm{STP}(m, l) = l_M$ if for some machine M and tape position A,

$$m = \#M \quad \text{and} \quad l = \#_t A.$$

For our purposes the value of STP when (m, l) does not satisfy the condition is immaterial. Later in this section we show that such a function exists, but first we use this fact to obtain a universal machine.

Now let TS (read "tape sequence") be the 3-function defined recursively as follows:

$$\mathrm{TS}(m, l, 1) = \mathrm{STP}(m, l),$$
$$\mathrm{TS}(m, l, k+1) = \mathrm{STP}(m, \mathrm{TS}(m, l, k)).$$

TS is computable, since STP is. Notice that if $A_1, A_2, \ldots$ is a sequence of tape positions determined by M, then $\#_{t+i} A_{i+1}$ is $\mathrm{TS}(\#M, \#_t A_1, i)$ for all $i \in \mathbf{N}^+$. Hence if $\mathrm{TS}(m, l, k) = \mathrm{TS}(m, l, k+1)$, then the sequence determined by M beginning with A_1 ends with the tape position whose number is $\mathrm{TS}(m, l, k)$.

Now let U_1 be a machine that prints out $\mu x(\mathrm{TS}(m, l, x) = \mathrm{TS}(m, l, x+1))$ when this is defined, and does not halt otherwise (an application of Theorem 3.23). Let U_2 compute TS. Let U' be the machine

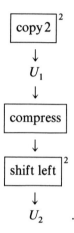

In a weak sense, U' is already a universal machine. For suppose M is a machine that yields the output B when the input is A. Then U' will yield the output $\#_s B$ when the input is $(\#M, \#_t A)$. However, U' has several shortcomings. First of all, A must be coded as $\#_t A$. In the second place, the output of U' obtained from the input $(\#M, \#_t A)$ is $\#_s B$ and not B. In order to do away with these deficiencies, we need machines to code and decode tape positions.

To decode the output tape position we use a machine

$$\boxed{\text{decode}}$$

that computes the function $D(n)$ that subtracts the location of the first 1 on the tape from 1 plus the location of the last 1:

$$D(n) = \left[1 + \left[\mu x \left[(x > 2) \wedge (\neg \mathrm{Prm}^2 x | n) \right. \right. \right.$$
$$\left. \left. \left. \wedge ((\neg \mathrm{Prm}(x+1)|n) \vee (\mathrm{Prm}^2(x+1)|n)) \right] \right] \right]$$
$$\dot{-} \mu x((x > 2) \wedge (\neg \mathrm{Prm}^2(x)|n)).$$

Clearly, $\boxed{\text{decode}}$ yields the output B when the input is $\#_t B$. (See Exercise 1.)

2.7 Universal Machines

The following computable functions will be useful in describing a machine that codes inputs:

$$f(n,1) = \prod_{i=1}^{n} \text{Prm}(i)$$

$$f\left(n, \prod_{i=1}^{m} \text{Prm}^{k_i}(i)\right) = \prod_{i=1}^{n} \text{Prm}(i) \cdot \text{Prm}^2(n+1) \cdot \prod_{i=1}^{m} \text{Prm}^{k_i}(i+n+1) \quad (2)$$

$$\text{when } \prod_{i=1}^{m} \text{Prm}^{k_i}(i+m+1) > 1.$$

It is easy to see that f is computable (Exercise 2). The usefulness of f in coding inputs can be seen by considering the input $A = (n_1, n_2, n_3)$. Then

$$f(n_1,(f(n_2,f(n_3,1))))$$

is $2^{d_1} \cdot 3^{d_2} \cdot 5^{d_3}$, where d_i is as in the definition of $\#_t(A)$.

We also need a function that will tack on the marker $(2,1)$, and this is given by

$$g\left(\prod_{i=1}^{t} \text{Prm}^{d_i}(i)\right) = 2^2 \cdot 3 \cdot \prod_{i=1}^{t} \text{Prm}^{d_i}(i+2). \quad (3)$$

Hence, in the example above, $g(f(n_1,f(n_2,f(n_3,1))))$ is $\#_t A$. It is easy to see that g is computable (Exercise 3).

Now let M_1 be any machine such that

$$M_1|(m,n_1,\ldots,n_k) \rightsquigarrow 0\left(m,n_1,\ldots,\overset{*}{n}_k,1\right).$$

Let M_f compute f, and let M_g compute g. Take M_2 to be the following machine:

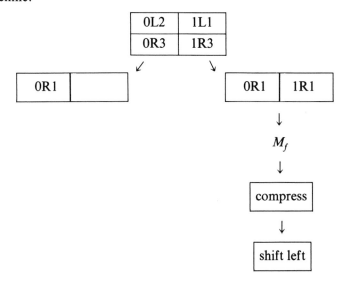

Now let boxed{code} be the machine

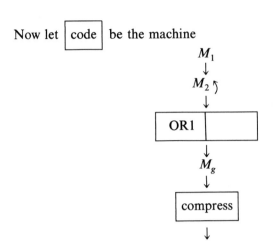

Now as our universal machine U we can take

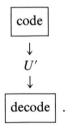

If M is any machine and $(n_1,\ldots,n_k)$ any tuple, then
$$M|(n_1,\ldots,n_k)=y \quad \text{iff} \quad U|(\# M, n_1,\ldots,n_k)=y.$$

We still have to show that STP, the successor tape function, is computable. To do this we require several functions, each of which is easily seen to be computable from its definition.

Recall that $\text{Exp}'(m,n)$ is computable, where $\text{Exp}'(m,n) = \mu x ((m^x \not\mid n) \vee (m=1))$. Now let
$$\text{Exp}(m,n) = \text{Exp}'(\text{Prm}(m),n) \dot{-} 1.$$

Then if the mth prime divides n, $\text{Exp}(m,n)$ is the largest power of $\text{Prm}(m)$ that divides n.

If m is the number of a machine and l is the number of a tape position, then the code number of the relevant row of the table is given by the function
$$\text{RR}(m,l) = \text{Exp}(\text{Exp}(2,l),m).$$

The code number of the relevant column is
$$\text{RC}(l) = \text{Exp}(\text{Exp}(1,l)+2,l).$$

(Recall that the 0-column is coded by 2 and the 1-column by 1.)

2.7 Universal Machines

The next-position function is

$$NP(m,l) = \text{Exp}(6 \dot{-} RC^2(m,l), RR(m,l)).$$

The next-state function is defined as

$$NS(m,l) = \begin{cases} \text{Exp}(2,l) & \text{if } \text{Exp}(1,l)=1 \text{ and } NP(m,l)=2, \\ \text{Exp}(7 \dot{-} RC^2(m,l), RR(m,l)) & \text{otherwise.} \end{cases}$$

The next scanned term is given by

$$NST(m,l) = \begin{cases} \text{Exp}(1,l)=1 & \text{if } NP(m,l)=1, \\ \text{Exp}(1,l) \dot{-} 1 & \text{if } NP(m,l)=2, \\ \text{Exp}(1,l) & \text{if } NP(m,l)=3. \end{cases}$$

The only term that might be altered in the successor tape is the scanned term whose location is $\text{Exp}(1,l)$, and the new value of this term will be the first or fourth entry in the relevant row of the machine, depending on the relevant column, i.e., if v is the new value and $j = \#v$, then $\text{Exp}(5 \dot{-} C^2(m,l), RR(m,l)) = j$. Terms other than the scanned term will not be altered, and we agree that no term will be altered if our machine dictates a move off the left end of the tape. In order to describe the kth term of the successor tape we use the following function:

$$T(k,m,l) = \begin{cases} j \text{ if } (k=\text{Exp}(1,l)) \wedge (\text{Exp}(5 \dot{-} RC^2(m,l), RR(m,l))=j) \\ \quad \wedge (k = \text{Ln}\,l \dot{-} 2) \wedge \neg(\text{Exp}(1,l)=1 \wedge NP(m,l)=2 \\ 2 \text{ if } k = \text{Ln}\,l \dot{-} 1 \\ \text{Exp}(k+2,l) \text{ otherwise.} \end{cases}$$

where $\text{Ln}\,l$ is the length of l, namely $\mu x(\neg \text{Prm}\,x|l) \dot{-} 1$.

Finally we obtain STP as

$$STP(m,l) = 2^{NST(m,l)} 3^{NS(m,l)} \prod_{i=1}^{\text{Ln}\,l+1} [\text{Prm}(i+2)]^{T(i,m,l)}.$$

EXERCISES FOR §2.7

1. Prove that the decode function $D(n)$ in (1) is computable.
2. Prove that the function f defined in (2) is computable.
3. Prove that the function g defined in (3) is computable.
4. Show that there is a machine M_1 such that $M_1|(m, n_1, \ldots, n_k) \stackrel{*}{\leadsto} (m, n_1, \ldots, n_k, 1)$.
5. Show that RR, RC, and NS are computable.
6. Show that NT and STP are computable.
7. Let $X = \{\#M : M \text{ is a machine}\}$. Show that X is computable.
8. Let $L = \{l : l \text{ is the number of a tape position}\}$. Show that L is computable.

2.8 Machine Enumerability

If f is a computable 1-function, then there is an effective procedure for enumerating its range. For suppose that M computes f. We can use M to compute $f(1)$, and then to compute $f(2)$, and then $f(3)$, and so on. In this way we begin an enumeration of the set $\{f(1),f(2),f(3),\ldots\}$ which is the range of f.

On the other hand, an effective procedure for enumerating a set X, say as the list $y_1, y_2, y_3, \ldots$, clearly yields an effective procedure for computing the 1-function $f(n) = y_n$. Accepting Turing's thesis, we conclude that f is machine computable, and so X is the range of a computable function. Thus we are led to make the following

Definition 8.1. A set $Y \subseteq \mathbf{N}^+$ is *machine enumerable* if and only if it is either the range of a computable 1-function or the empty set.

Notice that the range of a computable k-function g is also the range of the computable 1-function f defined by

$$f(x) = g(\mathrm{Exp}(1,x), \mathrm{Exp}(2,x), \ldots, \mathrm{Exp}(k,x)).$$

Hence the notion of machine enumerable is not broadened by substituting 'k-function' for '1-function' in the definition.

Is every machine enumerable set computable? Suppose Y is machine enumerable: say Y is the range of the computable 1-function f. We want to know if there is an effective procedure that will enable us to decide, given arbitrary y, whether or not $y \in Y$. We can, of course, begin to enumerate Y by computing $f(1), f(2), f(3)$, and so on. Then if y does belong to Y, it will appear in our list as some $f(n)$ after finitely many steps. But if $y \notin Y$, then we shall never know it by following this procedure. For after the first 100 computations, we have no guarantee that $y \neq f(101)$, and after the first 10,000 computations we have no guarantee that $y \neq f(10,001)$. We can never be certain that no future computation will yield y as some $f(n)$. Hence, this procedure does not constitute an effective method for determining membership in Y. Indeed, as we next see, there are sets that are machine enumerable but not computable.

EXAMPLE 8.2. Let $K = \{\#M : M$ is a machine which halts for the input $\#M\}$. In Example 6.4 we proved that K *is not computable*. However, K *is machine enumerable*. Our proof of this is based on the computability of the following relations and functions. Write $\mathrm{Row}(x)$ if x is the code number of a row of a machine, i.e.,

$\mathrm{Row}(x)$ iff $(\exists x_1, x_2, \ldots, x_6 < x)$

$[[x = \#_6(x_1, \ldots, x_6) \wedge (x_1 \dotdiv 1)(x_2 \dotdiv 2)(x_4 \dotdiv 1)(x_5 \dotdiv 4) = 1]$

$\vee [x = \#_3(x_1, x_2, x_3) \wedge (x_1 \dotdiv 1)(x_2 \dotdiv 2) = 1]$

$\vee [x = (7^{x_4} 11^{x_5} 13^{x_6}) \wedge (x_4 \dotdiv 1)(x_5 \dotdiv 2) = 1]].$

2.8 Machine Enumerability

The computability of $\text{Row}(x)$ gives us the computability of the set of code numbers of machines:

$$\text{Mach}(x) \quad \text{iff} \quad (\forall n < x)[\neg(\text{Prm}(n|x)) \lor \text{Row}(\text{Exp}(n,x)))].$$

The code number of the input x is given by

$$\text{In}(x) = 2^2 3^1 5^2 \prod_{i=1}^{x} \text{Prm}(i+3).$$

Now define

$$\text{Halt}(m,n,s) \quad \text{iff} \quad \text{Mach}(m) \land [\text{TS}(m,n,s) = \text{TS}(m,n,s+1)].$$

Let d be the number of any machine M that halts on the input $\#M$. Now define

$$f(k) = \begin{cases} m & \text{if } (\exists r,s < k)[2^r 3^s = k \land r = m \land \text{Halt}(m, \ln m, s)] \\ d & \text{otherwise.} \end{cases}$$

Then f is computable and its range is K. Thus K is a machine enumerable, non-computable set.

We extend our definition of machine enumerable to k-relations in a natural way as follows.

Definition 8.3. A k-relation R is *machine enumerable* if and only if $\{\#_k(n_1,\ldots,n_k) : R(n_1,\ldots,n_k)\}$ is.

Thus a k-relation R is machine enumerable if and only if there is a machine which enumerates the code numbers of the k-tuples that belong to R, or R is empty. In fact it is an easy but tedious exercise to modify such a machine so that it will print out the k-tuples themselves rather than the code numbers of the k-tuples. So by Turing's thesis, if there is a set of directions that will enable a man to enumerate the k-tuples belonging to the relation R, then R is machine enumerable, and conversely.

EXAMPLE 8.4. An example of a non-computable machine enumerable 2-relation is the relation R defined by

$R(m,n)$ if and only if m is the code number of a machine which halts for the input n.

R is machine enumerable, since the following function is computable:

$$F(r) = \begin{cases} 2^{\text{Exp}(1,r)} 3^{\text{Exp}(2,r)} & \text{if } (\exists m < r)(\exists n < r)(\exists s < r) \\ & (r = 2^m 3^n 5^s \land \text{Halt}(m,n,s)) \\ 2^{m^*} 3^{l^*} & \text{otherwise.} \end{cases}$$

Here l^* is the number of the 1-input 1, and m^* the number of some machine that halts on 1. From our discussion of the self-halting problem, we see that $R(m, \text{In}(m))$ is not computable, and so neither is $R(m,n)$.

In the next few theorems we explore the relationship between machine enumerability and computability.

Theorem 8.5. *Every computable k-relation is machine enumerable.*

PROOF: Let R be a computable k-relation. If R is empty, then it is machine enumerable by definition. In the case where R is not empty, choose a k-tuple $\bar{s} \in R$. Define f by

$$f(\bar{n}) = \begin{cases} \#_k(\bar{n}) & \text{if } R(\bar{n}), \\ \#_k(\bar{s}) & \text{otherwise.} \end{cases}$$

Clearly, f is computable and R is the range of f. Hence R is machine enumerable (making use of the remark following Definition 8.1). □

The examples above show that the converse of Theorem 8.5 is false.

Theorem 8.6. *R is computable if and only if both R and $\neg R$ are machine enumerable.*

PROOF: If R is computable, then so is $\neg R$ (by Theorem 3.17). So by the last theorem, both R and $\neg R$ are machine enumerable.

To prove the converse, assume that R and $\neg R$ are both machine enumerable k-relations. If R is empty or equal to $^k(\mathbf{N}^+)$, there is nothing to prove. If this is not the case, we choose computable 1-functions f and g such that $\text{Ran} f = \{\#(\bar{n}): R(\bar{n})\}$ and $\text{Ran} g = \{\#(\bar{n}): \neg R(\bar{n})\}$. Since every $\bar{n}$ is either in the relation R or $\neg R$, we know that for every $\bar{n}$ there is an m such that $f(m) = \#_k(\bar{n})$ or $g(m) = \#_k(\bar{n})$. Thus the following relation, which is easily seen to be R, is computable:

$$f(\mu m(f(m) = \#_k(\bar{n}) \vee g(m) = \#_k(\bar{n}))) = \#_k(\bar{n}). \qquad \square$$

Corollary 8.7. *If R is machine enumerable but not computable, then $\neg R$ is not machine enumerable.*

Hence, using the examples at the beginning of this section and taking complements, we obtain examples of non-machine enumerable relations.

We do not define machine enumerable functions. The reason is that machine enumerable functions are computable, as the following theorem shows.

Theorem 8.8. *Suppose R is a machine enumerable $(k+1)$-relation such that for every k-tuple $\bar{m}$ there is exactly one n such that $R(\bar{m}, n)$. Then R is computable, as is the k-function f defined by*

$$f(\bar{m}) = n \quad \text{if and only if} \quad R(\bar{m}, n).$$

PROOF. Let R and f be as in the hypothesis of the theorem. Then $f(\bar{m}) = \mu x(R(\bar{m}, x))$. Hence if R is computable, then so is f (by Theorem

3.23). To show that R is computable, let g be a computable 1-function whose range is $\{\#_{k+1}(\overline{m},n):R(\overline{m},n)\}$. Then $R(\overline{m},n)$ if and only if $g(\text{Exp}(1, \mu x(g(\text{Exp}(1, x)) = \#_k(\overline{m}) \cdot (\text{Prm}(k+1))^{\text{Exp}(2,x)}))) = \#_{k+1}(\overline{m}, n)$. Hence R is computable □

The next theorem pinpoints the relation between machine enumerability and computability even further. If S is a $(k+1)$-relation, we define the k-relation $(\exists x)S(\overline{m},x)$ to hold for $\overline{m}$ if and only if there is an n such that $S(\overline{m},n)$. We say that $(\exists x)S(\overline{m},x)$ is the relation resulting from $S(\overline{m},x)$ by applying the existential quantifier to x. We read '$\exists x$' as 'there is an x'.

Theorem 8.9. *R is a machine enumerable k-relation if and only if there is a computable $(k+1)$-relation S such that*

$$R(\overline{m}) \quad \text{if and only if} \quad (\exists x)S(\overline{m},x).$$

PROOF: Let R be a machine enumerable k-relation. If R is empty, then

$$R(\overline{m}) \quad \text{if and only if} \quad (\exists x)(\#_k(\overline{m}) \neq \#_k(\overline{m}) \wedge x \neq x),$$

where $\#_k(\overline{m}) \neq \#_k(\overline{m}) \wedge x \neq x$ is clearly computable. Suppose R is not empty, so that there is a computable 1-function f with $\text{Ran} f = \{\#_k(\overline{m}):R(\overline{m})\}$. Then

$$R(\overline{m}) \quad \text{if and only if} \quad (\exists x)(f(x) = \#_k(\overline{m})),$$

where $f(x) = \#_k(\overline{m})$ is a computable $(k+1)$-relation.

To prove the other direction of the theorem, we start with a computable $(k+1)$-relation S and define $R(\overline{m})$ if and only if $(\exists x)S(\overline{m},x)$. If S is empty, then R is empty and so is machine enumerable. If S is not empty, we choose $(\overline{m}^*, n^*) \in S$ and define

$$f(\overline{m},n) = \begin{cases} \#_k(\overline{m}) & \text{if } S(\overline{m},n), \\ \#_k(\overline{m}^*) & \text{otherwise.} \end{cases}$$

Then f is a computable function whose range is obviously $\{\#_k(\overline{m}):R(\overline{m})\}$. Hence R is machine enumerable, as we needed to show. □

Given a $(k+1)$-relation S we define a k-relation R by

$$R(\overline{m}) \quad \text{if and only if} \quad \text{for all } n, S(\overline{m},n).$$

We write $(\forall x)S(\overline{m},x)$ in place of $R(\overline{m})$ and say that $\forall x\, S(\overline{m},x)$ is the relation resulting from $S(\overline{m},x)$ by applying the universal quantifier to x. '$\forall x$' is read 'for every x'.

Corollary 8.10. *Let S_1 and S_2 be computable $(k+1)$-relations, and let R be a k-relation. Suppose that*

$$R(\overline{m}) \quad \text{if and only if} \quad (\forall x)S_1(\overline{m},x)$$

$$\quad \text{if and only if} \quad (\exists x)S_2(\overline{m},x).$$

Then R is computable.

PROOF: $\neg R(\bar{m})$ if and only if $(\exists x) \neg S_1(\bar{m}, x)$. Hence by the theorem, both R and $\neg R$ are machine enumerable. Therefore Theorem 8.6 applies, and so R is computable. □

We end this section with a result in the spirit of Theorem 3.18—a result that shows how machine enumerable relations can be combined to yield other machine enumerable relations.

Theorem 8.11.

i. *If R_1 and R_2 are machine enumerable k-relations, then so are $R_1 \cup R_2$ and $R_1 \cap R_2$.*
ii. *If R is a machine enumerable $(k+1)$-relation, then the following are machine enumerable k-relations:*

  (a) $(\exists x) R(\bar{y}, x)$,
  (b) $(\exists x \leqslant y_1) R(\bar{y}, x)$,
  (c) $(\forall x \leqslant y_1) R(\bar{y}, x)$.

PROOF OF I. By Theorem 8.9 there are computable relations S_1 and S_2 such that

$$R_1(\bar{m}) \quad \text{if and only if} \quad (\exists x) S_1(\bar{m}, x)$$

and

$$R_2(\bar{m}) \quad \text{if and only if} \quad (\exists x) S_2(\bar{m}, x).$$

Then $R_1(\bar{m}) \vee R_2(\bar{m})$ if and only if $(\exists x) S_1(\bar{m}, x) \vee (\exists x) S_2(\bar{m}, x)$ if and only if $(\exists x)(S_1(\bar{m}, x) \vee S_2(\bar{m}, x))$, where $S_1(\bar{m}, n) \vee S_2(\bar{m}, n)$ is computable. Hence $R_1 \cup R_2$ is machine enumerable by Theorem 8.9. Similarly, $R_1(\bar{m}) \wedge R_2(\bar{m})$ if and only if $(\exists x)(S_1(\bar{m}, \text{Exp}(1, x)) \wedge S_2(\bar{m}, \text{Exp}(2, x)))$, so $R_1 \cap R_2$ is machine enumerable. □

PROOF OF II. Let R be a machine enumerable $(k+1)$-relation. By Theorem 8.9 there is a computable $(k+2)$-relation S such that

$$R(\bar{y}, x) \quad \text{if and only if} \quad (\exists z) S(\bar{y}, x, z)$$

Hence,

$$(\exists x) R(\bar{y}, x) \quad \text{if and only if} \quad (\exists w)(S(\bar{y}, \text{Exp}(1, w), \text{Exp}(2, w))),$$

which (by Theorem 8.9) proves part ii(a). Also,

$$(\exists x \leqslant y_1) R(\bar{y}, x) \quad \text{if and only if} \quad (\exists z)(\exists x \leqslant y_1) S(\bar{y}, x, z),$$

which proves ii(b) (by an application of Theorem 8.9 and Theorem 3.21). Finally,

$$(\forall x \leqslant y_1) R(\bar{y}, x) \quad \text{if and only if} \quad (\exists w)(\forall x \leqslant y_1) S(\bar{y}, x, \text{Exp}(x, w)),$$

which proves ii(c) (by an application of Theorem 8.9 and Theorem 3.21). □

In a later section we shall see how computability theory is related to the problem of axiomatizing arithmetic, and we shall find some particularly interesting sets that are machine enumerable but not computable, and others that are not even machine enumerable.

EXERCISES FOR §2.8

1. Let f be a computable 1-function, and suppose that f is monotonic increasing, i.e., $f(n+1) \geq f(n)$ for all $n \in \mathbf{N}^+$. Show that $\mathrm{Ran} f$ is computable.

2. Let f and g be computable 1-functions with $f(n) \geq g(n)$ for all $n \in \mathbf{N}^+$. Suppose that g is 1-1 and that $\mathrm{Ran} g$ is computable. Show that $\mathrm{Ran} f$ is computable.

3. Let f be a computable 1-function and X a computable set. By $f^{-1}(X)$ we mean $\{x : f(x) \in X\}$. Must $f^{-1}(X)$ be computable?

4. What is the cardinality of the set of machine enumerable sets? What is the cardinality of the set of sets that are machine enumerable but not computable?

5. Show that every infinite computable set contains an infinite machine enumerable non-computable subset.

6. Show that every infinite machine enumerable set contains an infinite computable subset. (*Hint:* Use Exercise 1 above.)

7. Show that there is a machine enumerable 2-relation R that enumerates all machine enumerable 1-relations, in the sense that for any machine enumerable 1-relation S there is an m such that for all n

$$R(m,n) \quad \text{if and only if} \quad S(n).$$

8. Let $R(m,n)$ hold if and only if there are machines M_1, M_2 such that $m = \#M_1$, $n = \#M_2$, and M_1 and M_2 compute the same 1-functions, or one of the two machines does not compute a function. Is R machine enumerable?

2.9 An Alternate Definition of Computable Function

There are ways of defining the class of computable functions without mentioning machines, and some of these definitions are far more elegant than the one we have been working with. However, the notion of machine has considerable intuitive appeal and provides easier access to examples of non-computable functions that are of interest to those who deal with real computers.

In this section we present one of these alternate definitions of the class of computable functions and sketch a proof of the equivalence of the new definition with the one we have been using. The new definition will be of use in the next chapter.

Before giving the alternate formulation, we first show that any computable function f can be given a rather simple definition in terms of a machine M that computes it. Let $m = \#M$ and $\mathrm{In}_k(\bar{n})$ be the number of the k-input $\bar{n}$. Then

$$f(\bar{n}) = \text{Decode}\big[\text{TS}(m, \text{In}_k(\bar{n}), \text{Halt}(m, \text{In}_k(\bar{n}), s))\big]$$

$$= \text{Decode}\big[\text{TS}(m, \text{In}_k(\bar{n}), \mu s(R_=(\text{TS}(m, \text{In}_k(\bar{n}), s+1),$$

$$\text{TS}(m, \text{In}_k(\bar{n}), s)) = 1))\big]. \qquad (*)$$

Now let $\mathcal{C}$ be the set of computable functions. Suppose that $\mathcal{K}$ is any class of functions such that:

i. For each k the function In_k belongs to $\mathcal{K}$, and the functions Decode, TS, $R_=$, and $f(x) = x + 1$ belong to $\mathcal{K}$.
ii. If f is an r-function and belongs to $\mathcal{K}$, and if $g_1, \ldots, g_r$ are k-functions each belonging to $\mathcal{K}$, then the composition of f with $g_1, \ldots, g_r$ belongs to $\mathcal{K}$.
iii. if g is a $(k+1)$-function belonging to $\mathcal{K}$ and if for every k-tuple $\bar{n}$ there is an m such that $g(m, \bar{n}) = 1$, then the k-function f defined by

$$f(\bar{n}) = \mu x\big[\, g(x, \bar{n}) = 1\,\big]$$

belongs to $\mathcal{K}$.

Then $\mathcal{K}$ contains each function f defined as in (*) above, and so $\mathcal{K} \supseteq \mathcal{C}$.

Except for In_k we already know that each of the functions mentioned in i. belongs to $\mathcal{C}$. We now show that In_k also belongs to $\mathcal{C}$ for each k. First consider the function C defined by

$$C(x, y) = x \cdot \prod_{i=1}^{\text{Ln}\, y} \big[\text{Prm}(\text{Ln}\, x + i)\big]^{\text{Exp}(i, y)}.$$

So if $x = \#(n_1, \ldots, n_k)$ and $y = \#(m_1, \ldots, m_l)$, then $C(x, y) = \#(n_1, \ldots, n_k, m_1, \ldots, m_l)$. Clearly C is computable. Now we have

$$\text{In}_1(n_1) = 2^2 3^1 5^2 \prod_{i=1}^{n_1} \text{Prm}(3 + i)$$

and

$$\text{In}_{k+1}(n_1, \ldots, n_k, n_{k+1}) = C\bigg(C(\text{In}_k(n_1, \ldots, n_k), 2^2), \prod_{i=1}^{n_{k+1}} \text{Prm}(i)\bigg).$$

We see by induction on k that each of the In_k's belongs to $\mathcal{C}$.

Hence if $\mathcal{K}$ is the smallest set of functions satisfying conditions i, ii, and iii above, then $\mathcal{K} \subseteq \mathcal{C}$ (since $\mathcal{C}$ satisfies i, ii, and iii). Thus we have a new characterization of $\mathcal{C}$, namely, $\mathcal{C}$ is the smallest set $\mathcal{K}$ satisfying i, ii, and iii.

The definition as it stands is not particularly elegant or useful, because of the complexity of the functions mentioned in condition i and because the definition of some of them is motivated only within the framework of

2.9 An Alternate Definition of Computable Function

Turing machines. These shortcomings are bypassed by observing that each of the functions appearing in condition i was built up from those mentioned in Theorem 2.5 by composition, recursion, and the μ-operator.

Lemma 9.1. $\mathcal{C}$ is the smallest set $\mathcal{K}$ such that:

i. For each $k,d \in \mathbf{N}^+$ and $t \leq d$, $C_{k,d} \in \mathcal{K}$ and $P_{k,t} \in \mathcal{K}$. Also Sum, $R_=$, and Pred belong to $\mathcal{K}$.
ii. If $h(\bar{n}) = f(g_1(\bar{n}),\ldots,g_k(\bar{n}))$ and $f, g_1, \ldots, g_k \in \mathcal{K}$, then $h \in \mathcal{K}$.
iii. if h is defined by

$$h(1,\bar{n}) = g(\bar{n}),$$
$$h(m+1,\bar{n}) = f(m,\bar{n},h(m,\bar{n})),$$

and $g, f \in \mathcal{K}$, then $h \in \mathcal{K}$.
iv. If $g \in \mathcal{K}$ and for all $\bar{n}$ there is an m such that $g(m,\bar{n}) = 1$, and if $h(\bar{n}) = \mu x(g(x,\bar{n}) = 1)$, then $h \in \mathcal{K}$.

Next we will obtain a more elegant formulation of $\mathcal{C}$ that has only two closure conditions (closure under recursive definitions being omitted).

Definition 9.2. The set of *recursive functions*, Rec, is the smallest set $\mathcal{K}$ such that:

i. $+$, $\cdot$, and $R_>$ belong to $\mathcal{K}$, as does $P_{k,t}$ for each $k \in \mathbf{N}^+$ and $t \leq k$.
ii. If $h(\bar{n}) = f(g_1(\bar{n}),\ldots,g_k(\bar{n}))$ and $f, g_1, \ldots, g_k \in \mathcal{K}$, then $h \in \mathcal{K}$.
iii. If $g \in \mathcal{K}$, and for each $\bar{n}$ there is an m such that $g(m,\bar{n}) = 1$, and if $h(\bar{n}) = \mu x(g(x,\bar{n}) = 1)$, then $h \in \mathcal{K}$.

The main result of this section is

Theorem 9.3. $\mathcal{C} = \text{Rec}$.

As before, it is clear that $\text{Rec} \subseteq \mathcal{C}$, since each of the functions mentioned in condition i belong to $\mathcal{C}$, and $\mathcal{C}$ satisfies conditions ii and iii. By the lemma, if we show that $+$, $R_=$, Pred, and each $C_{k,d}$ and $P_{k,d}$ belong to Rec, and that Rec is closed under recursive definitions, we shall have $\text{Rec} \subseteq \mathcal{C}$ as needed. To see this we need the next lemma as well as some results from number theory.

Definition 9.4. A relation is recursive if its representing function is.

Lemma 9.5.

i. For every $k,d \in \mathbf{N}^+$, the constant function $C_{k,d}$ is recursive.
ii. $\text{Pred} \in \text{Rec}$.
iii. The equality relation is recursive.

iv. *If P and Q are recursive k-relations, then so are*

$$P \lor Q, \quad P \land Q, \quad \text{and} \quad \neg P.$$

v. *If P is a recursive $(k+1)$-relation then so are*

$$(\forall x < n_1)(Px, n_1, \ldots, n_k) \text{ and } (\exists x < n_1)(Px, n_1, \ldots, n_k).$$

PROOF: In each case, the proof follows easily from the form of the equation or equivalence given below that defines the function or relation in question.

i: $\quad C_{k,1}(\bar{n}) = \mu x(x < n_1 + n_1) = \mu x(R_<(x, P_{k,1}(\bar{n})) = 1).$

Now suppose $C_{k,j} \in \text{Rec}$. Then $C_{k,j+1} \in \text{Rec}$, since

$$C_{k,j+1}(\bar{n}) = \mu x(R_<(C_{k,j}(\bar{n}), x) = 1).$$

ii: $\quad \text{Pred}(n) = \mu x(n < x + 2) = \mu x(R_<(n, \text{Sum}(x, C_{1,2}(n))) = 1).$

iii: $\quad R_=(m, n) = R_<(R_<(m, n+1) \cdot R_<(n, m+1), 2).$

iv: $\quad R_{P \lor Q}(\bar{n}) = R_<(R_P(\bar{n}) + R_Q(\bar{n}), 4),$

$\quad\quad R_{P \land Q}(\bar{n}) = R_<(R_P(\bar{n}) + R_Q(\bar{n}), 3),$

$\quad\quad R_{\neg P}(\bar{n}) = R_<(1, R_P(\bar{n})).$

v: Suppose that P is recursive, and let F be the representing function of $(\forall x < n_1)(Px, n_1, \ldots, n_k)$. Then

$$F(\bar{n}) = {}^+ R_<(n_1, \mu x(\neg P(x, \bar{n}) \lor (x = n_1 + 1))).$$

Finally, note that $(\exists x \leq n_1)(Px, n_1, \ldots, n_k)$ iff $\neg(\forall x \leq n_1)(\neg Px, n_1, \ldots, n_k)$ and apply part iii. □

Recall that m and n are relatively prime if 1 is their greatest common divisor. We need several number theoretic facts whose proofs are short enough to be given here.

Theorem 9.6. *If m and n are relatively prime, then there are integers s and t such that $1 = sm + tn$.*

PROOF: Let u be the least positive integer that can be written in the form $sm + tn$ where s and t are integers (positive, negative, or 0). Notice that $u | m$, for if not, then $m = uk + l$ for some k and some l such that $0 < l < u$; hence $m = (sm + tn)k + l$, from which we get $l = (1 - sk)m - tkn$, contradicting the minimality of u. Similarly $u | n$. Since m and n are relatively prime, u must be 1. □

Now define the remainder function Rem as follows:

$$\text{Rem}(m, n) = \begin{cases} r & \text{if } m = nk + r \text{ for some } k \geq 1, n > r \geq 1, \\ 1 & \text{otherwise.} \end{cases}$$

2.9 An Alternate Definition of Computable Function 109

Rem is also defined by

$$\text{Rem}(m,n) = \mu x (\exists k < m (m = kn + x) \lor (x = 1 \land (\exists k < m(m = k \cdot n) \lor m < k))).$$

Using the lemma and this definition, we see that $\text{Rem} \in \text{Rec}$.

Theorem 9.7. (Chinese Remainder Theorem). *If $n_1,\ldots,n_k$ are relatively prime in pairs, and if $a_i < n_i$ for all $i \leq k$, then there is an m such that $\text{Rem}(m,n_i) = a_i$ for all $i \leq k$.*

PROOF: Let $z = n_1 \cdot n_2 \cdot \ldots \cdot n_k$ and let $z_i = z/n_i$. Since z_i and n_i are relatively prime, there are integers s_i and t_i such that $1 = s_i z_i + t_i n_i$. Hence $a_i = a_i s_i z_i + a_i t_i n_i$ or $a_i - a_i t_i n_i = a_i s_i z_i$. Dividing both sides by n_i we get $\text{Rem}(a_i s_i z_i, n_i) = a_i$. Let $m = a_1 s_1 z_1 + \ldots + a_k s_k z_k + lz$ where l is arbitrarily chosen so that $m > 0$. Then

$$\text{Rem}(m,n_i) = a_i \quad \text{for each } i \leq k. \qquad \square$$

Theorem 9.8. *For each finite sequence $a_1,\ldots,a_k$ there is an m and an n such that*

$$\text{Rem}(m, 1 + jn) = a_j$$

for each $j \leq k$.

PROOF: By the preceding theorem it is enough to find an n such that the numbers $1 + jn$ for $j \leq k$ are relatively prime in pairs and $1 + jn > a_j$. Let $n = b!$, where $b = \max\{a_1,\ldots,a_k,k\}$. Surely $1 + jn > a_j$ for each $j \leq k$. Suppose that some prime p divides both $1 + jn$ and $1 + in$ where $i < j \leq k$. Then p divides $(1+jn) - (1+in)$, i.e., p divides $(j-i)n$. Hence $p | j - i$ or $p | n$. But if $p | j - i$, then $p | n$, since $j - i < k < b$ and $n = b!$. So in any case, $p | n$. Along with our assumption that $p | 1 + jn$, we get $p | (1 + jn) - j(n)$, or $p | 1$ —a contradiction. Hence the $1 + jn$ are relatively prime in pairs for $j \leq k$. $\square$

Lemma 9.9. *Let g be a k-function and f a $(k+2)$-function such that g and f are recursive. Then the function h defined as follows is recursive:*

$$h(1, \bar{n}) = g(\bar{n}),$$
$$h(m+1, \bar{n}) = f(m, \bar{n}, h(m, \bar{n})).$$

PROOF: Using the lemma we see that the following relation H is recursive:

$$H(x, y, m, \bar{n}, s)$$

iff $\quad (g(\bar{n}) = \text{Rem}(x, 1+y))$

$\quad \land (\forall w < m)(\text{Rem}(x, 1+(w+1)y) = f(m, \bar{n}, \text{Rem}(x, 1+wy))$

$\quad \land \text{Rem}(x, 1+my) = s).$

Notice that $h(m,\bar{n}) = s$ iff there is an x and y such that $H(x,y,m,n,s)$. The lemma gives the recursiveness of the function

$$G(m,\bar{n}) = \mu x (\exists y < x)(\exists s < x) H(x,y,m,\bar{n},s).$$

This in turn gives the recursiveness of
$$K(m,\bar{n}) = \mu y (\exists s \leqslant G(m,\bar{n})) H(G(m,\bar{n}),y,m,\bar{n},s),$$
and hence the recursiveness of
$$h(m,\bar{n}) = Rm(G(m,\bar{n}), 1 + mK(m,\bar{n}))$$
as needed. □

The two lemmas above conclude the proof of Theorem 9.2.

Exercises for §2.9

1. The class of *primitive recursive functions*, Pr, is the least class Γ such that

    i. $C_{1,1}, R_=, \text{Succ} \in \Gamma$ [recall that $C_{1,1}(n) = 1$ for all n; $R_=(m,n) = 1$ if $m = n$, and $R_=(m,n) = 2$ if $m \neq n$; $\text{Succ } n = n + 1$],
    ii. Γ is closed under composition,
    iii. Γ is closed under definition by recursion.

    (a) Show that $\text{Succ}, \text{Prod} \in \text{Pr}$.
    (b) Show that there is some 2-function $F \in \text{Pr}$ such that for each 1-function $g \in \text{Pr}$ there is an m for which
    $$g(n) = F(m,n)$$
    whenever $n \in \text{Pr}$.
    (c) Conclude from (b) that $\text{Pr} \not\subseteq \text{Rec}$.

2. Let Ω be the least class of relations Δ such that

    i. $+, \cdot$ (as 3-relations) $\in \Delta$,
    ii. if $R, S \in \Delta$ and both are k-relations, then $R \cup S \in \Delta$ and $\neg R \in \Delta$,
    iii. if R is a k-relation in Δ and $p: k \to k$, then S is a k-relation in Δ, where S is defined by
    $$S(n_1,\ldots,n_k) \quad \text{iff} \quad R(n_{p(1)},\ldots,n_{p(k)}),$$
    iv. if R is a k-relation in Δ and
    $$S(n_1,\ldots,n_k) \quad \text{iff} \quad (\forall y \leqslant n_k) R(n_1,\ldots,n_{k-1},n_k),$$
    then $S \in \Delta$,
    v. if R_1, R_2 are $(k+1)$-relations in Δ, and if $S(n_1,\ldots,n_k)$ iff $(\exists y) R_1(n_1,\ldots,n_k,y)$ iff $(\forall y) R_2(n_1,\ldots,n_k,y)$, then $S \in \Delta$.

    Show that Ω is the set of recursive relations.

2.10 An Idealized Language

In his famous address of 1928 to the International Congress in Bologna, David Hilbert proposed a search for an adequate axiomatic foundation for the arithmetic of the natural numbers. His hope was to use such an axiomatization as a base for all of mathematics, since the other number

2.10 An Idealized Language

systems such as the rationals, the reals, the complex numbers, etc., can be defined in terms of the arithmetic of the natural numbers in such a way that their properties can be derived from those of the natural numbers. Unfortunately, in 1931, Gödel proved that this goal was not obtainable and with his incompleteness theorem pointed out startling limitations inherent in the axiomatic method. With Gödel's remarkable theorem as our goal, our first task is to define a language that is appropriate for an axiomatic development of number theory. This language, which we call L, can be viewed as a fragment of English, so formalized that the set of meaningful expressions is precisely defined, and the meaning of each assertion is unambiguous.

Since we shall be using the English language to discuss expressions of L, we make the symbols of L distinct from those of English in order to avoid confusion.

THE SYMBOLS OF L

Variables: $v_1, v_2, v_o, \ldots$.
Constant symbols: $\mathbf{1, 2, 3}, \ldots$.
Equality symbols: $\approx$.
Symbols for 'and', 'or', and 'not': $\wedge, \vee, \neg$.
Existential and universal quantifier symbols: $\exists, \forall$.
Symbols for the functions of addition and multiplication: $\oplus, \odot$.
Left and right parentheses: [,].

Definition 10.1a. An *expression* is a finite sequence of symbols of L.

For example,],], $\forall, \mathbf{9}, \exists, \forall$ is an expression. From now on we shall omit the commas between successive terms of expressions, so that this will be written $]]\forall\mathbf{9}\exists\forall$.

Definition 10.1b. The set of *terms*, Trm, is the smallest set X of expressions such that

i. $v_i \in X$ for all $i \in \mathbf{N}^+$,
ii. $\mathbf{i} \in X$ for all $i \in \mathbf{N}^+$,
iii. if $t_1, t_2 \in X$, then so are $[t_1 \oplus t_2]$ and $[t_1 \odot t_2]$.

Some examples of terms are v_{28}, $\mathbf{9}$, and $[[\mathbf{1} \oplus v_4] \odot [\mathbf{4} \odot \mathbf{3}]]$. To see that the last expression is a term, we first observe by conditions i and ii that $\mathbf{1}, v_4, \mathbf{4},$ and $\mathbf{3}$ are terms. Hence by condition iii, $[\mathbf{1} \oplus v_4]$ and $[\mathbf{4} \odot \mathbf{3}]$ are terms; call them t_1 and t_2 respectively. Hence by iii again $[t_1 \odot t_2]$ is a term as needed.

The expressions $\mathbf{3} \oplus \mathbf{4}$ and $\mathbf{2} \odot v_1]$ are not terms.

Without the parentheses, we could not distinguish between $[[\mathbf{2} \oplus \mathbf{4}] \odot \mathbf{3}]$ and $[\mathbf{2} \oplus [\mathbf{4} \odot \mathbf{3}]]$; i.e., the expression $\mathbf{2} \oplus \mathbf{4} \odot \mathbf{3}$ could be viewed both as $t_1 \oplus t_2$ and $t'_1 \odot t'_2$ where t_1, t_2, t'_1, t'_2 are respectively $\mathbf{2} \oplus \mathbf{4}, \mathbf{3}, \mathbf{2},$

and $4 \odot 3$. That this kind of ambiguity is avoided by the use of parentheses is the content of the following:

Theorem 10.2 (Unique Readability for Terms). *For each term t, exactly one of the following conditions hold*:

i. *there is a unique i such that $t = v_i$,*
ii. *there is a unique i such that $t = \mathbf{i}$,*
iii. *there is exactly one sequence t_1, t_2, s where $t_1, t_2 \in \mathrm{Trm}$ and $s \in \{\oplus, \odot\}$ such that $t = [t_1 s t_2]$.*

PROOF: We first observe that a proper initial segment of a term has fewer right parentheses than left parentheses, and that a proper end segment of a term has fewer left parentheses than right parentheses. For let Y be the set of all terms for which this is true. Clearly all variables and constant symbols belong to Y (these have no proper segments), and it is easily seen that $[t_1 \oplus t_2] \in Y$ and $[t_1 \odot t_2] \in Y$ if $t_1, t_2 \in Y$. Hence $Y = \mathrm{Trm}$. It follows that a term has as many right parentheses as left parentheses.

Now suppose that t is a term that is neither a variable or a constant symbol, say $t = [t_1 * t_2]$, where '$*$' is either '$\odot$' or '$\oplus$'. If t can also be written differently as $[s_1 \# s_2]$, then s_1 is a proper initial segment of t_1 or t_1 is a proper initial segment of s_1, which is impossible in view of our remarks about parentheses. □

Of course, our intention is to have the symbol **1** denote the number 1, the symbol **2** denote the number 2, and so on. In the most natural way we use terms other than the constant symbols to denote numbers also. For example, $[\mathbf{1} \oplus \mathbf{1}]$ denotes the number 2, while $[[[\mathbf{1} \oplus \mathbf{1}] \oplus \mathbf{1}] \oplus \mathbf{1}]$, $[\mathbf{2} \odot \mathbf{2}]$, $[\mathbf{1} \oplus [\mathbf{2} \oplus \mathbf{1}]]$ are three terms denoting the number 4.

On the other hand, terms having variables have no denotation. For example, $[\mathbf{3} \oplus v_9]$ has no denotation, since v_9 does not denote a specific number. However, if a particular number is assigned to the variable v_9, then we may think of $[\mathbf{3} \oplus v_9]$ as denoting a number. For example, if we assign the number 5 to v_9, then $[\mathbf{3} \oplus v_9]$ denotes 8. We now spell out these ideas more precisely.

Definition 10.3a. An *assignment* is a function whose domain is $\{v_1, v_2, v_3, \ldots\}$ and whose range is contained in $\mathbf{N}^+$.

Definition 10.3b. If z is an assignment and t is a term, we define $t\langle z \rangle$, the denotation of t under z, as follows:

i. If t is the variable v, then $t\langle z \rangle$ is $z(v)$.
ii. If t is $\mathbf{n}$, then $t\langle z \rangle = n$.
iii. If $t = [t_1 \oplus t_2]$, then $t\langle z \rangle = t_1\langle z \rangle + t_2\langle z \rangle$. If $t = [t_1 \odot t_2]$, then $t\langle z \rangle = t_1\langle z \rangle \cdot t_2\langle z \rangle$.

2.10 An Idealized Language

For example, let $t = [[v_3 \odot v_1] \oplus [2 \oplus v_9]]$, and let z be any assignment such that $z(v_1) = 5$, $z(v_3) = 2$, $z(v_9) = 3$. Then by clause i and clause ii, $v_1\langle z\rangle = 5$, $v_3\langle z\rangle = 2$, $v_9\langle z\rangle = 3$, and $2\langle z\rangle = 2$. Using clause iii, $[v_3 \odot v_1]\langle z\rangle = v_3\langle z\rangle \cdot v_1\langle z\rangle = 5 \cdot 2 = 10$, and $[2 \oplus v_9]\langle z\rangle = 2\langle z\rangle + v_9\langle z\rangle = 2 + 3 = 5$. Using clause iii again we get $t\langle z\rangle = [v_3 \odot v_1]\langle z\rangle + [2 \oplus v_9]\langle z\rangle = 10 + 5 = 15$.

Definition 10.4a. An *atomic formula* is an expression of the form $[t_1 \approx t_2]$ where t_1 and t_2 are terms.

Definition 10.4b. The set of all *formulas* (call it Fm) is the smallest set X of expressions such that

i. every atomic formula belongs to X,
ii. if $\varphi \in X$ then $[\neg \varphi] \in X$,
iii. if $\varphi \in X$ and $\psi \in X$, then both $[\varphi \wedge \psi]$ and $[\varphi \vee \psi]$ belong to X,
iv. if $\varphi \in X$ and v is a variable, then both $[\forall v \varphi]$ and $[\exists v \varphi]$ belong to X.

For example, the following expressions are formulas:
$[[v_3 \odot 3] \approx [[v_3 \oplus v_1] \oplus 9]]$, $[\exists v_3 [[[1 \oplus 1] \odot v_3] \approx v_1]]$, $[\forall v_4 [[v_4 \approx 2] \vee [\exists v_1 [[v_1 \oplus 1] \approx v_4]]]]$. Only the first of these is an atomic formula. To see that the last expression is a formula we observe by clause i that $[v_4 \approx 2]$ and $[[v_1 \oplus 1] \approx v_4]$ are (atomic) formulas. Hence by clause iv, $[\exists v_1 [[v_1 \oplus 1] \approx v_4]]$ is a formula. By clause iii, $[[v_4 \approx 2] \vee [\exists v_1 [[v_1 \oplus 1] \approx v_4]]]$ is a formula. Finally, by clause iv, $[\forall v_4 [[v_4 \approx 2] \vee [\exists v_1 [[v_1 \oplus 1] \approx v_4]]]]$ is a formula.

Theorem 10.5 (Unique Readability for Formulas). *If φ is a formula, then exactly one of the following conditions is true*:

i. *there are unique terms t_1, t_2 such that $\varphi = [t_1 \approx t_2]$*,
ii. *there is a unique formula ψ such that $\varphi = [\neg \psi]$*,
iii. *there is a unique triple ψ_1, ψ_2, s where ψ_1 and ψ_2 are formulas and s is either $\vee$ or $\wedge$ and $\varphi = [\psi_1 s \psi_2]$*,
iv. *there is a unique triple φ, v, ψ where Q is either $\forall$ or $\exists$, v is a variable, and ψ is a formula, and $\varphi = [Qv\psi]$*

A proof for Theorem 10.5 can be given along the lines of that for Theorem 10.2 (see Exercise 3).

Some of our formulas can be interpreted in the obvious way as assertions about arithmetic. For example, $[\forall v_1 [\exists v_2 [[[2 \odot v_2] \approx v_1] \vee [[2 \odot v_2] \approx [v_1 \oplus 1]]]]]$ is intended to assert that every number is either divisible by 2 or its successor is. The intended assertion of the formula $[\exists v_1 [\forall v_2 [\exists v_3 [[v_2 \oplus v_3] \approx v_1]]]]$ is that there is a largest natural number. On the other hand,

$[\exists v_1[\exists v_2[[\neg[v_1\approx 1]]\wedge[\neg[v_2\approx 1]]\wedge[[v_1\odot v_2]\approx v_3]]]]$ makes no assertion unless we assign a number to v_3, in which case the formula will be true if the number assigned is not a prime and false otherwise. We now make these notions more precise.

Let z be an assignment, v a variable, and $n\in\mathbf{N}^+$. Then by $z\binom{v}{n}$ we mean the assignment defined by

$$z\binom{v}{n}(u) = \begin{cases} z(u) & \text{if } u\neq v, \\ n & \text{if } u=v. \end{cases}$$

Definition 10.6. We say that the assignment z *satisfies the formula* φ and write $\mathbf{N}^+\vDash\varphi\langle z\rangle$ if either

i. for some $t_1, t_2 \in \text{Trm}$, φ is $[t_1 \approx t_2]$ and $t_1\langle z\rangle = t_2\langle z\rangle$,
ii. for some $\psi\in\text{Fm}$, φ is $[\neg\psi]$ and it is not the case that $\mathbf{N}^+\vDash\psi\langle z\rangle$,
iii. for some $\psi_1, \psi_2\in\text{Fm}$, φ is $[\psi_1\wedge\psi_2]$ and both $\mathbf{N}^+\vDash\psi_1\langle z\rangle$ and $\mathbf{N}^+\vDash\psi_2\langle z\rangle$,
iii. for some $\psi_1, \psi_2\in\text{Fm}$, φ is $[\psi_1\vee\psi_2]$ and either $\mathbf{N}^+\vDash\psi_1\langle z\rangle$ or $\mathbf{N}^+\vDash\psi_2\langle z\rangle$ (as usual, 'or' here is used in the inclusive sense),
iv. for some $\psi\in\text{Fm}$ and some variable v, φ is $[\forall v\psi]$, and for all n,

$$\mathbf{N}^+\vDash\psi\langle z\binom{v}{n}\rangle,$$

or

iv'. for some $\psi\in\text{Fm}$ and some variable v, φ is $[\exists v\psi]$, and there is at least one n such that

$$\mathbf{N}^+\vDash\psi\langle z\binom{v}{n}\rangle.$$

As an example, take $\varphi = [\forall v_1[\exists v_2[[2\odot v_2]\approx v_1]]]$ and let z be an arbitrary assignment. By clause iv, $\mathbf{N}^+\vDash\varphi\langle z\rangle$ only if for all n we have

$$\mathbf{N}^+\vDash\varphi\langle z\binom{v_1}{n}\rangle.$$

By iv', this is the case only if for all n there is an m such that

$$N^+\vDash[[2\odot v_2]\approx v_1]\langle z\binom{v_1}{n}\binom{v_2}{m}\rangle,$$

i.e., only if for all n there is an m such that $2\cdot m = n$. This is not the case if n is odd, and so z does not satisfy φ. Of course, since z is arbitrary in this example, we see that no z satisfies φ.

Definition 10.7.

i. If S is a sequence $s_1, s_2, \ldots, s_n$ and if $1\leq i\leq j\leq n$, then the *i,j-subsequence* of S is $s_i, s_{i+1}, \ldots, s_j$.
ii. ψ is the *i,j-subformula* of φ if ψ is a formula and the i,j-subsequence of φ. ψ is the *subformula* of φ if ψ is the i,j-subformula of φ for some i,j
iii. The symbol s *occurs at* i in φ if s is the ith term of the sequence φ.

2.10 An Idealized Language

iv. The variable v *is bound at* k in φ if v occurs at k in φ and for some $i < k < j$, the i,j-subsequence of φ is a subformula of the form $[\forall v \psi]$ or $[\exists v \psi]$.

v. A variable v that occurs at k in φ but is not bound there is said to occur *free at* k. v occurs *free* (*bound*) in φ if for some k, v occurs free (bound) at k.

vi. A formula φ is an *assertion* if no variable occurs free in it.

For example, let $\varphi = [\exists v_3[[\forall v_2[\neg[v_4 \approx v_2]]] \vee [v_3 \approx v_2]]]$. Then $\forall v_2[\neg[v_4 \approx v_2]]$ is the 6,15-subsequence of φ, while $[\neg[v_3 \approx v_2]]$ is the 8,15-subsequence of φ. The first is not a subformula but the second is. v_2 is bound at 7 and 13 but free at 21. v_3 is bound at both of its occurrences, and v_4 is free at its single occurrence. φ is not an assertion, but $[\forall v_2[\forall v_4 \varphi]]$ is.

Theorem 10.8.

i. *If $t \in \mathrm{Trm}$, and z and z' are assignments such that $z(v) = z'(v)$ for all variables v occurring in t, then*
$$t\langle z \rangle = t\langle z' \rangle.$$

ii. *If $\varphi \in \mathrm{Fm}$ and z and z' are assignments such that $z(v) = z'(v)$ for all variables v occurring free in φ, then*
$$\mathbf{N}^+ \vDash \varphi\langle z \rangle \text{ if and only if } \mathbf{N}^+ \vDash \varphi\langle z' \rangle.$$

PROOF OF I. Let X be the set of all terms t for which Theorem 10.8i is true, i.e., all terms t such that $t\langle z \rangle = t\langle z' \rangle$ whenever $z(v) = z'(v)$ for all variables v occurring in t. Clearly, every constant symbol and every variable belongs to X. Moreover, if t_1 and t_2 belong to X, then $[t_1 \oplus t_2]\langle z \rangle = t_1\langle z \rangle + t_2\langle z \rangle = t_1\langle z' \rangle + t_2\langle z' \rangle = [t_1 \oplus t_2]\langle z' \rangle$, and so $[t_1 \oplus t_2] \in X$. Similarly $[t_1 \odot t_2]$ belongs to X if $t_1, t_2 \in X$. Hence, by Definition 10.1b, $X = \mathrm{Trm}$ as we needed to show. □

PROOF OF II. Let X be the set of all formulas for which Theorem 10.8ii holds. Using part i, we see that the atomic formulas are in X.

Suppose that $\psi \in S$ and that z and z' are assignments that agree on the free variables that occur in ψ. Then $\mathbf{N}^+ \vDash \psi\langle z \rangle$ if and only if $\mathbf{N}^+ \vDash \psi\langle z' \rangle$. Noting that ψ and $[\neg \psi]$ have the same free variables, and using Definition 10.6ii, we see that the following statements are equivalent:

$\mathbf{N}^+ \vDash [\neg \psi]\langle z \rangle$.
It is not the case that $\mathbf{N}^+ \vDash \psi\langle z \rangle$.
It is not the case that $\mathbf{N}^+ \vDash \psi\langle z' \rangle$.
$\mathbf{N}^+ \vDash [\neg \psi]\langle z' \rangle$.

Hence, if $\psi \in X$, then $[\neg \psi] \in X$.

Now suppose that $\psi_1 \in X$ and $\psi_2 \in X$. Clearly, every variable that occurs free in ψ_1 or occurs free in ψ_2 also occurs free in $[\psi_1 \vee \psi_2]$. Now suppose that z and z' are assignments that agree on the variables occurring free in $[\psi_1 \vee \psi_2]$. Since $\psi_1, \psi_2 \in X$, we have that $\mathbf{N}^+ \vDash \psi_1 \langle z \rangle$ if and only if $\mathbf{N}^+ \vDash \psi_1 \langle z' \rangle$, and $\mathbf{N}^+ \vDash \psi_2 \langle z \rangle$ if and only if $\mathbf{N}^+ \vDash \psi_2 \langle z' \rangle$. Hence, using Definition 10.1b iii, we see that the following statements are equivalent:

$\mathbf{N}^+ \vDash [\psi_1 \vee \psi_2] \langle z \rangle$.
$\mathbf{N}^+ \vDash \psi_1 \langle z \rangle$ or $\mathbf{N}^+ \vDash \psi_2 \langle z \rangle$.
$\mathbf{N}^+ \vDash \psi_1 \langle z' \rangle$ or $\mathbf{N}^+ \vDash \psi_2 \langle z' \rangle$.
$\mathbf{N}^+ \vDash [\psi_1 \vee \psi_2] \langle z' \rangle$.

Hence, if $\psi_1, \psi_2 \in X$, then $[\psi_1 \vee \psi_2] \in X$. A similar argument shows that if $\psi_1, \psi_2 \in X$, then $[\psi_1 \wedge \psi_2] \in X$.

Now suppose $\psi \in X$, and let z and z' be assignments that agree on all variables occurring free in $[\exists v \psi]$. Suppose that $\mathbf{N}^+ \vDash [\exists v \psi] \langle z \rangle$. By Definition 10.6iv' this means that for some assignment w agreeing with z on all variables except possibly v, we have $\mathbf{N}^+ \vDash \psi \langle w \rangle$. Now let $w'(v) = w(v)$ and $w'(u) = z'(u)$ for all $u \neq v$. Since we are assuming that $\psi \in X$, and since a variable that occurs free in ψ either is v or occurs free in $[\exists v \psi]$, we see that $\mathbf{N}^+ \vDash \psi \langle w' \rangle$. Hence, by Definition 10.6iv' again, $\mathbf{N}^+ \vDash [\exists v \psi] \langle z' \rangle$. Thus if $\psi \in X$, so is $[\exists v \psi]$. Similarly, one shows that if $\psi \in X$, then $[\forall v \psi] \in X$.

Hence, by Definition 10.4a, we see that $X = \mathrm{Fm}$, as needed to prove the theorem. □

Recall that an assertion is a formula without free variables. By Theorem 10.8ii, if φ is an assertion, then $\mathbf{N}^+ \vDash \varphi \langle z \rangle$ either for all z or for no z. Since the assignment is immaterial, we shall write $\mathbf{N}^+ \vDash \varphi$ and say that φ *is true* in the first case, and write $\mathbf{N}^+ \nvDash \varphi$ and say that φ *is false* in the second case.

To facilitate the writing and reading of formulas of L, we shall omit those occurrences of parentheses that are not needed to avoid ambiguity, and we shall write $+$ and $\cdot$ instead of $\oplus$ and $\odot$. For example, we might write

$$\forall v_1 \exists v_2 [v_1 \approx v_2 \wedge v_2 \cdot c_3 \approx v_4 \wedge \exists v_5 [v_3 + v_5 \approx v_4]] \tag{1}$$

instead of

$$\left[\forall v_1 \left[\exists v_2 \left[[[v_1 \approx v_2] \wedge [[v_2 \odot v_3] \approx v_4]] \wedge [\exists v_5 [[v_3 \oplus v_5] \approx v_4]] \right] \right] \right] \tag{2}$$

[So according to the context, (1) might be an expression that is not a formula, or an abbreviation for the formula (2).]

When in a mathematical discussion a statement of the form 'If A then B' or of the form 'A implies B' is made, it is intended to mean that B is true if A is true, but if A is false then B may be either true or false. So the statement 'If A then B' has the same meaning as 'Either A is false or B is true'. It is handy to have a counterpart of such statements in L, so we may abbreviate formulas of the form '$[\neg\varphi]\vee\psi$' by '$\varphi\rightarrow\psi$' and read these as 'If φ then ψ' or as 'φ implies ψ'. Similarly, a mathematical statement of the form 'A if and only if B' means that either A and B are both true or both false, i.e., if A then B, and if B then A. The formalized counterpart of this in L is '$[\varphi\rightarrow\psi]\wedge[\psi\rightarrow\varphi]$', which we further abbreviate as '$\varphi\leftrightarrow\psi$' and read '$\varphi$ if and only if ψ'.

Definition 10.9. The variable v *is free at k for the term τ* in the formula φ if

i. v occurs free at k in φ, and
ii. if u is any variable occurring in τ, and φ' is the result of replacing v at the kth place in φ by u, then u is free at k in φ'.

For example, if v is v_2, τ is $[v_1+v_2]\cdot v_1$, and φ is $\exists v_3[v_2\approx v_3]\wedge\forall v_1[v_2+v_1\approx v_1]$, then v_2 is free for τ in its first occurrence but not in its second.

We say that v *is free for τ in φ* if every free occurrence of v in φ is free for τ in φ.

If we let φ,τ_1,τ_2 be respectively $[\exists v_1[v_1+v_2\approx 3]]\wedge[\forall v_3[v_1\cdot v_2\approx v_3]]$, $v_1\cdot v_2$, and $v_3\cdot v_4$, then v_1 is free for τ_1 in φ but not τ_2, v_2 is not free for τ_1 or τ_2, and v_3 is free for τ_1 and τ_2.

When $v_{i_1},v_{i_2},\ldots,v_{i_k}$ is a list of the variables occurring free in φ and $i_1<i_2<\cdots<i_k$, we may indicate this by writing $\varphi(v_{i_1},v_{i_2},\ldots,v_{i_k})$. Then if v_{i_j} is free for t_j in φ for each $j\leq k$, we write $\varphi(t_1,t_2,\ldots,t_k)$ for the result of replacing each free occurrence of v_{i_j} by t_j. For example, if φ is $\forall v_3[v_2+3\approx v_3]\wedge\exists v_2[v_2\approx v_3]$, we may write $\varphi(v_2,v_3)$ to indicate that v_2,v_3 are the only variables occurring free in φ. Then $\varphi(v_2+8,v_4)$ is the formula $\forall v_3[[v_2+8]+3\approx v_3]\wedge\exists v_2[v_2\approx v_4]$.

If v_{i_j} is not free for t_j in $\varphi(v_{i_1},\ldots,v_{i_k})$ for each $j\leq k$, then by $\varphi(t_1,\ldots,t_k)$ we mean the formula that is obtained from φ as follows. Let $u_1,u_2,\ldots,u_l$ be a list of the bound variables occurring in φ, and let $u'_1,u'_2,\ldots,u'_l$ be distinct variables not occurring in φ or in any of the t_j's. Let φ' be the result of replacing each occurrence of u_i by u'_i for $i\leq l$. (By Theorem 10.8ii, we have that for all assignments z, $\mathbf{N}^+\vDash\varphi\langle z\rangle$ if and only if $\mathbf{N}^+\vDash\varphi'\langle z\rangle$.) Notice that v_{i_j} is free for t_j in φ'. Then by $\varphi(t_1,\ldots,t_k)$ we mean $\varphi'(t_1,\ldots,t_k)$. For example, if $\varphi(v_2,v_4)$ is $\exists v_1\forall v_3[v_1+v_2\approx v_3\cdot v_4]$, then $\varphi(v_1,v_2+v_3)$ is $\exists v_5\forall v_6[v_5+v_1\approx v_6\cdot[v_2+v_3]]$.

This completes our description of L and attendant semantic notions. Definitions 10.1b and 10.4b can be regarded as giving the complete rules of grammar for L, while Definition 10.6 provides a complete dictionary.

The set of true assertions is then a well-defined set. Is there a decision procedure for determining membership in this set? I.e., is there a machine-like way of determining which assertions of L are true? In the next section we investigate the expressive power of L and §2.12 will be devoted to this decision problem.

EXERCISES FOR §2.10

1. Show that $[[[v_1+3]\cdot v_4]+[v_1\cdot 2]]$ is a term.
2. Show that $[\forall v_1[\forall v_2[\exists v_3[\exists v_4[[v_1+v_3]\approx[v_2+v_4]]]]]]$ is a formula.
3. Prove Theorem 10.5.
4. Find an assertion of L that is true just in case there are finitely many twin prime pairs. [The pair (p,q) is a twin prime pair if p and q are primes and $q=p+2$.]
5. It is not known if there are infinitely many twin prime pairs. Why doesn't Definition 10.6 along with Exercise 4 yield a solution?
6. Find an assignment that makes the following formula true: $\exists v_2\exists v_3[[v_1+v_1]+[v_2\cdot v_2]\approx[v_3+v_3]]$.
7. Let φ be $\forall v_2[v_1\approx v_2]\vee\exists v_3\forall v_1[v_3\approx[v_2+v_1]]$. Let τ_1 be $[v_1\cdot v_3]$, and let τ_2 be $[v_2+v_4]$. What is $\varphi(\tau_1,\tau_2)$?

2.11 Definability in Arithmetic

In this section we show that the expressive power of L is adequate to provide a definition of each computable function.

Definition 11.1.

i. Let R be a k-relation. Say that φ *defines* R if $R=\{(n_1,\ldots,n_k): \mathbf{N}^+\vDash\varphi(n_1,\ldots,n_k)\}$.
ii. The k-function f is *defined by* φ if the $(k+1)$-relation $f(\bar{n})=y$ is defined by φ.
iii. A function or relation is *arithmetical* if there is some formula that defines it.

For example, let **Div** be the formula $\exists v_3[v_1\cdot v_3\approx v_2]$. Then clearly **Div** defines the 2-relation $x|y$.

As another example, Prime is defined by the following formula:

$$\forall v_2[\mathbf{Div}(v_2,v_1)\rightarrow[[v_2\approx 1]\vee[v_2\approx v_1]]]\wedge[v_1\not\approx 1].$$

Theorem 11.2. *Every computable function is arithmetical.*

PROOF: Let Arth be the set of arithmetical functions. By Theorem 9.1 it is enough to prove that Arth $\supseteq$ Rec. Recalling the definition of Rec the proof breaks up into three cases as follows:

i. for all k,t the function $P_{k,t}$ belongs to Arth; also $+,\cdot,R_<\in$ Arth. Indeed, $+$ is defined $v_1+v_2=v_3$, $\cdot$ is defined by $v_1\cdot v_2\approx v_3$, $R_<$ is defined by

$$[\exists v_4[v_1+v_4\approx v_2]\wedge[v_3\approx 1]]\vee[\neg\exists v_4[v_1+v_4\approx v_2]\wedge[v_3\approx 2]],$$

and $P_{k,t}$ by $[v_1\approx v_1]\wedge\cdots\wedge[v_k\approx v_k]\wedge[v_{k+1}\approx v_t]$.

2.11 Definability in Arithmetic

ii. Arth is closed under composition. For suppose that the k-function f is defined by ψ and that the l-functions $g_1, \ldots, g_k$ are defined by $\varphi_1, \ldots, \varphi_k$ respectively. Then the composition of f with the g's is defined by

$$(\exists v_{l+2} \cdots \exists v_{l+k+2})[\varphi_1(v_1, \ldots, v_l, v_{l+2}) \wedge \cdots \wedge \varphi_k(v_1, \ldots, v_l, v_{l+k+2})$$
$$\wedge \psi(v_{l+2}, \ldots, v_{l+k+2}, v_{l+1})].$$

iii. The set of arithmetical functions is closed under the restricted μ-operator. For suppose that g is an arithmetical $(k+1)$-function with φ defining g. Suppose further that for all k-tuples $\bar{n}$ there is an m such that

$$g(\bar{n}, m) = 1.$$

Then the k-function f whose value at $\bar{n}$ is $\mu x(g(\bar{n}, x) = 1)$ has as a defining formula

$$\varphi(v_1, \ldots, v_k, v_{k+1}, 1) \wedge (\forall v_{k+2})(v_{k+1} \leq \neg \varphi(v_1, \ldots, v_k, v_{k+1}, 1)),$$

where $v_i \leq v_j$ is $(v_i \approx v_j) \vee (\exists v_l)(v_i + v_l \approx v_j)$.

In view of Definition 9.1, this completes the proof that Arth $\supseteq$ Rec. □

Theorem 11.3. *All machine enumerable relations (and hence all computable relations) are arithmetical.*

PROOF: We first show that the computable relations are arithmetical. To say that the k-relation S is computable means that its representing function R_S is computable. By Theorem 11.2, R_S is arithmetical. Let φ define R_S. Then clearly, S is defined by

$$\varphi(v_1, v_2, \ldots, v_k, 1).$$

Hence all computable relations are arithmetical. □

Now let S' be a machine enumerable k relation. By Theorem 8.9 there is a computable $(k+1)$-relation S such that for all k-tuples $\bar{n}$

$$S'(n_1, \ldots, n_{k-1}) \quad \text{if and only if} \quad (\exists n_k) S(n_1, \ldots, n_{k-1}, n_k).$$

As we have just observed, there is an χ that represents S. Hence S' is represented by

$$(\exists v_k) \chi(v_1, \ldots, v_{k-1}, v_k).$$

The converse of Theorem 11.3 is false. There are sets that are arithmetical but not machine enumerable. In fact, if X is any machine enumerable non-computable set (cf. Example 8.2), then $\mathbf{N}^+ - X$ is not machine enumerable, by Corollary 8.7. However, $\mathbf{N}^+ - X$ is arithmetical, since by Theorem 11.3 there is a formula φ that defines X, and so clearly $\neg \varphi$ defines $\mathbf{N}^+ - X$.

Since there are only countably many formulas, it follows that there are only countably many arithmetical sets. Hence the vast majority of sets are not arithmetical. In the next section we shall exhibit an important example of a set that is not arithmetical.

EXERCISES FOR §2.11

1. Find an arithmetical formula φ that defines the function
   $f(m,n)$ is the greatest divisor of m and n.

2. Give an example of an arithmetical non-computable function.

3. Is there an arithmetical 2-function F such that for every arithmetical 1-function g there is some m for which
   $$F(m,n) = g(n)$$
   for all n?

2.12 The Decision Problem for Arithmetic

Let $S(m)$ be the statement 'm is the number of a machine that halts on the input m'. Is there an effective procedure for deciding for which m the statement $S(m)$ is true? If such a decision procedure existed, then there would be an effective way of determining membership in the set $\{m: S(m)\}$. By Turing's thesis, this set would then be computable, but as we have seen in Example 6.4, it is not computable. Hence there is no algorithm for deciding for which m the sentence $S(m)$ is true.

What does this have to do with arithmetic? First of all, $S(m)$ can be viewed as a statement about arithmetic. Indeed, there is a formula $\varphi(v_1)$ of L such that the assertion $\varphi(\mathbf{m})$ is true iff $S(m)$ is true. To see this let $X = \{m: S(m)\}$. In Example 8.2 we proved that X is machine enumerable. Hence, by Theorem 11.3, $X \in \text{Arth}$. This means that there is a formula $\varphi(v_1)$ such that

$$\mathbf{N}^+ \vDash \varphi(\mathbf{m}) \quad \text{iff} \quad m \in X.$$

Thus we may consider $\varphi(\mathbf{m})$ to be the formalization in L of the statement $S(m)$; it is in this sense that $S(m)$ is a statement about arithmetic.

Thus we see that *there is no effective procedure that will work for all m for determining the truth of the assertion $\varphi(\mathbf{m})$ where φ is as above.*

From this we conclude that *there is no decision procedure for determining which assertions of L are true.* In other words, if T is the set of true assertions of L, then no algorithm exists for determining membership in T. For such an algorithm, when applied to an assertion of the form $\varphi(\mathbf{m})$, φ as above, would enable us to decide whether or not $\varphi(\mathbf{m})$ is true.

Moreover, this argument applies to any language whose set of assertions contains the assertions $\varphi(\mathbf{m})$. Thus we see that *there is no decision procedure for arithmetic.*

Indeed, we can say more: *There is no effective way of listing the true assertions of L.* In fact, if Z is any set of assertions in any language extending L, and Z contains all assertions $\varphi(\mathbf{m})$ as well as all assertions $\neg \varphi(\mathbf{m})$, then no machinelike procedure exists for enumerating the true assertions of Z. For if such a procedure existed, then one could check

2.12 The Decision Problem for Arithmetic

whether or not $\varphi(\mathbf{m})$ is true by enumerating the true assertions in Z until $\varphi(\mathbf{m})$ is reached or $\neg\varphi(\mathbf{m})$ is reached.

Of course one fully expects that a reasonable numbering of the assertions of L (or the assertions of some richer language) would enable us to argue that the set of numbers of true sentences is not machine enumerable, and indeed this is the case. However, instead of doing this directly, we shall prove a much stronger result of Tarski, namely, that "truth is not arithmetical." More precisely, we shall associate with each assertion σ of L a number $\#\sigma$, and then prove that the set $\{\#\sigma: N \vDash \sigma\}$ is not arithmetical, and so by Theorem 11.3 not machine enumerable. However, one should notice that non-arithmetical sets are, in a sense, quite plentiful. For L has countably many symbols, and so only countably many formulas, since each formula is a finite sequence of symbols (see Theorem 4.12 in Part I). Since each arithmetical set has a defining formula in L, there are at most countably many arithmetical subsets of $\mathbf{N}^+$. Hence 2^w subsets of $\mathbf{N}^+$ are not arithmetical (cf. Theorem 5.3 in Part I).

Of course, there must be some restrictions on the numbering $\#$, for otherwise nothing can be said about $\{\#\sigma: N \vDash \sigma\}$. Indeed, if X is any infinite subset of $\mathbf{N}^+$ such that $\mathbf{N}^+ - X$ is also infinite, then there is a $1-1$ correspondence between the assertions of L and $\mathbf{N}^+$ under which the true assertions of L are exactly those that correspond to elements of X. Since X can be computable, machine enumerable but not computable, arithmetical but not machine enumerable, or not arithmetical, we see that arbitrary numberings of the assertions of L are meaningless; such numberings will tell us nothing about the set of true assertions.

To assure a meaningful relation between σ and $\#\sigma$, we should restrict our attention to numberings that are effective in the sense that there is a decision procedure by which given an assertion σ we can find its number, and given a number we can decide if it corresponds to an assertion, and if so, to what assertion. Given such a numbering $\#$, the computability of $\{\#\sigma: N \vDash \sigma\}$ implies a decision procedure for membership in $\{\sigma: N \vDash \sigma\}$, and conversely; machine enumerability of $\{\#\sigma: N \vDash \sigma\}$ implies the existence of an algorithm for enumerating $\{\sigma: N \vDash \sigma\}$, and conversely.

Now suppose that $\#$ is a numbering of the assertions of L in the sense just described. Let σ be an assertion, and let c be the largest k such that $\mathbf{k}$ occurs in σ (if such a k exists). Let $s(\sigma, n)$ be the result of replacing c throughout σ by $\mathbf{n}$. If no constant symbol occurs in σ, then $s(\sigma, n)$ is just σ. Now define the substitution function S as follows:

$$S(m, n) = \# s(\sigma, n) \quad \text{if } m = \#\sigma \quad \text{for some } \sigma.$$

Assuming that $\#$ is effective, it is clear that there is an algorithm for computing S, and that S can be extended to an effectively computable function whose domain consists of all pairs of positive integers. These considerations are intended to show that the assumptions in the next theorem are natural. In fact, in the hypotheses of the theorem we assume only that S is an arithmetical extension, and this assumption is not as strong as the assumption that S is computable (see Theorem 11.2 and the last two paragraphs of §2.11).

Lemma. *There is no arithmetical 2-relation R such that for each arithmetical 1-relation P there is an m for which*

$$Rmn \text{ iff } Pn$$

for all $n \in \mathbf{N}^+$.

PROOF: This is a straightforward diagonal argument. If such an R exists, then R is represented by a formula, say $\varphi v_0, v_1$. Then $\neg \varphi v_0 v_0$ represents $\neg Rnn$, and so $\neg Rnn$ is an arithmetical 1-relation. But then there is an m_0 such that $Rm_0 n$ iff $\neg Rnn$ for all n, by our assumption on R. Taking $n = m_0$ gives $Rm_0 m_0$ iff $\neg Rm_0 m_0$, a contradiction. Hence no such R exists. □

Theorem 12.1 (Tarski). *Let $\#$ be a 1-1 function from assertions to positive integers such that the substitution function S (defined above) has an arithmetical extension. Let $\mathcal{T} = \{\#\sigma: \mathbf{N}^+ \vDash \sigma\}$. Then $\mathcal{T}$ is not arithmetical, and so $\mathcal{T}$ is neither computable or machine enumerable.*

PROOF: We argue by contradiction. Assume that the substitution function has an arithmetical extension defined by $\mathbf{S}$. Assume also that $\mathcal{T}$ is arithmetical, say $\mathcal{T}$ is defined by $\mathfrak{T}$. Now define Rm,n to hold iff $\mathbf{N}^+ \vDash \exists v_1(\mathfrak{T}(v_1) \wedge \mathbf{S}(\mathbf{m}, \mathbf{n}, v_1))$. R is an arithmetical 2-relation. Moreover, if P is an arithmetical 1-relation (say P is defined by φ), then Pn iff $R(\#\varphi', n)$, where φ' is $\varphi(\mathbf{k})$. k is the first l such that l does not occur in φ. But this is impossible by the lemma. Hence $\mathcal{T}$ is not arithmetical, as needed. The fact that $\mathcal{T}$ is not machine enumerable (and hence not computable) follows from Theorem 11.3. □

Can the hypotheses of the theorem be satisfied; i.e., is there a function $\#$ that numbers the assertions of L in such a way that S is arithmetical? The answer is yes, as we shall now show, following a procedure analogous to that in §2.7, where machines, tape positions, computations, etc., were coded as numbers. In fact we display a $\#$ such that S is computable.

First assign to each symbol s of L a number $\#'(s)$ as indicated below:

$$s: +, \cdot, \approx, \wedge, \vee, \neg, \forall, \exists, [,], v_i, \mathbf{n}$$

$$\#'(s): 1, 2, 3, 4, 5, 6, 7, 8, 9, 10, 9+2i, 10+2n.$$

If $s_1 s_2 \ldots s_n$ is a sequence of symbols of L, we let

$$\#(s_1 s_2 \ldots s_n) = \prod_{i=1}^{n} (\text{Prm}(i))^{\#'(s_i)}.$$

The numbering $\#$ is defined for all expressions of L, and so $\#\sigma$ is defined for all assertions σ. Clearly this numbering is effective in the sense discussed above, and so we have every reason to suspect that S has a computable extension and hence an arithmetical one.

2.12 The Decision Problem for Arithmetic

To prove this, we first consider the 3-function f defined as follows:

$$f(k,l,m) = \begin{cases} 1 & \text{if } m=1, \\ p_1^{n'_1} p_2^{n'_2} \cdots p_r^{n'_r} & \text{if } m = p_1^{n_1} p_2^{n_2} \cdots p_r^{n_r}, \text{ and the } p\text{'s} \\ & \text{are distinct primes, and } n'_i = n_i \\ & \text{when } n_i \neq k, \text{ and } n'_i = l \text{ when } n_i = k. \end{cases}$$

f is computable, as we see by writing it as

$$f(k,l,m) = \mu x (\forall y \leqslant m)((\text{Exp}(y,m) \neq k \wedge \text{Exp}(y,x) = \text{Exp}(y,m))$$
$$\vee (\text{Exp}(y,m) = k \wedge \text{Exp}(y,x) = l)).$$

Now we need a computable function that will pick out the constant with the largest index occurring in a sequence:

$$q(n) = \mu x (\forall y \leqslant n)(\forall z \leqslant n)(\text{Exp}(y,n) \neq 10 + 2(x+z))).$$

We can now define

$$S'(m,n) = f(10 + 2g(m), 10 + 2(n), m).$$

Clearly S' is computable and hence an arithmetical extension of S, as was to be shown.

EXERCISES FOR §2.12

1. If we replace $\mathbf{N}^+$ by $\mathbf{I}$, the set of integers $\{0, 1, -1, 2, -2, \ldots\}$, in Definition 10.6 and take our assignments $z \in {}^{\text{vbl}}\mathbf{I}$, then we obtain the notion $\mathbf{I} \vDash \varphi \langle z \rangle$. Replacing $\mathbf{N}^+$ by $\mathbf{I}$ in Definition 11.1 gives us the definition of Arth'.
   (a) Is $\{\#\sigma: \mathbf{I} \vDash \sigma\} \in \text{Arth'}$?
   (b) Is $\{\#\sigma: \mathbf{I} \vDash \sigma\}$ computable?
   (c) Is $\{\#\sigma: \mathbf{N}^+ \vDash \sigma\} \in \text{Arth'}$?
   (*Hint:* If n is an integer, then $n \in \mathbf{N}^+$ iff $\mathbf{I} \vDash (\exists s, t, u, v)[s^2 + t^2 + u^2 + v^2 \approx \mathbf{n}]$.)

2. Let $\text{sq}(n) = n^2$ for all $n \in \mathbf{N}^+$. Alter L by replacing $\cdot$ with sq. Let X be the set of assertion true in $\mathbf{N}^+$ in the new language. Show that X is not computable. [*Hint:* $r = m \cdot n$ if $r + r = \text{sq}(m+n) - \text{sq}(m) - \text{sq}(n)$. Now show that there is a computable function taking assertions $\sigma \in L$ into assertion σ' in the new language such that σ is true iff σ' is true.]

3. Let $\text{Sq}(n)$ be the relation on the integers 'n in a square', i.e., $\text{Sq}(n)$ iff $n = m^2$ for some $m \in \mathbf{I}$. Replace $\cdot$ in L by Sq, getting a language L', and define term, formula, satisfaction for L'-formulas accordingly. Show that the set of true L'-assertions is not computable. [*Hint:* Refer to the hint in Exercise 2. Also notice that since $(x+1)^2 = x^2 + 2x + 1$, we have $y = x^2$ iff $\text{Sq}(y) \wedge \exists z (\text{Sq}(z) \wedge y < z \wedge \neg (\exists u)(\text{Sq}(u) \wedge y < u \wedge u < z) \wedge z \approx y + x + x + 1)$. Since $(x+y)^2 = x^2 + 2xy + y^2$, we have $z = x \cdot y$ iff $x^2 + z + z + y^2 = (x+y)^2$.]

2.13 Axiomatizing Arithmetic

In the last section we saw that no decision procedure exists for determining which assertions of L are true; indeed, we proved that the true assertions cannot be effectively enumerated. Now we investigate the possibility of summarizing the true statements of arithmetic by means of an axiom system.

In the beginning of the usual high school geometry course one is given a set of statements about points and lines called axioms. In this context a proof is a finite sequence of statements, each being an axiom or a "logical consequence" of preceding statements. A theorem is the last statement of a proof. With this example in mind, we make the following definition.

Definition 13.1.

i. An *axiom* is an assertion of L.
ii. A k-premise *rule of inference* is a $(k+1)$-tuple $(\sigma_1,\ldots,\sigma_k,\sigma_{k+1})$, where each σ_i is an L-assertion and $\sigma_i \neq \sigma_j$ for $1 \leq j \leq k$. σ_{k+1} is the conclusion of the rule, and the σ_i's for $i \leq k$ are the premises.
iii. An *axiom system* is a set $\mathcal{S} = \mathcal{C} \cup \mathcal{B}$ where $\mathcal{C}$ is a set of axioms and $\mathcal{B}$ is a set of rules of inference. If we want to make explicit mention of the axioms and rules, we write $\mathcal{S}(\mathcal{C},\mathcal{B})$ instead of $\mathcal{S}$.
iv. A finite sequence $(\rho_1,\rho_2,\ldots,\rho_n)$ is an $\mathcal{S}(\mathcal{C},\mathcal{B})$-*proof* if for each ρ_i either
 (a) $\rho_i \in \mathcal{C}$, or
 (b) There is a k-premise rule $(\sigma_1,\ldots,\sigma_k,\rho_i)$ in $\mathcal{B}$ such that for all $l \leq k$, $\sigma_l \in \{\rho_j : j < i\}$.
v. An $\mathcal{S}$-*theorem* is the last line of an $\mathcal{S}$-proof. We write $\vdash_\mathcal{S} \sigma$ if σ is an $\mathcal{S}$-theorem.

What requirements should an axiom system meet in order that it may provide a fully adequate summary of the assertions true in $\mathbf{N}^+$? Ideally, every assertion true in $\mathbf{N}^+$ should be provable in $\mathcal{S}$, and every $\mathcal{S}$-theorem should be true in $\mathbf{N}^+$, i.e., $\vdash_\mathcal{S} \sigma$ iff $\mathbf{N}^+ \vDash \sigma$ for every L-assertion σ. Of course, these conditions are met by the axiom system $\mathcal{S}_1(\mathcal{C}_1,\mathcal{B}_1)$ where $\mathcal{C}_1 = \{\sigma : \mathbf{N}^+ \vDash \sigma\}$ and $\mathcal{B}_1$ is empty. Every theorem σ has a 1-term $\mathcal{S}_1$-proof, namely (σ). Clearly, there is no effective method for deciding whether a given sequence of L-assertions is an $\mathcal{S}_1$-proof, for such a method when applied to 1-term sequences would provide a test for truth in $\mathbf{N}^+$.

Axiom systems for which there is no decision procedure for determining which sequences are proofs are of little interest. For an axiom system $\mathcal{S}$ to be of some use, one should be able to decide if an alleged $\mathcal{S}$-proof is in fact an $\mathcal{S}$-proof. As we show in Theorem 13.7, a system that is effectively given in the sense of Definition 13.2 below has this property.

Definition 13.2. Let $\mathcal{S}(\mathcal{C},\mathcal{B})$ be an axiom system.

i. $\mathcal{S}$ is *effectively given* if there is a decision procedure for determining membership in the sets $\mathcal{C}$ and $\mathcal{B}$.

2.13 Axiomatizing Arithmetic   125

ii. S is *correct* for $\mathbf{N}^+$ if for all L-assertions σ, $\vdash_S \sigma$ implies $\mathbf{N}^+ \vDash \sigma$.
iii. S is *complete* for $\mathbf{N}^+$ if for all L-assertions σ, $\mathbf{N}^+ \vDash \sigma$ implies $\vdash_S \sigma$.

EXAMPLE 13.3. S_1 (described above) is complete and correct for $\mathbf{N}^+$, but not effectively given.

EXAMPLE 13.4. Let $\mathcal{Q} = \{1 = 1\}$ and $\mathcal{B}_2 = \{(1 = 1, \sigma): \sigma \text{ an } L\text{-assertion}\}$. Let $S_2 = S_2(\mathcal{Q}_2, \mathcal{B}_2)$. Then S_2 is effectively given and complete for $\mathbf{N}^+$, but not correct for $\mathbf{N}^+$.

EXAMPLE 13.5. Let $S_3 = S_3(\mathcal{Q}_3, \mathcal{B}_3)$, where $\mathcal{Q}_3 = \{\neg[1 = 1]\}$ and $\mathcal{B}_3 = \{(\neg[1 = 1], \sigma): \sigma \text{ is an } L\text{-assertion}\}$. Then S_3 is effectively given and complete for $\mathbf{N}^+$, but not correct for $\mathbf{N}^+$.

EXAMPLE 13.6. Let $S_4 = S_4(\mathcal{Q}_4, \mathcal{B}_4)$, where $\mathcal{Q}_4$ is the set of all atomic assertions that are true in $\mathbf{N}^+$, and $\mathcal{B}_4$ is the set of all sequences $(\sigma_1, \ldots, \sigma_k, [\sigma_1 \wedge \ldots \wedge \sigma_k] \vee \rho)$ where $k = \mathbf{N}^+$ and $\sigma_1, \ldots, \sigma_k, \rho$ are L-assertions. Then S_4 is effectively given and correct for $\mathbf{N}^+$, but is not complete for $\mathbf{N}^+$. Indeed, $\neg[1 = 2]$ is true in $\mathbf{N}^+$ but is not an S_4 theorem, for $\neg[1 = 2]$ is clearly not an axiom or a conclusion of an S_4-rule.

Lemma. *There is an effective procedure for enumerating the set of finite sequences of L-assertions.*

PROOF: Let # be the numbering described at the end of §2.12. If n is of the form $\prod_{i=1}^{k}(\mathrm{Prm}(i))^{\#\sigma_i}$ for some $\sigma_1, \ldots, \sigma_k$, we let the nth member of the enumeration be $(\sigma_1, \ldots, \sigma_n)$. Otherwise the nth member is $(1 = 1)$. □

Theorem 13.7. *Let S be effectively given. Then*

i. *there is an effective way of deciding which finite sequences of L-assertions are S-proofs;*
ii. *there is an effective procedure for enumerating the set of S-theorems.*

PROOF OF i. Let X be the set of all finite sequences of the form $(\sigma_1, \ldots, \sigma_k, \sigma_{k+1})$ where σ_{k+1} is the conclusion of an S-rule whose premises belong to $\{\sigma_1, \ldots, \sigma_k\}$. We claim that X is decidable. For given a sequence $(\sigma_1, \sigma_2, \ldots, \sigma_k, \sigma_{k+1})$, we can find the finite set Y of those sequences $(\rho_1, \ldots, \rho_l, \sigma_{k+1})$ having the property that $l \leq k$ and each ρ_i is some σ_j for $j \leq k$. Since S is effectively given, we can decide if any member of Y belongs to $\mathcal{B}$. Since $(\sigma_1, \ldots, \sigma_k, \sigma_{k+1}) \in X$ iff $Y \cap X \neq 0$, this gives an effective test for membership in X.

Now let $(\delta_1, \ldots, \delta_m)$ be a sequence of L-assertions. Clearly, this sequence is an S-proof iff for each $n \leq m$ either

i. δ_n is an S-axiom, or
ii. $(\delta_1, \ldots, \delta_n) \in X$ (X as above).

We have just seen that there is an effective test for membership in X. Also there is an effective test for membership in the set of axioms of $\mathcal{S}$. Applying these tests for $m=1,2,\ldots,n$ enables us to decide if $(\delta_1,\ldots,\delta_m)$ is an $\mathcal{S}$-proof or not. □

PROOF OF ii. If there are no axioms in the system $\mathcal{S}$, then the set of all $\mathcal{S}$-theorems is empty and there is nothing to prove. So suppose that σ is an axiom of $\mathcal{S}$. By the lemma, there is an effective enumeration of the finite sequences of L-assertions. Let $\bar{\sigma}_1,\bar{\sigma}_2,\ldots$ be such an enumeration with $\bar{\sigma}_i=(\sigma_{i,1},\ldots,\sigma_{i,k_i},\sigma_{i,k_i+1})$. We enumerate the theorems of $\mathcal{S}$ by letting σ_n be σ if $\bar{\sigma}_n$ is not a proof, and letting $\sigma_n=\sigma_{n,k_n+1}$ otherwise. Clearly, this is an effective enumeration. □

Theorem 13.8. *No axiom system $\mathcal{S}$ can satisfy all three of the following conditions*:

i. $\mathcal{S}$ *is effectively given*,
ii. $\mathcal{S}$ *is correct for* $\mathbf{N}^+$,
iii. $\mathcal{S}$ *is complete for* $\mathbf{N}^+$.

PROOF: Suppose $\mathcal{S}$ is effectively given, correct for $\mathbf{N}^+$, and complete for $\mathbf{N}^+$. By the preceding theorems, there is an effective way of enumerating $\{\sigma:\vdash_\mathcal{S}\sigma\}$. Also $\{\sigma:\vdash_\mathcal{S}\sigma\}=\{\sigma:\mathbf{N}^+\vDash\sigma\}$. But this contradicts the fact that no effective enumeration exists for $\{\sigma:\mathbf{N}^+\vDash\sigma\}$, as shown in Theorem 12.1. □

As one expects, the last two theorems can be placed in the context of machine enumerability. This presupposes a numbering $\#$ of the L-assertions.

Definition 13.2′. Let $\#$ be a numbering of the assertions of L. Say that $\mathcal{S}(\mathcal{C},\mathcal{B})$ is $\#$-*effectively given* if both of the following sets are computable:

i. $\{\#\sigma:\sigma\in\mathcal{C}\}$,
ii. $\{\Pi_{i=1}^{k+1}(Prm(i))^{\#\sigma_i}:(\sigma_1,\ldots,\sigma_k,\sigma_{k+1})\in\mathcal{B}\}$.

Theorem 13.7′. *Let $\mathcal{S}$ be $\#$-effectively given. Then*

i. $\{\Pi_{i=1}^n(\mathrm{Prm}(i))^{\#\sigma_i}:(\sigma_1,\ldots,\sigma_n)$ *is an $\mathcal{S}$-proof*$\}$ *is computable.*
ii. $\{\#\sigma:\vdash_\mathcal{S}\sigma\}$ *is machine enumerable.*

PROOF OF I. Our argument is the obvious modification of that given in Theorem 13.7. Let $\mathcal{S}=\mathcal{S}(\mathcal{C},\mathcal{B})$, $\mathcal{C}^\#=\{\#\sigma:\sigma\in\mathcal{C}\}$, $\mathcal{B}^\#=\{\Pi_{i=1}^{k+1}(\mathrm{Prm}(i))^{\#\sigma_i}:(\sigma_1,\ldots,\sigma_k,\sigma_{k+1})\in\mathcal{B}\}$. Let $X^\#$ be the set of all numbers of the form $\Pi_{i=1}^{l+1}(\mathrm{Prm}(i))^{n_i}$, where n_{l+1} is the number of the conclusion of some $\mathcal{S}$-rule $(\sigma_1,\ldots,\sigma_k,\sigma_{k+1})$ such that $\{\#\sigma_1,\ldots,\#\sigma_k\}\subseteq\{n_1,\ldots,n_l\}$. $X^\#$ is computable, since $n\in X^\#$ iff $\operatorname{Seq}n\wedge(\exists y\leqslant n^n)(\mathcal{B}^\#y\wedge\operatorname{Exp}(\operatorname{Ln}y,y)=\operatorname{Exp}(\operatorname{Ln}n,n)\wedge(\forall z<\operatorname{Ln}y)(\exists w<\operatorname{Ln}n)(\operatorname{Exp}(z,y)=\operatorname{Exp}(w,n)))$. It follows that n is the

2.13 Axiomatizing Arithmetic   127

number of an $\mathfrak{S}$-proof iff
$$\operatorname{Seq} n \wedge (\forall x \leqslant \operatorname{Ln} n) \left(\mathcal{Q}^{\#}(\operatorname{Exp}(x,n)) \vee \mathcal{B}^{\#} \left(\prod_{i=1}^{x} (\operatorname{Prm}(i))^{\operatorname{Exp}(i,n)} \right) \right).$$
This proves part i.   □

PROOF OF II. We have this immediately if $\{\sigma: \vdash_S \sigma\}$ is empty. So suppose $\vdash_S \sigma^*$. Define
$$f(n) = \begin{cases} \operatorname{Exp}(\operatorname{Ln} n, n) & \text{if } n \text{ is the number of an } \mathfrak{S}\text{-proof,} \\ \#\sigma & \text{otherwise.} \end{cases}$$
From part i we see that f is computable, and clearly the range of f is $\{\#\sigma: \vdash_S \sigma\}$. Hence $\{\#\sigma: \vdash_S \sigma\}$ is machine enumerable as asserted in part ii.

Theorem 13.8'. *Let $\#$ be a 1-1 function from assertions to positive integers such that the substitution function S (defined in §2.12) has an arithmetical extension. Then no axiom system $\mathfrak{S}$ can satisfy all three of the following conditions*:

i. $\mathfrak{S}$ is $\#$-effectively given,
ii. $\mathfrak{S}$ is correct for $\mathbf{N}^+$,
iii. $\mathfrak{S}$ is complete for $\mathbf{N}^+$.

PROOF: Suppose $\#$ is as specified and that $\mathfrak{S}$ satisfies i, ii, and iii. Condition i allows us to use Theorem 13.7' to conclude that $\{\#\sigma: \vdash_S \sigma\}$ is machine enumerable. By ii and iii, $\{\#\sigma: \vdash_S \sigma\} = \{\#\sigma: \mathbf{N}^+ \vDash \sigma\}$. But then $\{\#\sigma: \mathbf{N}^+ \vDash \sigma\}$ is machine enumerable, contradicting Theorem 12.1.   □

Definition 13.9. σ is *undecidable* with respect to $\mathfrak{S}$ if neither $\vdash_S \sigma$ nor $\vdash_S \neg \sigma$.

Given a function $\#$ from assertions to positive integers, we define the *diagonal function* D by
$$D(\#\sigma) = \# \neg S(\sigma, \#\sigma),$$
where S is defined in §2.12. If we can effectively find $\#\sigma$ from σ and σ from $\#\sigma$, then surely there is an algorithm for computing D. For example, the numbering described near the end of §2.12 has this property. In the next theorem we assume that D has an arithmetical extension, an assumption much weaker than the assumption that D is computable.

Notice that if $\#$ is as above and $\mathfrak{S}$ is $\#$-effectively given and correct for $\mathbf{N}^+$ then Theorem 13.8' assures the existence of assertions that are undecidable with respect to $\mathfrak{S}$. The next theorem displays such an assertion.

Let $\varphi(u)$ be a formula and n the least number k such that k does not occur in φ. By φ^0 we mean the result of replacing each free occurrence of u in φ by $\mathbf{n}$.

Recall that every machine enumerable set is definable in $\mathbf{N}^+$ (Theorem 11.3). We use this fact in the following stronger version of Theorem 13.8'.

Theorem 13.10. *Let $\#$ be a 1-1 function from assertions to positive integers such that the diagonal function has an arithmetical extension. Let the axiom system $\mathcal{S}$ be $\#$-effectively given and $\mathbf{N}^+$-correct. Let $\psi_\mathcal{S}$ and $\mathbf{D}$ define $\{\#\sigma:\vdash_\mathcal{S}\sigma\}$ and the diagonal function respectively. Let $\varphi(v_2)$ be $\exists v_1[\mathbf{D}(v_2,v_1)\wedge\psi_\mathcal{S}(v_1)]$, and let $\sigma=\varphi(\#\varphi^0)$. Then σ is not decidable with respect to $\mathcal{S}$.*

PROOF: Assume the hypotheses of the theorem and suppose $\vdash_\mathcal{S}\sigma$. Then by the correctness of $\mathcal{S}$, $\mathbf{N}^+\vDash\sigma$. Hence for some k, $\mathbf{N}^+\vDash D(\#\varphi^0,k)$ and $\mathbf{N}^+\vDash\psi_\mathcal{S}(k)$. Since D defines the diagonal function, k must be $\#\neg\varphi(\#\varphi^0)$, i.e., $k=\#\neg\sigma$. Hence $\mathbf{N}^+\vDash\psi_\mathcal{S}(\#\neg\sigma)$. Since $\psi_\mathcal{S}$ defines $\{\rho:\vdash_\mathcal{S}\rho\}$, $\vdash_\mathcal{S}\neg\sigma$. This and our assumption $\vdash_\mathcal{S}\sigma$ contradict the correctness of $\mathcal{S}$. Thus $\vdash_\mathcal{S}\sigma$ is impossible.

On the other hand, suppose that $\vdash_\mathcal{S}\neg\sigma$. Since $\psi_\mathcal{S}$ defines $\{\rho:\vdash_\mathcal{S}\rho\}$, we have $\mathbf{N}^+\vDash\psi_\mathcal{S}(\#\neg\sigma)$. We also have $\mathbf{N}^+\vDash D(\#\varphi^0,\#\neg\sigma)$. Hence $\mathbf{N}^+\vDash\exists v_1[\mathbf{D}(\#\varphi^0,v_1)\wedge\psi_\mathcal{S}(v_1)]$, i.e., $\mathbf{N}^+\vDash\sigma$. But $\mathcal{S}$ is correct, so that our assumption $\vdash_\mathcal{S}\neg\sigma$ implies $\mathbf{N}^+\vDash\neg\sigma$—a contradiction. Therefore, $\neg\sigma$ is not a theorem of $\mathcal{S}$. Hence, with what we have above, σ is undecidable with respect to $\mathcal{S}$. □

Thus there is no axiomatization in L that provides a completely satisfactory summary of the true assertions about $\mathbf{N}^+$.

Theorem 13.8 (or 13.10) is Gödel's incompleteness theorem, so called because it says that every effectively given correct axiomatization is incomplete. This is one of the most outstanding results of twentieth century mathematics, for this theorem caused a profound alteration in views held about mathematics and science in general. No longer can mathematics be thought of as an idealized science that can be formalized using self-evident axioms and rules of inference in such a way that all things true are provable. Any correct formalization whose proofs can be checked effectively must admit undecidable assertions.

In our chapter on set theory we mentioned that the axiomatic set theory ZFC provides an axiomatic foundation for all of classical mathematics. In view of the incompleteness theorem, we should qualify this statement, since Gödel's result can be extended to cover axiomatic set theory (as well as any other axiomatic system strong enough to code an adequate notion of proof). Virtually all mathematicians believe that ZFC is correct. If this is so, since ZFC is effectively given, the incompleteness theorem tells us that ZFC is incomplete, and the analog of Theorem 13.10 will produce an undecidable sentence.

EXERCISES FOR §2.13

1. Let $\mathcal{S}$ be a correct effectively given axiomatization for $\mathbf{N}^+$, and let φ define the relation $\{\#M: M(\#M)=1\}$.
   (a) Show that there is an M^* such that for all machines M
   $$M^*(\#M) = 2 \quad \text{if } \vdash_{\mathcal{S}} \varphi(\#\mathbf{M})$$
   $$M^*(\#M) = 1 \quad \text{if } \vdash_{\mathcal{S}} \neg \varphi(\#\mathbf{M})$$
   (See Theorem 3.23.) Convince yourself that given a machine that enumerates the axioms and rules of $\mathcal{S}$, a machine M^* satisfying the above conditions can be found explicitly.
   (b) Let σ be $\varphi(\#M^*)$. Prove that σ is undecidable.

2. Prove that the σ of Exercise 1 is false.

3. Let $\mathcal{S}$ and σ be as in Exercise 1. Let $\mathcal{S}'$ be the result of adding σ to the set of axioms of $\mathcal{S}$.
   (a) Is $\mathcal{S}$ effectively given?
   (b) Is $\mathcal{S}$ correct?
   (c) Is $\mathcal{S}$ complete?

2.14 Some Directions in Current Research

A closer look at the incompleteness theorem, especially a version like that in Exercise 1 of the last section in which a specific example of an undecidable assertion σ is presented, leads to Gödel's second incompleteness theorem. Roughly speaking, this result states that a sufficiently strong consistent axiomatization does not yield a proof of its own consistency. A detailed proof of this is rather messy, and so we content ourselves with the barest elements of an outline. Let $\mathcal{S}$ be effectively given. Let $\text{Prf}_{\mathcal{S}}(n)$ define the relation 'n is an assertion and there is an $\mathcal{S}$ proof of n'. (Notice that this relation is machine enumerable.) Let $f(n)$ be a definable function that maps $\#\psi$ to $\#\neg\psi$ for each formula ψ. As the formal counterpart of our assertion that $\mathcal{S}$ is consistent (i.e., free of contradiction) we take $\text{Cons}_{\mathcal{S}}$: $\neg \exists v [\text{Prf}_{\mathcal{S}} v \wedge \text{Prf}_{\mathcal{S}} f(v)]$. Let σ be the undecidable assertion of Exercises 1 and 2 of the last section. Exercise 2 can be answered in outline as follows:

i. Suppose σ is true, i.e., suppose $\mathbf{N}^+ \vDash \sigma$.
ii. Since σ is true, $M^*(\#M^*) = 1$, so by definition of M^*, $\vdash_{\mathcal{S}} \neg \sigma$. By the correctness of $\mathcal{S}$, $\mathbf{N}^+ \vDash \neg \sigma$.
iii. But then $\mathbf{N}^+ \vDash \sigma \wedge \neg \sigma$—a contradiction. Hence $\mathbf{N}^+ \vDash \neg \sigma$.

Instead of assuming the correctness of $\mathcal{S}$ in step ii, let us suppose that $\mathcal{S}$ is strong enough so that when $\mathbf{N}^+ \vDash \varphi(\mathbf{n})$, then $\vdash_{\mathcal{S}} \varphi(\mathbf{n})$ where φ is the formula described in exercise 1 of §2.13. This is not an outlandish requirement, and many known axiom systems have this feature. Now the above argument can be modified as follows:

i. If σ is true, then $\vdash_S \sigma$.
ii. If σ is true, then $\vdash_S \neg \sigma$.
iii. But S is consistent, so not $[\vdash_S \sigma$ and $\vdash_S \neg \sigma]$. Hence $\neg \sigma$ is true.

Now suppose that the axioms and rules of S are strong enough to enable one to give the above argument inside of S. Again, this is not an outrageous requirement. Many of the interesting axiomatizations of number theory have been made up to be as strong as possible and satisfy this requirement. Suppose also that $\vdash_S \text{Cons}_S$. Now we can internalize the above argument:

i. $\vdash_S \sigma \to \text{Prf}_S \# \sigma$
ii. $\vdash_S \sigma \to \text{Prf}_S \# \neg \sigma$
iii. $\vdash_S \text{Cons}_S$ from which we get $\vdash_S \neg [\text{Prf}_S \# \sigma \wedge \text{Prf}_S \# \neg \sigma]$. Hence $\vdash_S \neg \sigma$. But $\nvdash_S \neg \sigma$, and so we can not have $\vdash_S \text{Cons}_S$.

A more dramatic way of stating the second incompleteness theorem might be: Provided S is strong enough, any proof of the consistency of S must use methods not formalizable within S, and hence the proof of consistency of S is as much in doubt as the consistency of S.

The undecidability of arithmetic is one of the most important and profound discoveries of logic. However, there are many structures that are of interest to mathematicians, and one can ask if their theories are decidable. The general notion of a first order theory will be discussed in the last chapter, but the intuitive idea is straightforward. For example, the reals (with the usual addition, multiplication, $<$, and $0, 1$) can be thought of as a structure $\mathcal{R} = (\mathbf{R}, +, \cdot, <, 0, 1)$. The first order language appropriate for this structure has binary function symbols $+, \cdot$, a binary relation symbol $<$, and constant symbols $\mathbf{0, 1}$. The notion of formula, satisfaction, and truth are defined in the obvious way by modifying Definitions 10.4, and 10.6 appropriately. Now one can ask about the decidability of the set of assertions that are true in $\mathcal{R}$.

Surprisingly, the set of assertions true in $\mathcal{R}$ is decidable. In particular this means that the natural numbers cannot be defined by a formula of this language. Similarly the theory of the complex numbers with plus and times is decidable. On the other hand, the set of assertions true in the rationals with $+$ and $\cdot$ is undecidable. The theory of Abelian groups (i.e., the set of assertions in the language of group theory that are true in every Abelian group) is decidable, but the theory of groups is not. The theory of linear orderings, the theory of well orderings, and the theory of Boolean algebras are decidable. The theory of an equivalence relation is decidable, but the theory of two equivalence relations is not. The theory of a symmetric reflexive relation is not decidable either. The theory of a single unary function is decidable, but the theory of two such or of one binary function is not. On the other hand, the theory of two successor functions or even the theory of countably many successor functions is decidable. Such results are most surprising when a seemingly weak theory turns out to be undecidable or a seemingly strong theory turns out to be decidable.

Of course it is not necessary to limit oneself to first order languages when considering questions of decidability. For example, one can consider a language extending L that not only has variables ranging over elements but also has variables ranging over relations. This language is called the second order calculus. Although the set of first order assertions true in $(\mathbf{N}, <)$ is a decidable set, the set of second order assertions true in $(\mathbf{N}, <)$ is not. A remarkable theorem along these lines proved by Rabin is that the monadic second order theory of countably many successor functions is decidable. (The second order variables of the monadic second order calculus are restricted to range over sets of elements of the structure in question and not over relations of arbitrary arity.)

Proceeding in another direction, one can restrict attention to assertions that have some special form. For example, the set of Diophantine equations that are solvable in $\mathbf{N}$ is undecidable. A Diophantine equation has the form

$$Q(x_0, \ldots, x_k) = 0,$$

where $Q(x_0, \ldots, x_k)$ is a polynomial with integer coefficients in the variables $x_0, \ldots, x_k$ and we ask for $n_0, \ldots, n_n \in \mathbf{N}$ such that $Q(n_0, \ldots, n_k) = 0$. This problem was first raised by Hilbert in 1900 and finally solved by Matijasevič in 1970. Thus the decision problem for $\mathbf{N}$ has a negative solution even if we restrict our attention to just the assertions $\exists x_0, \ldots, x_k [Q(x_0, \ldots, x_k) = 0]$. What Matijasevič shows is that there is a formula $E(u, v)$ of the form $\exists x_0, \ldots, x_k [Q(x_0, \ldots, x_k, u, v) = 0]$ where $Q(x_0, \ldots, x_k, u, v) = 0$ is diophantine and for every machine enumerable set X there is a n for which $X = \{x : \mathbf{N} \models E(x, n)\}$. Hence taking X to be any machine enumerable undecidable set gives us a set of assertions $\{E(x, n) : \mathbf{N} \models E(x, n)\}$ which is undecidable.

As a corollary to this result, we get a polynomial $P(\bar{x}, u, v)$ with coefficients from $\mathbf{N}$ and domain $^{k+2}\mathbf{N}$ such that for each machine enumerable X there is an n such that X is the set of positive integers in the range of P. For if $E(u, n)$ is the formula for X mentioned above, then we can take $P(\bar{x}, u, n)$ to be $u[1 - Q^2(\bar{x}, u, n)]$. So in particular, there is a polynomial P in several variables that enumerates the primes in the sense that the positive portion of the range of P is the set of prime numbers. This came as a surprise to those working in number theory.

Here is another surprising consequence. In the next section we shall describe an effectively given axiomatization for set theory, called ZFC, that is complete for classical mathematics in the sense that all mathematical notions can be defined and all theorems of classical mathematics can be proved in this axiomatic framework. This axiomatization is thought to be consistent by virtually all mathematicians, but cannot be proved so within its own framework, by the second incompleteness theorem. Being effectively given, the set $\mathcal{T}$ of theorems of ZFC is effectively enumerable (as in Theorem 13.7). So the theorems provable within the framework of classical mathematics can be enumerated effectively. Assign numbers to formulas of set theory in the way we assigned numbers to formulas of arithmetic at the end of §2.12, except assign 0 to each assertion of the form $\sigma \wedge \neg \sigma$. Let $\mathcal{T}^{\#}$ be the set of numbers of elements of $\mathcal{T}$. $\mathcal{T}^{\#}$, being effectively enumerable, is the non-negative part of the range of a polynomial P. We believe that P does not have a solution for 0 in integers, since we believe ZFC is consistent. On the other hand we cannot prove that P has no solution in integers within the classical framework of mathematics, since this would be tantamount to a proof of consistency of ZFC within ZFC.

The problem of deciding whether an arbitrarily given Diophantine equation has a solution in the rationals remains open.

One of the most active areas of investigation in the theory of computers in recent years is the theory of computational complexity. The notion of machine computability as an abstraction of our intuitive notion of algorithm or man computability is very generous in that no restriction is placed on amount of tape or number of steps required in a given computation as long as these quantities are finite. However, length of computation is a crucial consideration in the operation of real computers. A problem may be computable in the theoretical sense, but non-computable in the practical sense that given any program for the problem, unavoidable calculations will require a totally unacceptable amount of time.

A remarkable example of this was discovered by Fischer and Rabin. Let L^- be the set of L-assertions without occurrences of $\cdot$ or constant symbols. Let T_1 be the set of L^--assertions true in $\mathbf{N}^+$ under addition, and let T_2 be the set of L^--assertions true in the reals under addition. It is known that T_1 and T_2 are decidable. However, Fischer and Rabin prove that they are not decidable in the practical sense. For T_1 they show that there is a constant c such that for any machine M there is an n_0 such that for each $n > n_0$ there is an L^--assertion of length n which takes at least $2^{2^{cn}}$ steps by M to decide. For T_2 the situation is analogous except that the decision takes at least 2^{dn} steps. Thus the number of steps required for a decision is growing much, much faster than the length of the question and soon exceeds, say, the number of atoms in the universe. The methods used to prove these results have had many applications to questions of computational complexity in mathematical structures other than the reals or natural numbers under addition.

There is a beautiful theorem of P. Young which states, loosely speaking, that there is a recursively enumerable set X such that given any recursive enumeration $(x_1, x_2, x_3, \ldots)$ of X, there is another enumeration $(y_1, y_2, y_3, \ldots)$ of X and a machine M that will enumerate X in the order $(y_1, y_2, y_3, \ldots)$ significantly faster than any machine can enumerate X in the order $(x_1, x_2, x_3, \ldots)$.

2.14 Some Directions in Current Research

Another bizarre result of complexity theory is that there is no effectively given sequence of computable functions $f_1, f_2, f_3, \ldots$ such that f_{i+1} is more complicated than f_i in the sense that any program for the computation of f_{i+1} is larger than the minimal program for f_i. In other words, if f is a computable 2-function, and $f_i(n) = f(i, n)$ for all i, n, and cf_i is, say, the least number of the form $\# M$ where M computes f_i, then $cf_{i+1} \leqslant cf_i$ for some i's. So in particular, letting $f_{i+1}(n) = n^{f_i(n)}$ with $f_1(x)$ the identity function does not yield a sequence of functions of increasing complexity.

As a consequence of this we have the following strange corollary. Given any axiomatic framework for mathematics, there are only finitely many machines M for which there is a proof that M computes a function that cannot be computed by a simpler machine (simpler, say, in the sense of the size of $\# M$).

There is a particularly notorious open question in this domain. Consider the class P of problems that can be done in polynomial bounded time. A problem in this class can be thought of as a recursive set X for which there is a machine M and a polynomial p such that $M(n) = R_x(n)$ (where R_x is the representing function of X) and the computation $M(n)$ takes fewer than $p(n)$ steps. P is the class of deterministic, polynomial bounded problems. The class NP of non-deterministic polynomial bounded problems consists of those recursive sets X for which there is a non-deterministic machine M and a polynomial p such that M has some computation of length $\leqslant p(n)$ that *accepts* n if and only if $n \in X$. A non-deterministic machine differs from our machines in that the next-state function s maps elements of $\{0, 1\} \times \mathbf{N}^+$ into finite subsets of the set of states. Definition 2.3 is modified so that a successor tape position $M(t)$ can have any marker of the form $(j + p(a_j, k), u)$ where $u \in s(a_j, k)$. No one has yet been able to show that $P \neq \mathrm{NP}$.

Beside the attempts to classify the computable functions into a hierarchy according to a computational complexity, considerable success has been achieved in classifying non-computable functions according to their relative complexity. If we add the stipulation that $g \in K$ to the definition of Rec given in Definition 9.1, we obtain the set of functions recursive with respect to g, Rec^g. Intuitively, $f \in \mathrm{Rec}^g$ means that there is a canonical procedure for computing f that may require at any given step of a computation some value of g. If $f \in \mathrm{Rec}^g$, we write $f \leqslant g$. Clearly if $f \in \mathrm{Rec}$, then $f \leqslant g$ for all g. If we restrict our attention to representing functions of effectively enumerable sets, then it is not difficult to find an h such that $f \leqslant h$ for all such f's. More work is required to show that each of our effectively enumerable non-computable sets given in the examples of §2.8 has a representing function that is such a maximal h. Post's problem asks if there are any functions g that are neither maximal or minimal with respect to $\leqslant$. After many years the question was settled affirmatively by Friedberg and Mučnik. If we identify f and g whenever $f \leqslant g$ and $g \leqslant f$, then the resulting equivalence classes are partially ordered by $\leqslant$ with a least element and a maximal element. Using the methods developed by

Friedberg and Mučnik, much was learned about this ordering. For example, any countable partially ordered structure can be embedded in it.

Notions of computability have been developed for objects of higher type, such as functions that map functions to functions.

Analogs of computability have been proposed by Kreisel for functions defined on some ordinal α with range in α. This generalization grew into a rich theory developed by Barwise and others.

Formal language theory is another direction that has received considerable attention. An alphabet is a set Σ of symbols. Σ^* is the set of all finite sequences of terms in Σ. Σ^* is called the set of words on Σ. A language is a subset of some Σ^*. Languages can be specified by syntactic conditions or by a process. As an example of the latter kind, suppose we fix a word $w \in \Sigma^*$ and a function $f : \Sigma \to \Sigma^*$. Now define a function $F: \Sigma^* \to \Sigma^*$ as follows. If $\sigma \in \Sigma^*$, say $\sigma = a_1 a_2 \ldots a_n$ then $F\sigma = f(a_1)f(a_2)\ldots f(a_n)$, i.e., each a_i is replaced by the word $f(a_i)$ and the resulting concatenation of symbols is $F\sigma$. Let L be the orbit of F on w, that is, $L = \{w, Fw, F^2 w, \ldots\}$. Then L is a language, and languages obtained in this way are simple examples of Lindenmeyer languages. Lindenmeyer is a botanist who first proposed the study of these languages, and the cause was subsequently championed by Rozenberg. These languages have been used to model simple morphogenic processes in biology and are of interest to computer scientists as examples of parallel programming.

Many other interesting classes of languages have been described, and the relationships among them are an active current area of research.

PART III
An Introduction to Model Theory

3.1 Introduction

For the remainder of the text we turn our attention to that branch of logic called model theory. Here we consider formal languages with enough expressive power to formulate a large class of notions that arise in many diverse areas of mathematics. Within our idealized language we shall be able to describe different kinds of orderings, groups, rings, fields, and other commonly studied mathematical notions.

If an assertion σ is a true statement about a mathematical structure $\mathfrak{A}$, then $\mathfrak{A}$ is said to be a model of σ. The main concern of model theory is the relation between assertions in some formal language and their models. For example, we shall describe a language strong enough to capture many of the important properties of the real number field but weak enough so that the sentences true in this structure also have a non-standard model, a model in which there are infinitesimaly small numbers and infinitely large numbers. Within such non-standard models one can justify a development of the calculus along lines close to that of Newton's original conception. For example $\lim_{n\to\infty} a_n = a$ would mean that $|a_n - a|$ is infinitesimal whenever n is infinite.

Where model theory is developed within the context of algebra, the latter subject undergoes considerable unification and generalization, while model theory benefits from the examples, methods, and problems of algebra. This interaction has been particularly fruitful in the areas of Boolean algebra and the theory of groups.

The interaction of set theory and model theory has given tremendous impetus to both, and each has contributed techniques and theorems to the other that were used to solve famous problems of long standing.

In fact, over the past several decades, the connections between model theory, computable function theory, set theory, and infinitary combinatorics have become more and more closely knit. There are areas in which the symbiosis is so strong that any division between them is bound to be artificial. In the last section we shall briefly indicate some of the more recent directions taken by model theory and hint at the growing interrelation between the various branches of logic.

3.2 The First Order Predicate Calculus

We now expand the language L, introduced in §2.10, by adding new function symbols, relation symbols, and constant symbols. In the expanded language we shall be able to make assertions about groups, rings, fields, orderings, and other objects of mathematical interest. Our new language (which we also call L, or the first order predicate calculus) and the new definitions of assignment and satisfaction are natural extensions of the corresponding notions previously introduced in §2.10.

As before, the language L can be viewed as a formalized fragment of mathematical English.

THE SYMBOLS OF L

Variables: $v_0, v_1, v_2, \cdots$.
For each ordinal α, a constant symbol c_α.
Equality symbol: $\approx$.
Symbols for 'and', 'or', and 'not': $\wedge, \vee, \neg$.
Existential and universal quantifier symbols: $\exists, \forall$.
Left and right parentheses: [,].
For each $n \in \mathbf{N}^+$ and each ordinal α, an n-function symbol $f_{n,\alpha}$.
For each $n \in \mathbf{N}^+$ and each ordinal α, an n-relation symbol $R_{n,\alpha}$.

Definition 2.1a. A set whose elements are either constant symbols, function symbols, or relation symbols is called a *type*. An *expression* is a finite sequence of symbols. If φ is an expression then the *type of* φ, written $\tau(\varphi)$, is the set of constant symbols, function symbols, and relation symbols occurring in φ. If s is a type, then φ *is of type s* if $\tau(\varphi) \subseteq s$.

EXAMPLE. $]]v_9 \forall c_\omega R_{3,9} f_{84,2}[$ is an expression φ such that $\tau(\varphi) = \{c_\omega, R_{3,9}, f_{84,2}\}$. φ is of type $\tau(\varphi)$ and of any type containing $\tau(\varphi)$.

Definition 2.1b. Let s be a type. The set of *terms* of type s, Trm_s, is the smallest set X of expressions such that

i. $v_i \in X$ for all $i \in \omega$,
ii. $c_\alpha \in X$ for all $c_\alpha \in s$,
iii. if $t_1, \ldots, t_n \in X$ and $f_{n,\alpha} \in s$, then $f_{n,\alpha} t_1, \ldots, t_n \in X$.

3.2 The First Order Predicate Calculus

EXAMPLE. Suppose that v is a variable, c a constant symbol, f a 1-function symbol, and q a 2-function symbol. Then c and v are terms by conditions i and ii; so fc is a term by iii, and so $gfcv$ is a term (of type $\{g,f,c\}$) by iii also.

As before (Theorem 10.2 in part II), unique readability for terms is easy but tedious to prove, so we leave both the statement and the proof as an exercise.

Definition 2.1c. An *atomic formula* is an expression of the form $[t_1 = t_2]$ or of the form $[R_{n,\alpha} t_1, \ldots, t_n]$ where the t_i's are terms.

Definition 2.1d. Let s be a type. The set Fm_s, *the formulas of type s*, is the smallest set X such that

i. every atomic formula of type s belongs to X,
ii. if $\varphi \in X$ then $[\neg \varphi] \in X$,
iii. if $\varphi \in X$ and $\psi \in X$, then $[\varphi \vee \psi] \in X$ and $[\varphi \wedge \psi] \in X$,
iv. if $\varphi \in X$ and v is a variable, then $[\exists v \varphi] \in X$ and $[\forall v \varphi] \in X$.

If Σ is a set of formulas, then $\tau\Sigma$, the type of Σ, is $\bigcup \{\tau(\sigma) : \sigma \in \Sigma\}$.

EXAMPLE. Let f be a 1-function symbol, R a 2-relation symbol, u and v variables, and c a constant symbol. Then we can check that

$$\left[\forall u \left[\exists v \left[\left[R u f v \right] \vee \left[f v \approx c \right] \right] \right] \right] \tag{*}$$

is a formula. By Definition 2.1b, u, fv, and c are terms. Hence $[Rufv]$ and $[fv \approx c]$ are formulas by Definition 2.1d i, and so by Definition 2.1d iii we get that $[[Rufv] \vee [fv \approx c]]$ is a formula. By Definition 2.1d iii used twice we see that (*) is a formula.

We shall make the usual notational abuses in the name of readability when writing down formulas, such as omitting or inserting brackets or commas (see discussion preceding Definition 10.9 in Part II). For example, we may write

$$\forall u \exists v [R(u, f(v)) \vee f(v) \approx c]$$

in place of (*).

The obvious analog of Theorem 10.5 in Part II (unique readability for formulas) holds. We leave the precise statement of the theorem and its proof as an exercise.

The notions of free occurrence, bound occurrence, free for, and assertion carry over from Definition 10.7 of Part II without change.

In Part I we remarked that all of mathematics can be developed within an axiomatic framework such as the Zermelo-Fraenkel axiomatization or some extension of it, and model theory is no exception. To do so all objects under discussion would have to be sets whose existence followed from the axioms. For example, we could take v_n, c_α, $f_{n,\alpha}$, $R_{n,\alpha}$ to be $(1,n)$, $(2,\alpha)$, $(3,n,\alpha)$, $(4,n,\alpha)$ respectively, and $\approx$, $\vee$, $\wedge$, $\neg$, $\exists$, $\forall$, $[$, $]$ to be $(0,0)$, $(0,1),\ldots,(0,3)$ respectively. Expressions, terms, and formulas could then be defined as sequences in an appropriate set theoretic way (see Part I), and unique readability would then be proved from the axioms of set theory. Indeed, everything that follows could be developed formally within the axiom system ZFC. However, within the present context, this would be a tedious exercise in needless rigor. On the other hand, it is important to realize that this can be done, and in several problems that arise in logic such a formalization of model theory is given explicitly or at least assumed.

EXERCISES FOR §3.2

1. Revise Theorem 10.2 of Part II to obtain a unique readability theorem for terms as defined in Definition 2.1 above, and give a proof of the new theorem.
2. Obtain a unique readability theorem for formulas as defined in Definition 2.1d by revising Theorem 10.5 of Part II, and give a proof of the theorem.

3.3 Structures

In Part I we defined a structure to be an ordered pair (A,e) where $A \neq \emptyset$ and e is a binary relation on A. We now extend the notion of structure so as to encompass a great variety of constructs that are of interest to mathematicians.

Definition 3.1. Let s be a type. By a *structure of type s* we mean a function $\mathfrak{A}$ whose domain is $s \cup \{\emptyset\}$ satisfying the following requirements:

i. $\mathfrak{A}(\emptyset)$ is a non-empty set.
ii. $\mathfrak{A}(c_\alpha) \in \mathfrak{A}(\emptyset)$ for each $c_\alpha \in s$.
iii. $\mathfrak{A}(R_{n,\alpha})$ is an n-relation on $\mathfrak{A}(\emptyset)$ for each $R_{n,\alpha} \in s$.
iv. $\mathfrak{A}(f_{n,\alpha})$ is an n-function on $\mathfrak{A}(\emptyset)$ for each $f_{n,\alpha} \in s$.

So $\mathfrak{A}$ maps constant symbols into constants in $\mathfrak{A}(\emptyset)$, n-ary relation symbols into n-relations on $\mathfrak{A}(\emptyset)$, and n-ary function symbols into n-functions on $\mathfrak{A}(\emptyset)$.

We shall write $|\mathfrak{A}|$ instead of $\mathfrak{A}(\emptyset)$ and call it the universe of the structure $\mathfrak{A}$. We may also write $c_\alpha^\mathfrak{A}$, $f_{n,\alpha}^\mathfrak{A}$, and $R_{n,\alpha}^\mathfrak{A}$ in place of $\mathfrak{A}(c_\alpha)$, $\mathfrak{A}(f_{n,\alpha})$, and $\mathfrak{A}(R_{n,\alpha})$ respectively. If S is a symbol, then $S^\mathfrak{A}$ is called the denotation of S in $\mathfrak{A}$. Write $\tau\mathfrak{A} = s$ if s is the type of $\mathfrak{A}$.

3.3 Structures

With this notation, structures of the kind described in §I12 can be thought of as functions $\mathfrak{A}$ with domain $\{\varnothing, R_{2,\alpha}\}$, so we have $\mathfrak{A} = (|\mathfrak{A}|, R_{2,\alpha}^{\mathfrak{A}})$ $=(A, e^{\mathfrak{A}})$. Here α is fixed but arbitrary.

If $\mathfrak{A}$ is a structure with universe A and $\{H_0, H_1, H_\alpha, \cdots\} = \tau\mathfrak{A}$, then we may write $(A, H_0^{\mathfrak{A}}, H_1^{\mathfrak{A}}, \ldots, H_\alpha^{\mathfrak{A}}, \cdots)$ instead of $\mathfrak{A}$. Or it may be convenient to write $\mathfrak{A} = (A, R_i^{\mathfrak{A}}, f_n^{\mathfrak{A}}, c_k^{\mathfrak{A}})_{i \in I, j \in J, k \in K}$ if the type of $\mathfrak{A}$ is $\{R_i : i \in I\} \cup \{f_j : j \in J\} \cup \{c_k : k \in K\}$, where the R_i's are relation symbols, the f_j's are function symbols, and the c_k's are constant symbols.

We shall use capital German letters $\mathfrak{A}$, $\mathfrak{B}$, $\mathfrak{C}$, etc. to denote structures.

EXAMPLE 3.2. We can consider $(N^+, +, \cdot, 1, 2, \cdots)$ a structure $\mathfrak{A}$ of type $\{f_{2,0}, f_{2,1}, c_0, c_1, \cdots\}$, where $|\mathfrak{A}| = N^+$, $f_{2,0}^{\mathfrak{A}} = +$, $f_{2,1}^{\mathfrak{A}} = \cdot$, $c_i^{\mathfrak{A}} = i+1$.

EXAMPLE 3.3. A group can be thought of as a structure $\mathfrak{G} = (G, *)$ of type $\{f_{2,0}\}$, where $|\mathfrak{G}| = G$ and $f_{2,0}^{\mathfrak{G}} = *$, $*$ being a binary group operation. Alternatively, it may be convenient to regard a group as a structure $\mathfrak{G} = (G, *, \text{In}, c)$ where $\tau\mathfrak{G} = \{f_{2,0}, f_{1,0}, c_0\}$ and $f_{2,0}^{\mathfrak{G}} = *$ ($*$ the binary group operation), $f_{1,0}^{\mathfrak{G}} = \text{In}$ (In the unary inverse operation), $c_0^{\mathfrak{G}} = c$ (c the identity element).

EXAMPLE 3.4. An ordered ring or field may be viewed as a structure $\mathfrak{A} = (A, \oplus, \odot, <)$ with $\mathfrak{A}$ of the same type as $(N^+, +, \cdot, <)$, say $\tau\mathfrak{A} = \{f_{2,0}, f_{2,1}, R_{2,0}\}$. Or one might find it convenient to have the type be $\{f_{2,\alpha}, f_{2,\beta}, R_{2,\gamma}\}$ for some (α, β, γ) different than $(0, 1, 0)$ with $\alpha \neq \beta$.

EXAMPLE 3.5. A partial ordering $\leq$ on a set A may be viewed as a structure $\mathfrak{A}$ of type $\{R_{2,\alpha}\}$ where $|\mathfrak{A}| = A$ and $R_{2,\alpha}^{\mathfrak{A}} = \leq$.

EXAMPLE 3.6. With a little hanky-panky, vector spaces can be regarded as structures. Suppose we have a vector space with $(V, \oplus)$ the underlying group of vectors, $(F, +, \cdot)$ the field of scalars, and $\odot$ the multiplication of a vector by a scalar. We can regard the vector space as the structure $\mathfrak{C} = (V \cup F, V, F, R_\oplus, R_+, R_\cdot, R_\odot)$ where V, F are unary relations, $R_\oplus = \{(x,y,z) : x \oplus y = z\}$, $R_+ = \{(x,y,z) : x+y=z\}$, $R_\cdot = \{(x,y,z) : x \cdot y = z\}$ and $R_\odot = \{(x,y,z) : x \odot y = z\}$. Note that we have used relations rather than functions for $\oplus$, $+$, $\cdot$, and $\odot$ in $\mathfrak{C}$; this is because clause iv of Definition 3.1 demands that function symbols $f_{n,\alpha}$ denote functions in $\mathfrak{C}$ that are defined for all n-tuples of $\mathfrak{C}$.

EXAMPLE 3.7. Structures of the empty type are allowed, as for example (N). We also allow two different symbols to have the same denotation, as in the structure $(A, c_0^{\mathfrak{A}}, c_1^{\mathfrak{A}})$ where $A = N$, $c_0^{\mathfrak{A}} = c_1^{\mathfrak{A}} = 3$. Since structures are functions, $\mathfrak{A} = \mathfrak{B}$ iff $\text{Dom}\,\mathfrak{A} = \text{Dom}\,\mathfrak{B}$ and for all x in the common domain, $\mathfrak{A}(x) = \mathfrak{B}(x)$. Thus if $\mathfrak{A} = (A, f_{1,0}^{\mathfrak{A}})$ and $\mathfrak{B} = (B, f_{1,1}^{\mathfrak{B}})$, then $\mathfrak{A} \neq \mathfrak{B}$ even if $A = B$, $f_{1,0}^{\mathfrak{A}} = f_{1,1}^{\mathfrak{B}}$.

Definition 3.8. We say that $\mathfrak{A}$ is a *substructure* of $\mathfrak{B}$, or that $\mathfrak{B}$ is an *extension* of $\mathfrak{A}$, and write $\mathfrak{A} \subseteq \mathfrak{B}$, if

i. $\tau\mathfrak{A} = \tau\mathfrak{B}$,
ii. $|\mathfrak{A}| \subseteq |\mathfrak{B}|$,
iii. $c_\alpha^\mathfrak{A} = c_\alpha^\mathfrak{B}$ for all $c_\alpha \in \tau\mathfrak{A}$,
iv. $f_{n,\alpha}^\mathfrak{A} \bar{a} = f_{n,\alpha}^\mathfrak{B} \bar{a}$ for every $f_{n,\alpha} \in \tau\mathfrak{A}$ and all $\bar{a} \in {}^n|\mathfrak{A}|$, and
v. $R_{n,\alpha}^\mathfrak{A} \bar{a}$ iff $R_{n,\alpha}^\mathfrak{B} \bar{a}$ for every $R_{n,\alpha} \in \tau\mathfrak{A}$ and all $\bar{a} \in {}^n|\mathfrak{A}|$.

If $\mathfrak{B}$ has a substructure with universe X, then that substructure will be called the *restriction* of $\mathfrak{B}$ to X and denoted by $\mathfrak{B}|X$.

Notice that if $\mathfrak{A}$ is a subgroup of $\mathfrak{B}$ then $\mathfrak{A} \subseteq \mathfrak{B}$. A similar statement holds for rings, fields, lattices, etc. The converse, however, is not always true, but depends on the way in which the group (ring, field, etc.) is realized as a structure. For example, $(\mathbf{N}^+, +)$ is a substructure of the integers under addition, but is not a subgroup. On the other hand, if $\mathfrak{C} = (G, \cdot, {}^{-1})$ is a group ($^{-1}$ being the inverse operation of the group), then any substructure of $\mathfrak{C}$ is a subgroup of $\mathfrak{C}$ and conversely.

The notation '$\subseteq$' for substructure is a bit misleading in that $\mathfrak{A} \subseteq \mathfrak{B}$ does not mean that as functions, $\mathfrak{B}$ extends $\mathfrak{A}$. Extension in the sense of functions is considered in the following definition.

Definition 3.9. If $s \subseteq \tau\mathfrak{B}$, then $\mathfrak{A}$ is the *reduct* of $\mathfrak{B}$ to s, written $\mathfrak{A} = \mathfrak{B} \restriction s$, if

i. $\tau\mathfrak{A} = s$,
ii. $|\mathfrak{A}| = |\mathfrak{B}|$,
iii. $S^\mathfrak{A} = S^\mathfrak{B}$ for all $S \in s$.

If $\mathfrak{A} = \mathfrak{B} \restriction s$ for some type s then $\mathfrak{B}$ is called an *expansion* of $\mathfrak{A}$. When convenient we will write $\mathfrak{B} = (\mathfrak{A}, S^\mathfrak{B})_{S \in \tau\mathfrak{B} - s}$.

EXAMPLE Let $\mathfrak{B} = (\mathbf{N}^+, +, \cdot, <, 1, 2) = (\mathbf{N}^+, f_{2,0}^\mathfrak{B}, f_{2,1}^\mathfrak{B}, R_{2,0}^\mathfrak{B}, c_0^\mathfrak{B}, c_1^\mathfrak{B})$, $\mathfrak{A} = (\mathbf{N}^+, \cdot, <, 2) = (\mathbf{N}^+, f_{2,1}^\mathfrak{A}, R_{2,0}^\mathfrak{A}, c_1^\mathfrak{A})$, and $s = \{f_{2,1}, R_{2,0}, c_1\}$. Then $\mathfrak{B} \restriction s = \mathfrak{A}$ and $\mathfrak{B} = (\mathfrak{A}, +, 1)$.

Definition 3.10. Let $\tau\mathfrak{A} \subseteq \tau\mathfrak{B}$. A function g on $|\mathfrak{A}|$ into $|\mathfrak{B}|$ is an *injection* if

i. g is 1-1,
ii. $g(c^\mathfrak{A}) = c^\mathfrak{B}$ for all constant symbols $c \in \tau\mathfrak{A}$,
iii. $R^\mathfrak{A} a_1, \ldots, a_n$ iff $R^\mathfrak{B} g(a_1), \ldots, g(a_n)$ for all $a_1, \ldots, a_n \in |\mathfrak{A}|$ and all n-relation symbols $R \in \tau\mathfrak{A}$,
iv. $g(f^\mathfrak{A} a_1, \ldots, a_n) = f^\mathfrak{B} g(a_1), \ldots, g(a_n)$ for all $a_1, \ldots, a_n \in |\mathfrak{A}|$ and all n-function symbols $f \in \tau\mathfrak{A}$.

If in addition g is onto $|\mathfrak{B}|$, then g is an *isomorphism* of $\mathfrak{A}$ onto $\mathfrak{B}$, in which case we write $\mathfrak{A} \cong_g \mathfrak{B}$. Or we may write $\mathfrak{A} \cong \mathfrak{B}$ if explicit mention of the isomorphism is unnecessary.

3.3 Structures

Notice that this definition coincides completely with the use of 'injection' and 'isomorphism' in algebra. Several other notions from abstract algebra will be generalized in the problems and in the sections that follow.

EXERCISES FOR §3.3

1. Show that if $\mathfrak{A} \subseteq \mathfrak{B} \subseteq \mathfrak{C}$, then $\mathfrak{A} \subseteq \mathfrak{C}$.

2. Show that '$\cong$' is an equivalence relation by showing that for all structures $\mathfrak{A}, \mathfrak{B}, \mathfrak{C}$,
   (a) $\mathfrak{A} \cong \mathfrak{A}$,
   (b) $\mathfrak{A} \cong \mathfrak{B}$ implies $\mathfrak{B} \cong \mathfrak{A}$, and
   (c) $\mathfrak{A} \cong \mathfrak{B}$ and $\mathfrak{B} \cong \mathfrak{C}$ implies $\mathfrak{A} \cong \mathfrak{C}$.

3. Give examples of structures $\mathfrak{A}, \mathfrak{A}', \mathfrak{B}$, and $\mathfrak{B}'$ such that
$$\mathfrak{A} \cong \mathfrak{B}' \subseteq \mathfrak{B} \cong \mathfrak{A}' \subseteq \mathfrak{A}$$
but $\mathfrak{A} \not\cong \mathfrak{B}$.

4. Suppose that $\mathfrak{A}_i \subseteq \mathfrak{A}_{i+1}$ for $i = 0, 1, 2, \ldots$. Let $\mathfrak{A}$ be that structure of type $\tau\mathfrak{A}_0$ such that
$$|\mathfrak{A}| = \bigcup_{i \in \mathbf{N}} |\mathfrak{A}_i|,$$
$$R^{\mathfrak{A}} = \bigcup_{i \in \mathbf{N}} R^{\mathfrak{A}_i} \quad \text{for each } R \in \tau\mathfrak{A},$$
$$F^{\mathfrak{A}} = \bigcup_{i \in \mathbf{N}} f^{\mathfrak{A}_i} \quad \text{for each } f \in \tau\mathfrak{A},$$
$$c^{\mathfrak{A}} = c^{\mathfrak{A}_i} \quad \text{for each } c \in \tau\mathfrak{A}.$$
Clearly each $\mathfrak{A}_i$ is a substructure of $\mathfrak{A}$. Find such a substructure chain $\mathfrak{A}_1 \subseteq \mathfrak{A}_2 \subseteq \mathfrak{A}_3 \subseteq \ldots$ such that each $\mathfrak{A}_i$ is isomorphic to $(\mathbf{N}^+, <)$ but $\mathfrak{A}$ is not.

5. Let A be the set of all integral powers of 3, $A = \{3^j : j \in \mathbf{I}\}$. Show that $(A, \cdot) \cong (I, +)$.

6. A function h mapping $|\mathfrak{A}|$ into $|\mathfrak{B}|$ is a homomorphism if $\tau\mathfrak{A} = \tau\mathfrak{B}$ and whenever $c, R, f \in \tau\mathfrak{A}$, then
   i. $hc^{\mathfrak{A}} = c^{\mathfrak{B}}$,
   ii. $hf^{\mathfrak{A}}(a_1, \ldots, a_n) = f^{\mathfrak{B}}(ha_1, \ldots, ha_n)$,
   iii. $R^{\mathfrak{A}} a_1, \ldots, a_n$ iff $R^{\mathfrak{B}} ha_1, \ldots, ha_n$.
   Let k be a positive integer and let $+_k$ and $\cdot_k$ be addition and multiplication modulo k. Let $h(n) = n \bmod a$. Show that h is a homomorphism from $(I, +, \cdot)$ onto $(\{0, 1, \ldots, k\}, +_k, \cdot_k)$.

7. An equivalence relation C on $|\mathfrak{A}|$ is a congruence relation on $\mathfrak{A}$ if whenever $a_i C b_i$ for $i \leq n$ and R is an n-relation in $\tau\mathfrak{A}$ and f is an n-function in $\tau\mathfrak{A}$, then
$$Ra_1, \ldots, a_n \quad \text{iff} \quad Rb_1, \ldots, b_n$$
and
$$fa_1, \ldots, a_n \, C \, fb_1, \ldots, b_n.$$
Let h be a homomorphism on $\mathfrak{A}$ to $\mathfrak{B}$. Let $C = \{(a, b) : a, b \in \mathfrak{A}, h(a) = h(b)\}$. Show that C is a congruence relation.

8. Let C be a congruence relation on $\mathfrak{A}$. For each $a \in |\mathfrak{A}|$ we let $\overline{a} = \{b \in |\mathfrak{A}| : b\,C\,a\}$. The quotient structure $\mathfrak{A}$ modulo C is that structure $\mathfrak{B}$ of type $\tau\mathfrak{A}$ with

$$|\mathfrak{B}| = \{\overline{a} : a \in |\mathfrak{A}|\},$$
$$c^{\mathfrak{B}} = \overline{c^{\mathfrak{A}}} \quad \text{for each } c \in \tau\mathfrak{A},$$
$$f^{\mathfrak{B}}\overline{a_1},\ldots,\overline{a_n} = \overline{f^{\mathfrak{A}}a_1,\ldots,a_n} \quad \text{for each } f \in \tau\mathfrak{A},$$
$$R^{\mathfrak{B}}\overline{a_1},\ldots,\overline{a_n} \text{ iff } R^{\mathfrak{A}}a_1,\ldots,a_n \quad \text{for each } R \in \tau\mathfrak{A}.$$

Show that if $h(a) = \overline{a}$ for each $a \in |\mathfrak{A}|$, then h is a homomorphism from $\mathfrak{A}$ onto $\mathfrak{B}$.

3.4 Satisfaction and Truth

The purpose of this section is to extend the notions of satisfaction and truth as defined in Section 2.10 to our more general language. The extension is made in the obvious way, in correspondence to our usual use of the symbols in mathematics. After this is done, several examples are given to illustrate the expressive power of L.

Definition 4.1. An *assignment* to $\mathfrak{A}$ is a function z with $\text{Dom}\, z = \{v_i : i \in \mathbf{N}\}$ and $\text{Rng}\, z \subseteq |\mathfrak{A}|$. If z is an assignment to $\mathfrak{A}$, u a variable, and $a \in |\mathfrak{A}|$, then $z\binom{u}{a}$ is the assignment z' defined as follows:

$$z'(v) = z(v) \quad \text{for all variables } v \neq u,$$
$$z'(u) = a.$$

Letting Vbl be the set of variables, we can use the notation of Part I and write '$z \in {}^{\text{Vbl}}|\mathfrak{A}|$' when z is an assignment to $\mathfrak{A}$.

Definition 4.2. Let t be a term of type $\subseteq \tau\mathfrak{A}$ and let $z \in {}^{\text{Vbl}}|\mathfrak{A}|$. We define $t^{\mathfrak{A}}\langle z \rangle$ (by induction on the length of t) as follows:

i. $v_n^{\mathfrak{A}}\langle z \rangle = z(v_n)$,
ii. $c_{\alpha}^{\mathfrak{A}}\langle z \rangle = c_{\alpha}^{\mathfrak{A}}$,
iii. $(f_{n,\alpha}(t_1,\ldots,t_n))^{\mathfrak{A}}\langle z \rangle = f_{n,\alpha}^{\mathfrak{A}}(t_1^{\mathfrak{A}}\langle z \rangle,\ldots,t_n^{\mathfrak{A}}\langle z \rangle)$.

Definition 4.3. Let φ be a formula of type $\mathfrak{A}$ and let $z \in {}^{\text{Vbl}}|\mathfrak{A}|$. We say that z *satisfies the formula* φ *in* $\mathfrak{A}$, written $\mathfrak{A} \models \varphi\langle z \rangle$, if either

i. $\varphi = [t_1 \approx t_2]$ and $t_1^{\mathfrak{A}}\langle z \rangle \approx t_2^{\mathfrak{A}}\langle z \rangle$,
i'. $\varphi = [R_{n,\alpha}t_1,\ldots,t_n]$ and $R_{n,\alpha}^{\mathfrak{A}}t_1^{\mathfrak{A}}\langle z \rangle,\ldots,t_n\langle z \rangle$,
ii. $\varphi = [\neg \psi]$ and it is not the case that $\mathfrak{A} \models \psi\langle z \rangle$,
iii. $\varphi = [\psi_1 \wedge \psi_2]$ and both $\mathfrak{A} \models \psi_1\langle z \rangle$ and $\mathfrak{A} \models \psi_2\langle z \rangle$,
iii'. $\varphi = [\psi_1 \vee \psi_2]$ and either $\mathfrak{A} \models \psi_1\langle z \rangle$ or $\mathfrak{A} \models \psi_2\langle z \rangle$,
iv. $\varphi = [\exists v_n \psi]$ and for some $a \in |\mathfrak{A}|$, $\mathfrak{A} \models \psi\langle z\binom{v_n}{a}\rangle$, or
iv'. $\varphi = [\forall v_n \psi]$ and for all $z \in |\mathfrak{A}|$, $\mathfrak{A} \models \psi\langle z\binom{v_n}{a}\rangle$.

3.4 Satisfaction and Truth

Theorem 4.4. *For all* $\mathfrak{A}, z, \varphi$,

i. $\mathfrak{A} \vDash \forall v_n \varphi \langle z \rangle$ *iff* $\mathfrak{A} \vDash \neg [\exists v_n \neg \varphi] \langle z \rangle$, $\mathfrak{A} \vDash \exists v_n \varphi \langle z \rangle$ *iff* $\mathfrak{A} \vDash \neg [\forall v_n \neg \varphi] \langle z \rangle$;
ii. $\mathfrak{A} \vDash [\varphi \wedge \psi] \langle z \rangle$ *iff* $\mathfrak{A} \vDash \neg [\neg \varphi \vee \neg \psi] \langle z \rangle$, $\mathfrak{A} \vDash [\varphi \vee \psi] \langle z \rangle$ *iff* $\mathfrak{A} \vDash \neg [\neg \varphi \wedge \neg \psi] \langle z \rangle$.

The proof consists of an easy unwinding of the above definition, and is left as an exercise (see Exercise 1).

One may view Theorem 4.4 as saying that each quantifier can be defined in terms of the other and '$\neg$', and that each of the symbols '$\vee$' and '$\wedge$' can be defined in terms of the other and '$\neg$'. In other words, L suffers no loss in expressive power if we delete from its list of symbols '$\forall$' and '$\wedge$' (or '$\exists$' and '$\vee$', or '$\forall$' and '$\vee$', or '$\exists$' and '$\wedge$'), and if we delete from Definition 4.3 clauses iii and iv' (or, respectively, clauses iii' and iv, or iii' and iv', or iii and iv). In addition to the added succinctness in describing L, this approach makes Definition 4.3 easier to use in that fewer clauses need be checked (see the proof of Theorem 4.5). On the other hand, the choice of symbols for L as given seems natural and is a bit more convenient in the statement of several theorems. So from now on the formulation of L that is used in a given discussion will be chosen on the basis of convenience.

The definitions of bound occurrence of a variable, free occurrence, assertion, 'v is free for t in φ', etc., are exactly as before (see Part II, Definitions 10.7 and 10.9i). The meanings of $\varphi(v_{i_1}, \ldots, v_{i_k})$ and $\varphi(t_1, \ldots, t_k)$ carry over to our extended language (see end of §2.10). *We shall often find it convenient to write* $\varphi(t_1^{\mathfrak{A}} \langle z \rangle, \ldots, t_k^{\mathfrak{A}} \langle z \rangle)$ *instead of* $\mathfrak{A} \vDash \varphi(t_1, \ldots, t_k) \langle z \rangle$—even though this is an ambiguous convention, since z might be an assignment to two structures $\mathfrak{A}$ and $\mathfrak{B}$ with $t_i^{\mathfrak{A}} \langle z \rangle = t_i^{\mathfrak{B}} \langle z \rangle$ for $i = 1, 2, \ldots, k$ and $\mathfrak{A} \vDash \varphi(t_1, \ldots, t_k) \langle z \rangle$ but $\mathfrak{B} \vDash \neg \varphi(t_1, \ldots, t_k) \langle z \rangle$. However, we shall use this convention only where such confusion is unlikely.

Theorem 4.5.

i. *Let t be a term, and let z and z' be assignments to $\mathfrak{A}$ such that $z(v) = z'(v)$ for all variables v not occurring in t. Then $t^{\mathfrak{A}} \langle z \rangle = t^{\mathfrak{A}} \langle z' \rangle$.*
ii. *Let φ be a formula, and z and z' assignments to $\mathfrak{A}$ such that $z(v) = z'(v)$ for all variables v occurring free in φ. Then $\mathfrak{A} \vDash \varphi \langle z \rangle$ iff $\mathfrak{A} \vDash \varphi \langle z' \rangle$.*

PROOF: Exactly like that for Theorem 10.8i in Part II. □

The theorem implies that if σ is an assertion, i.e., a formula without free variables, then $\mathfrak{A} \vDash \sigma \langle z \rangle$ for all assignments z to $\mathfrak{A}$, or no assignment satisfies σ in $\mathfrak{A}$. Thus we write $\mathfrak{A} \vDash \sigma$ if there is an assignment that satisfies σ in $\mathfrak{A}$, and we say that σ is true in $\mathfrak{A}$ or that $\mathfrak{A}$ satisfies σ. We say that σ is valid and write $\vDash \sigma$ if for all $\mathfrak{A}$ of type $\supseteq \tau \sigma$, $\mathfrak{A} \vDash \sigma$. For example, $\forall x [Rx \vee \neg Rx]$ is valid.

Definition 4.6.

i. *The theory of* $\mathfrak{A}$, abbreviated $\text{Th}\,\mathfrak{A}$, is $\{\sigma: \mathfrak{A} \vDash \sigma\}$. If $\mathcal{K}$ is a class of structures, then $\text{Th}\,\mathcal{K} = \{\sigma: \sigma \in \text{Th}\,\mathfrak{A} \text{ for all } \mathfrak{A} \in \mathcal{K}\}$. $\mathfrak{A}$ is *elementarily equivalent* to $\mathfrak{B}$, in symbols $\mathfrak{A} \equiv \mathfrak{B}$, if $\text{Th}\,\mathfrak{A} = \text{Th}\,\mathfrak{B}$.

ii. If Σ is a set of sentences, then $\text{Mod}\,\Sigma$, *the class of models of* Σ, is the class of all $\mathfrak{A}$ such that $\mathfrak{A} \vDash \sigma$ for all $\sigma \in \Sigma$. We write $\mathfrak{A} \in \text{Mod}\,\sigma$ instead of $\mathfrak{A} \in \text{Mod}\{\sigma\}$. A class of structures $\mathcal{K}$ is an *elementary class* if $\mathcal{K} = \text{Mod}\,\Sigma$ for some set of sentences Σ.

Throughout the rest of the section we shall use ρ and σ to denote assertions, φ and ψ to denote formulas, Δ, Γ, and Σ to denote sets of assertions, and z to denote assignments.

We now give several examples of elementary classes, first giving the class $\mathcal{K}$ and then a set of assertions Σ such that $\text{Mod}\,\Sigma = \mathcal{K}$. In Examples 4.7 through 4.10, $\tau\mathcal{K} = \{\leqslant\}$, where $\leqslant$ is some binary relation symbol. As before, $\varphi \rightarrow \psi$ abbreviates $\neg\varphi \vee \psi$, and $\varphi \leftrightarrow \psi$ abbreviates $(\neg\varphi \vee \psi) \wedge (\neg\psi \vee \varphi)$.

EXAMPLE 4.7. Partially ordered structures:

$$\forall v_0 [v_0 \leqslant v_0],$$
$$\forall v_0 v_1 [[v_0 \leqslant v_1 \wedge v_1 \leqslant v_0] \rightarrow v_0 \approx v_1],$$
$$\forall v_0 v_1 v_2 [[v_0 \leqslant v_1 \wedge v_1 \leqslant v_2] \rightarrow v_0 \leqslant v_2].$$

EXAMPLE 4.8. Linearly ordered structures: The assertions of Example 4.7 along with

$$\forall v_0 v_1 [v_0 \leqslant v_1 \vee v_1 \leqslant v_0].$$

EXAMPLE 4.9. Densely ordered structures: The assertions of Example 4.8 along with

$$\exists v_0 v_1 [v_0 \not\approx v_1] \wedge [\forall v_0 v_1 [[v_0 < v_1] \rightarrow \exists v_2 [v_0 < v_2 \wedge v_2 < v_1]]].$$

Of course, $v_0 < v_1$ abbreviates $v_0 \leqslant v_1 \wedge v_0 \not\approx v_1$. The rationals and the reals along with the usual orderings are examples of densely ordered structures.

EXAMPLE 4.10. Discretely ordered structures: The assertions of Example 4.8 along with

$$\forall v_0 [\exists v_1 [v_1 < v_0] \rightarrow \exists v_2 [v_2 < v_0 \wedge \forall v_3 [v_3 < v_0 \rightarrow v_3 \leqslant v_2]]],$$
$$\forall v_0 [\exists v_1 [v_0 < v_1] \rightarrow \exists v_2 [v_0 < v_2 \wedge \forall v_3 [v_0 < v_3 \rightarrow v_2 \leqslant v_3]]].$$

The integers are discretely ordered by the usual 'less than' relation.

3.4 Satisfaction and Truth   145

EXAMPLE 4.11. Groups are structures $(A, *^{\mathfrak{A}})$ satisfying

$$\forall v_0 v_1 v_2 [v_0 * [v_1 * v_2] \approx [v_0 * v_1] * v_2],$$
$$\exists v_0 \forall v_1 [v_0 * v_1 \approx v_1 * v_0 \approx v_1],$$
$$\forall v_0 \exists v_1 \forall v_2 [[v_0 * v_1] * v_2 \approx [v_1 * v_0] * v_2 \approx v_2].$$

Here, $*$ is $f_{2,\alpha}$ for some convenient α, and we write $u * v$ instead of $f_{2,\alpha} uv$. Similar notational devices will be used throughout the chapter without mention.

Alternatively, we could consider a group to be a structure $(A, *^{\mathfrak{A}}, '^{\mathfrak{A}}, e^{\mathfrak{A}})$ satisfying

$$\forall v_0 v_1 v_2 [v_0 * [v_1 * v_2] \approx [v_0 * v_1] * v_2],$$
$$\forall v_0 [v_0 * i \approx i * v_0 \approx v_0],$$
$$\forall v_0 v_1 [v_0 * v_0' \approx i \approx v_0' * v_0].$$

It is not difficult to show that the groups in the first sense are exactly the groups in the second sense reducted to the type $\{*\}$.

A group is Abelian if

$$\forall v_0 v_1 [v_0 * v_1 \approx v_1 * v_0]$$

EXAMPLE 4.12. Rings are structures $\mathfrak{A} = (A, *^{\mathfrak{A}}, \circ^{\mathfrak{A}})$ where $(A, *^{\mathfrak{A}})$ is an Abelian group, i.e., $\mathfrak{A}$ satisfies the assertions in Example 4.11, and also $\mathfrak{A}$ satisfies

$$\forall v_0 v_1 v_2 [v_0 \circ [v_1 \circ v_2] \approx [v_0 \circ v_1] \circ v_2],$$
$$\forall v_0 v_1 v_2 [[v_0 \circ [v_1 * v_2] \approx [v_0 \circ v_1] * [v_0 \circ v_2]]$$
$$\wedge [[v_1 * v_2] \circ v_0 \approx [v_1 \circ v_0] * [v_2 \circ v_0]]].$$

Of course each alternative formulation of a group as an elementary class yields an alternative formulation of a ring. For example, we can view rings as structures $\mathfrak{A} = (A, *^{\mathfrak{A}}, ^{-1\mathfrak{A}}, i^{\mathfrak{A}}, \circ^{\mathfrak{A}})$ where $(A, *^{\mathfrak{A}}, ^{-1\mathfrak{A}}, i^{\mathfrak{A}})$ is an Abelian group according to our second formulation of a group, and $\mathfrak{A}$ satisfies the above two assertions.

EXAMPLE 4.13. Fields are rings satisfying the additional assertions

$$\exists v_0 \forall v_1 [v_0 \circ v_1 \approx v_1 \circ v_0 \approx v_1],$$
$$\forall v_0 [v_0 * v_0 \approx v_0 \vee \exists v_1 \forall v_2 [[v_0 \circ v_1] \circ v_2 \approx [v_1 \circ v_0] \circ v_2 \approx v_2]].$$

Again, there are alternative formulations.

These examples by no means exhaust the elementary classes that are mathematically interesting. On the other hand, as we show in §3.5, there are many classes of interest that are not elementary classes.

In our next example, we use L to give a precise statement of the axioms of Zermelo–Fraenkel set theory (with regularity). In particular, the ambiguities in our statement of the axiom of replacement as given in Part I are avoided. Here e is a binary relation symbol whose intended interpretation is $\in$.

EXAMPLE 4.14 (The Axioms of Zermelo–Fraenkel Set Theory).

i. Extensionality:
$$\forall v_0 v_1 [\forall v_2 [v_2 e v_0 \leftrightarrow v_2 e v_1] \rightarrow v_0 \approx v_1].$$

ii. Null set:
$$\exists v_0 \forall v_1 [\neg v_1 e v_0].$$

iii. Pairing:
$$\forall v_0 v_1 \exists v_2 \forall v_3 [v_3 e v_2 \leftrightarrow [v_3 \approx v_0 \lor v_3 \approx v_1]].$$

iv. Union:
$$\forall v_0 \exists v_1 \forall v_2 [v_2 e v_1 \leftrightarrow \exists v_3 [v_2 e v_3 \land v_3 e v_0]].$$

v. Power set:
$$\forall v_0 \exists v_1 \forall v_2 [v_2 e v_1 \leftrightarrow \forall v_3 [v_3 e v_2 \rightarrow v_3 e v_0]].$$

vi. Replacement schema: For each formula φ of L with free variables $v_0, v_1, \ldots, v_n$ the following is an axiom:
$$\forall v_2 \cdots v_n [\forall v_0 \exists v_{n+1} \forall v_1 [\varphi \leftrightarrow v_1 \approx v_{n+1}]]$$
$$\rightarrow \forall v_{n+1} \exists v_{n+2} \forall v_1 [v_1 \in v_{n+2} \leftrightarrow \exists v_0 [v_0 e v_{n+1} \land \varphi]].$$

vii. Infinity:
$$\exists v_0 [\emptyset e v_0 \land \forall v_1 [v_1 e v_0 \rightarrow v_1 \cup \{v_1\} e v_0]].$$

Here $\emptyset e v_0$ is an abbreviation for $\exists v_1 \forall v_2 [\neg v_2 e v_1 \land v_1 e v_0]$, and $v_1 \cup \{v_1\} \in v_0$ is an abbreviation for $\exists v_2 [\forall v_3 [v_3 \in v_2 \leftrightarrow v_3 \approx v_1 \lor v_3 e v_1] \land v_2 e v_0].$

viii. Regularity:
$$\forall v_0 [\forall v_1 [v_1 \not{e} v_0] \lor \exists v_1 [v_1 e v_0 \land \forall v_2 [v_2 e v_0 \rightarrow v_2 \not{e} v_1]]].$$

In addition to the Zermelo–Fraenkel axioms, the other axioms discussed in §1.11, such as the axiom of choice and the generalized continuum hypotheses, can also be formulated in L. This we leave as an exercise.

Our last two examples are number theories. The second is an attempt to realize the Paeno axioms for arithmetic within L. In contrast to the second example, the first involves only a single assertion, and it is this property which we will need in §3.12 when considering the possibility of an algorithmic test for validity.

EXAMPLE 4.15 (The Formalization Q). Our formulation is in the type $\{\stackrel{\cap}{+}, \stackrel{\cap}{\cdot}, S, 0\}$ (where S denotes the successor function):

$$\forall v_0 v_1 [[Sv_0 \approx Sv_1 \to v_0 \approx v_1]$$
$$\wedge [0 \not\approx Sv_0]$$
$$\wedge [0 \not\approx v_0 \to \exists v_1 [Sv_1 \approx v_0]]$$
$$\wedge [0 \stackrel{\cap}{+} v_0 \approx v_0]$$
$$\wedge [v_0 \stackrel{\cap}{+} Sv_1 \approx S(v_0 \stackrel{\cap}{+} v_1)]$$
$$\wedge [v_0 \stackrel{\cap}{\cdot} 0 \approx 0]$$
$$\wedge [v_0 \stackrel{\cap}{\cdot} Sv_1 \approx v_0 \cdot v_1 \stackrel{\cap}{+} v_0].$$

EXAMPLE (A Fragment of Peano's Arithmetic). Adjoin to the single assertion of Q above the infinite list of assertions in the following induction schema: All assertions of the form

$$\forall u_0, \ldots, u_{n-1} [[[\varphi(u_0, \ldots, u_{n-1}, 0)$$
$$\wedge \forall u_n [\varphi(u_0, \ldots, u_{n-1}, u_n) \to \varphi(u_0, \ldots, u_{n-1}, Su_n)]]$$
$$\to \forall u_n [\varphi(u_0, \ldots, u_{n-1}, u_n)]],$$

where $u_0, \ldots, u_n$ is a list of the free variables in φ.

The induction axiom for Peano's axiomatization asserts: For all subsets X of $\mathbf{N}$, if $0 \in X$ and if $n+1 \in X$ whenever $n \in X$, then $X = \mathbf{N}$. In particular, if $X = \{n : \mathbf{N} \vDash \varphi(n)\}$, we obtain a typical assertion of the induction schema listed above. We cannot state the induction axiom itself in L since this requires quantifying over arbitrary sets ("for all subsets of $\mathbf{N}$, if ...").

In the remainder of this section we discuss several theorems relating the notions of substructures and isomorphism to the language L.

Definition 4.16. Say that an assertion is *simple* if it has one of the following forms: $d_1 \approx d_2$, $d_1 \not\approx d_2$, $Rd_1 \ldots d_n$, $\neg Rd_1 \ldots d_n$, $fd_1, \ldots, d_n = d$, $fd_1 \ldots d_n \not\approx d$, where R is a relation symbol, f a function symbol, and the d's constant symbols. If $\mathfrak{B}$ is an expansion of $\mathfrak{A}$ such that for all $a \in |\mathfrak{A}|$ there is a $c \in \tau \mathfrak{B}$ with $c^{\mathfrak{B}} = a$, then the set of all simple sentences true in $\mathfrak{B}$ is called a *diagram* of $\mathfrak{A}$. Even though a structure has many diagrams, there will be no harm in speaking of 'the diagram of $\mathfrak{A}$' and writing $\mathfrak{D}\mathfrak{A}$.

For example, if $\mathfrak{A} = (\mathbf{N}, +, \cdot, 0, 1)$, where $c_0^{\mathfrak{A}} = 0$ and $c_0^{\mathfrak{A}} = 1$, then we can take $\mathfrak{B} = (\mathfrak{A}, c_{n+2}^{\mathfrak{B}})_{n \in \omega}$, where $c_{n+2}^{\mathfrak{B}} = n+2$ (or alternatively $c_{n+2}^{\mathfrak{B}} = n$; the fact that several symbols may denote the same element is of no consequence). Then $c_5 + c_3 \approx c_8$, $\neg[c_8 < c_5]$, $\neg[c_7 \cdot c_3 \approx c_{10}]$ are all members of $\mathfrak{D}\mathfrak{A}$.

Theorem 4.17. $\mathfrak{A}$ *is isomorphic to a substructure of $\mathfrak{B}$ if and only if some expansion of $\mathfrak{B}$ is a model of $\mathfrak{D}\,\mathfrak{A}$ and $\tau\mathfrak{A} = \tau\mathfrak{B}$.*

PROOF: Suppose that g is an isomorphism on $\mathfrak{A}$ onto $\mathfrak{B}' \subseteq B$. Let $\mathfrak{A}^+$ be an expansion of $\mathfrak{A}$ such that each $a \in |\mathfrak{A}|$ is a constant $c_a^{\mathfrak{A}^+}$ in $\mathfrak{A}^+$. Let $\mathfrak{B}^+$ be $(\mathfrak{B}, c_a^{\mathfrak{B}^+})_{a \in |\mathfrak{A}|}$ where $c_a^{\mathfrak{B}^+} = g(c_a^{\mathfrak{A}^+})$. For each $R, f \in \tau \mathfrak{A}$ we have $R^{\mathfrak{A}} a_1, \ldots, a_n$ iff $R^{\mathfrak{B}} g(a_1), \ldots, g(a_n)$ iff $R^{\mathfrak{B}^+} c_{a_1}^{\mathfrak{B}^+}, \ldots, c_{a_n}^{\mathfrak{B}^+}$, and $f^{\mathfrak{A}^+} a_1, \ldots, a_n = a_0$ iff $f^{\mathfrak{B}} g(a_1), \ldots, g(a_n) = g(a_0)$ iff $f^{\mathfrak{B}^+} c_{a_1}^{\mathfrak{B}^+}, \ldots, c_{a_n}^{\mathfrak{B}^+}$. Also, $c_{a_1}^{\mathfrak{A}^+} = c_{a_2}^{\mathfrak{A}^+}$ iff $g(c_{a_1}^{\mathfrak{A}^+}) = g(c_{a_2}^{\mathfrak{A}^+})$ iff $c_{a_1}^{\mathfrak{B}^+} = c_{a_2}^{\mathfrak{B}^+}$. Hence $\mathfrak{B}^+ \in \text{Mod}\,\mathfrak{D}\,\mathfrak{A}$.

For the converse, suppose that $\mathfrak{B}^+$ is an expansion of $\mathfrak{B}$ and $\mathfrak{B}^+ \in \text{Mod}\,\mathfrak{D}\,\mathfrak{A}$. For each $a \in |\mathfrak{A}|$ let c_a be the symbol in $\tau\mathfrak{D}\,\mathfrak{A}$ denoting a. We claim that the function g, defined by $g(a) = c_a^{\mathfrak{B}^+}$, is an isomorphism on $\mathfrak{A}$ onto Ran g and that Ran $g \subseteq \mathfrak{B}$.

i. Ran g is the universe of a substructure of $\mathfrak{B}$: We must show that Ran g is closed under $f^{\mathfrak{B}}$ for all $f \in \tau\mathfrak{B}$. Let $b_1, \ldots, b_n \in \text{Ran}\,g$, say $b_i = c_{a_i}^{\mathfrak{B}^+}$. Let $a = f^{\mathfrak{A}} a_1, \ldots, a_n$. Then $c_a \approx f c_{a_1}, \ldots, c_{a_n} \in \mathfrak{D}\,\mathfrak{A}$ and so is true in $\mathfrak{B}^+$, i.e., $c_a^{\mathfrak{B}^+} = f^{\mathfrak{B}^+} c_{a_1}^{\mathfrak{B}^+}, \ldots, c_{a_n}^{\mathfrak{B}^+}$. Thus $c_a^{\mathfrak{B}^+} = f^{\mathfrak{B}} b_1, \ldots, b_n$ and $c_a^{\mathfrak{B}^+} \in \text{Ran}\,g$, as needed.

ii. g is 1-1: If $a_1 \neq a_2$, then $c_{a_1} \not\approx c_{a_2} \in \mathfrak{D}\,\mathfrak{A}$, so $\mathfrak{B}^+ \vDash c_{a_1} \neq c_{a_2}$ and $c_{a_1}^{\mathfrak{B}^+} \not\approx c_{a_2}^{\mathfrak{B}^+}$, i.e., $g(a_1) \neq g(a_2)$.

iii. "g preserves relations: $R^{\mathfrak{A}} a_1, \ldots, a_n$ iff $R c_{a_1}, \ldots, c_{a_n} \in \mathfrak{D}\,\mathfrak{A}$ iff $\mathfrak{B}^+ \vDash R c_{a_1}, \ldots, c_{a_n}$ iff $R^{\mathfrak{B}^+} c_{a_1}^{\mathfrak{B}^+}, \ldots, c_{a_n}^{\mathfrak{B}^+}$ iff $R^{\mathfrak{B}} g(a_1), \ldots, g(a_n)$.

iv. "g preserves functions": $f^{\mathfrak{A}} a_1, \ldots, a_n = a$ iff $f c_{a_1}, \ldots, c_{a_n} \approx a \in \mathfrak{D}\,\mathfrak{A}$ iff $\mathfrak{B}^+ \vDash f c_{a_1}, \ldots, c_{a_n} \approx c_a$ iff $f^{\mathfrak{B}^+} c_{a_1}^{\mathfrak{B}^+}, \ldots, c_{a_n}^{\mathfrak{B}^+} = c_a^{\mathfrak{B}^+}$ iff $f^{\mathfrak{B}} g(a_1), \ldots, g(a_n) = g(a)$. Hence $g(f^{\mathfrak{A}} a_1, \ldots, a_n) = f^{\mathfrak{B}} g(a_1), \ldots, g(a_n)$. Thus we see that g is an isomorphism onto a substructure of $\mathfrak{B}$. □

Theorem 4.18. *If $\mathfrak{A}$ is isomorphic to a substructure of $\mathfrak{B}$, then there is a $\mathfrak{C} \cong \mathfrak{B}$ such that $\mathfrak{A} \subseteq \mathfrak{C}$.*

PROOF: Let $\mathfrak{A} \cong_g \mathfrak{B}' \subseteq \mathfrak{B}$. Let h be a 1-1 function on $|\mathfrak{B}|$ such that $h(b) = g^{-1}(b)$ if $b \in \text{Ran}\,g$, and $h(b) \notin |\mathfrak{A}|$ otherwise. Now define $\mathfrak{C}$ as follows:

$$|\mathfrak{C}| = \text{Ran}\,h,$$

$$c^{\mathfrak{C}} = c^{\mathfrak{A}} \quad \text{for every } c \in \tau\mathfrak{A},$$

$$R^{\mathfrak{C}} d_1, \ldots, d_n \text{ iff } R^{\mathfrak{B}} h^{-1}(d_1), \ldots, h^{-1}(d_n) \quad \text{for all } R \in \tau\mathfrak{A},$$

$$f^{\mathfrak{C}} d_1, \ldots, d_n = h\big(f^{\mathfrak{B}} h^{-1}(d_1), \ldots, h^{-1}(d_n)\big) \quad \text{for all } f \in \tau\mathfrak{A}.$$

Clearly, $\mathfrak{B} \cong_h \mathfrak{C}$ and $\mathfrak{A} \subseteq \mathfrak{C}$, as needed. □

3.4 Satisfaction and Truth

We shall use this theorem, often without explicit mention, to excuse the writing of $\mathfrak{A} \subseteq \mathfrak{B}$ when we mean that $\mathfrak{A}$ is isomorphic to a substructure of $\mathfrak{B}$.

Theorem 4.19. *If $\mathfrak{A} \cong \mathfrak{B}$, then $\mathfrak{A} \equiv \mathfrak{B}$. In fact, if $\mathfrak{A} \cong_g \mathfrak{B}$, then for all formulas φ and all $z \in {}^{\mathrm{Vbl}}|\mathfrak{A}|$, $\mathfrak{A} \models \varphi\langle z \rangle$ iff $\mathfrak{B} \models \varphi \langle g \circ z \rangle$.*

PROOF: Let g be an isomorphism from $\mathfrak{A}$ to $\mathfrak{B}$. We first show that for all $z \in {}^{\mathrm{Vbl}}|\mathfrak{A}|$ and all terms t,

$$g(t^{\mathfrak{A}}\langle z \rangle) = t^{\mathfrak{B}}\langle g \circ z \rangle. \tag{*}$$

For t variable or a constant symbol we have

$$g(v_n^{\mathfrak{A}}\langle z \rangle) = g(z(v_n)) = v_n^{\mathfrak{B}}\langle g \circ z \rangle,$$
$$g(c_\alpha^{\mathfrak{A}}\langle z \rangle) = gc_\alpha^{\mathfrak{A}} = c_\alpha^{\mathfrak{B}}\langle g \circ z \rangle.$$

Now suppose (*) is true for all terms $t_1, \ldots, t_n$, and let t be $ft_1, \ldots, t_n$. Then

$$g((ft_1, \ldots, t_n)^{\mathfrak{A}}\langle z \rangle) = g(f^{\mathfrak{A}} t_1^{\mathfrak{A}}\langle z \rangle, \ldots, t_n^{\mathfrak{A}}\langle z \rangle)$$
$$= f^{\mathfrak{B}} g(t_1^{\mathfrak{A}}\langle z \rangle), \ldots, g(t_n^{\mathfrak{A}}\langle z \rangle)$$
[since g is an isomorphism]
$$= f^{\mathfrak{B}} t_1^{\mathfrak{B}}\langle g \circ z \rangle, \ldots, t_n^{\mathfrak{B}}\langle g \circ z \rangle$$
[since (*) holds for $t_1, \ldots, t_n$]
$$= (ft_1, \ldots, t_n)^{\mathfrak{B}}\langle g \circ z \rangle \ [\text{by Definition 4.2}].$$

This proves (*).

We now show by induction on formulas that for all $z \in {}^{\mathrm{Vbl}}|\mathfrak{A}|$ and all formulas φ

$$\mathfrak{A} \models \varphi\langle z \rangle \quad \text{iff} \quad \mathfrak{B} \models \varphi\langle g \circ z \rangle. \tag{**}$$

We first prove (**) for φ atomic:

i. $\mathfrak{A} \models [t_1 \approx t_2]\langle z \rangle$ iff $t_1^{\mathfrak{A}}\langle z \rangle = t_2^{\mathfrak{A}}\langle z \rangle$
    iff $g(t_1^{\mathfrak{A}}\langle z \rangle) = g(t_2^{\mathfrak{A}}\langle z \rangle)$ (since g is an isomorphism)
    iff $t_1^{\mathfrak{B}}\langle g \circ z \rangle = t_2^{\mathfrak{B}}\langle g \circ z \rangle$ [by (*)] iff $\mathfrak{B} \models [t_1 \approx t_2]\langle g \circ z \rangle$.
ii. $\mathfrak{A} \models Rt_1, \ldots, t_n\langle z \rangle$ iff $R^{\mathfrak{A}} t_1^{\mathfrak{A}}\langle z \rangle, \ldots, t_n^{\mathfrak{A}}\langle z \rangle$
    iff $R^{\mathfrak{B}} g(t_1^{\mathfrak{A}}\langle z \rangle), \ldots, g(t_n^{\mathfrak{A}}\langle z \rangle)$ (since g is an isomorphism)
    iff $R^{\mathfrak{B}} t_1^{\mathfrak{B}}\langle g \circ z \rangle, \ldots, t_n^{\mathfrak{B}}\langle g \circ z \rangle$ [by (*)]
    iff $\mathfrak{B} \models [Rt_1, \ldots, t_n]\langle g \circ z \rangle$ (by Definition 4.3).

Next we show that if (**) is true for φ_1 and φ_2 and all z, then it holds for $\neg \varphi_1, \varphi_1 \wedge \varphi_2$, and $\exists v \varphi_1$:

iii. $\mathfrak{A} \models [\neg \varphi_1]\langle z \rangle$ iff not $\mathfrak{A} \models \varphi_1 \langle z \rangle$ iff not $\mathfrak{B} \models \varphi_1 \langle g \circ z \rangle$ (by the induction assumption) iff $\mathfrak{B} \models [\neg \varphi_1] \langle g \circ z \rangle$.

iv. $\mathfrak{A}[\varphi_1 \wedge \varphi_2]\langle z\rangle$ iff $\mathfrak{A}\vDash\varphi_1\langle z\rangle$ and $\mathfrak{A}\vDash\varphi_2\langle z\rangle$ iff $\mathfrak{B}\vDash\varphi_1\langle g\circ z\rangle$ and $\mathfrak{B}\vDash\varphi_2\langle g\circ z\rangle$ (by our induction assumption) iff $\mathfrak{B}\vDash[\varphi_1\wedge\varphi_2]\langle g\circ z\rangle$.

v. $\mathfrak{A}\vDash\exists v\varphi_1\langle z\rangle$ iff for some $a\in|\mathfrak{A}|$,

$$\mathfrak{A}\vDash\varphi_1\left\langle z\binom{v}{a}\right\rangle$$

iff for some $a\in|\mathfrak{A}|$,

$$\mathfrak{B}\vDash\varphi_1\left\langle g\circ\left(z\binom{v}{a}\right)\right\rangle \text{ [since we assume (**) for } \varphi_1]$$

iff for some $a\in|\mathfrak{A}|$,

$$\mathfrak{B}\vDash\varphi_1\left\langle (g\circ z)\binom{v}{g(a)}\right\rangle$$

iff $\mathfrak{B}\vDash\exists v\varphi_1\langle g\circ z\rangle$.

In the last equivalence, the implication from left to right is immediate from the definition of satisfaction, while the implication from right to left uses both the definition of satisfaction and the fact that g is onto.

This completes the proof of the second clause of the theorem. The first clause follows immediately from the second. □

The converse of this theorem is false, as we shall see in the next section.

EXERCISES FOR §3.4

1. Prove Theorem 4.4.

2. If $\mathfrak{A}$ is finite and $\tau\mathfrak{A}$ is finite then there is a σ such that $\mathfrak{B}\vDash\sigma$ and $\tau\mathfrak{B}=\tau\mathfrak{A}$ iff $\mathfrak{A}\cong\mathfrak{B}$.

3. Suppose that no variable other than v occurs free in φ or ψ. Show that $\vDash\forall v[\varphi\to\psi]\to[\forall v\varphi\to\forall v\psi]$, but for some φ and ψ $\nvDash[\forall v\varphi\to\forall v\psi]\to\forall v[\varphi\to\psi]$ and $\nvDash\neg[[\forall v\varphi\to\forall v\psi]\to\forall v[\varphi\to\psi]]$.

4. Suppose that ψ is an assertion and that v is not free in ψ. Show that $\vDash[\exists v\varphi\to\psi]\leftrightarrow\forall v[\varphi\to\psi]$.

5. Show that two structures are isomorphic iff they have identical diagrams.

6. Given a structure $\mathfrak{A}$, describe a set of simple assertions $\Sigma\subseteq\mathcal{D}\mathfrak{A}$ such that any structure $\mathfrak{B}$ contains a homomorphic image of $\mathfrak{A}$ iff $\Sigma\subseteq\mathcal{D}\mathfrak{B}$.

3.5 Normal Forms

Different assertions can have the same meaning. For example, $\forall v_1\exists v_2\forall v_3[Rv_1v_3\leftrightarrow v_3\approx v_2]$ and $\forall v_1\exists v_2[Rv_1v_2]\wedge\forall v_1v_2v_3[Rv_1v_2\wedge Rv_1v_3\to v_2\approx v_3]$ both assert that R is a function. When investigating the properties of the class of models of σ it is often helpful to find some assertion equivalent to σ that has some convenient form. As we shall see in §3.9, the

3.5 Normal Forms

algebraic properties of Mod σ and the form of assertions equivalent to σ are closely related.

Definition 5.1.

i. If φ is a formula with free variables $u_0, \ldots, u_{n-1}$, then $\forall u_0, \ldots, u_{n-1}\varphi$ is a *universal closure* of φ.
ii. Say that φ is *valid*, and write $\vDash\varphi$ if it has a valid universal closure. (Recall that an assertion σ is valid if $\mathfrak{A}\vDash\sigma$ for all $\mathfrak{A}$ of type $\supseteq \tau\sigma$.)

Of course, if φ has a valid universal closure, then any universal closure of φ is valid.

iii. We say that φ and ψ are *equivalent* if $\vDash\varphi\leftrightarrow\psi$.

Unwinding the definitions, we see that $\vDash\varphi\leftrightarrow\psi$ iff for all $\mathfrak{A}$ of type $\tau(\varphi\leftrightarrow\psi)$ and for all $z\in{}^{\mathrm{Vbl}}|\mathfrak{A}|$,

$$\mathfrak{A}\vDash\varphi\langle z\rangle \quad \text{iff} \quad \mathfrak{A}\vDash\psi\langle z\rangle.$$

From this it is clear that logical equivalence is an equivalence relation on the class of formulas; i.e., for all formulas φ, ψ, ξ:

$$\vDash\varphi\leftrightarrow\varphi,$$
$$\vDash\varphi\leftrightarrow\psi \quad \text{implies} \quad \vDash\psi\leftrightarrow\varphi,$$
$$\vDash\varphi\leftrightarrow\psi \text{ and } \vDash\psi\leftrightarrow\xi \quad \text{implies} \quad \vDash\varphi\leftrightarrow\xi.$$

Theorem 5.2. *If $\vDash\varphi\leftrightarrow\varphi'$ and ξ' results from ξ by replacing an occurrence of φ with φ', then $\vDash\xi\leftrightarrow\xi'$.*

PROOF: Let $\vDash\varphi\leftrightarrow\varphi'$ and say that ξ has the property (*) if whenever ξ' is obtained from ξ by replacing an occurrence of φ with φ', then $\vDash\xi\leftrightarrow\xi'$.

Case i. Every atomic formula has property (*): Indeed, if ξ is atomic and ξ' is obtained from ξ by replacing an occurrence of φ with φ', then $\xi=\varphi$ and $\xi'=\varphi'$.

Case ii. Suppose $\xi=\neg\psi$ and (*) holds for ψ. Let ξ' be obtained from ξ by replacing an occurrence of φ with φ'. If $\xi=\varphi$, then $\xi'=\varphi'$ and we are done. If $\xi\neq\varphi$, then $\xi'=\neg\psi'$, where ψ' is obtained from ψ by replacing φ by φ'. By assumption, $\vDash\psi\leftrightarrow\psi'$, from which it follows that $\vDash\xi\leftrightarrow\xi'$.

Case iii. Suppose $\xi=\psi_1\wedge\psi_2$, where both ψ_1 and ψ_2 satisfy (*). Again, we need only consider the case where $\xi\neq\varphi$. Then if ξ' is obtained from ξ by replacing an occurrence of φ with ψ, then ξ' has the form $\psi'_1\wedge\psi'_2$, where ψ'_i is ψ_i or is obtained from ψ_i by replacing an occurrence of φ with φ'. By assumption $\vDash\psi_i\leftrightarrow\psi'_i$ for $i=1,2$. Thus for all $\mathfrak{A}$ of the appropriate type and all $z\in{}^{\mathrm{Vbl}}|\mathfrak{A}|$, we have $\mathfrak{A}\vDash\xi\langle z\rangle$ iff $\mathfrak{A}\vDash\psi_1\langle z\rangle$, and $\mathfrak{A}\vDash\psi_2\langle z\rangle$ iff $\mathfrak{A}\vDash\psi'_1\langle z\rangle$, and $\mathfrak{A}\vDash\psi'_2\langle z\rangle$ iff $\mathfrak{A}\vDash\psi'_1\wedge\psi'_2\langle z\rangle$ if $\mathfrak{A}\vDash\xi\langle z\rangle$. Hence $\vDash\xi\leftrightarrow\xi'$.

Case iv. Suppose $\xi=\exists v\psi$ and ψ satisfies (*). The case $\xi=\varphi$ being immediate, we suppose that $\xi'=\exists v\psi'$, where ψ' results from ψ by replacing an occurrence of φ in ψ with φ'. Then $\vDash\psi\leftrightarrow\psi'$ by assumption, so

$$\mathfrak{A}\vDash\xi\langle z\rangle \quad \text{iff} \quad \text{for some } a\in|\mathfrak{A}|,\ \mathfrak{A}\vDash\psi\left\langle z\binom{v}{a}\right\rangle$$

$$\text{iff} \quad \text{for some } a\in|\mathfrak{A}|,\ \mathfrak{A}\vDash\psi'\left\langle z\binom{v}{a}\right\rangle$$

$$\text{iff} \quad \mathfrak{A}\vDash\xi'\langle z\rangle.$$

Thus all formulas have the property (*), which proves the theorem. □

Lemma 5.3. *Let s be an expression and u a variable that does not occur in s. Let s' be the result of replacing every occurrence in s of the variable v by u. Then for all $\mathfrak{A}$ of type $\supseteq \tau s$ and all $z\in {}^{Vbl}|\mathfrak{A}|$:*

a. *if s is a term, then*

$$s^{\mathfrak{A}}\langle z\rangle = s'^{\mathfrak{A}}\left\langle z\binom{u}{z(v)}\right\rangle.$$

b. *If s is a formula, then*

$$\mathfrak{A}\vDash s\langle z\rangle \quad \text{iff} \quad \mathfrak{A}\vDash s'\left\langle z\binom{u}{z(v)}\right\rangle.$$

PROOF: In what follows, if r is an expression then r' denotes the result of replacing v by u throughout r. Also, we abbreviate $z\binom{u}{z(v)}$ by z'.

PROOF OF a is by induction on terms:

Case i. s is a variable. If $s\neq v$ then $s=s'$ and $s^{\mathfrak{A}}\langle z\rangle = z(v) = z'(v) = s'^{\mathfrak{A}}\langle z'\rangle$. If $s=v$ then $s'=u$ and $s^{\mathfrak{A}}\langle z\rangle = z(v) = z'(u) = s'^{\mathfrak{A}}\langle z'\rangle$.

Case ii. s is a constant symbol. Then $s=s'$ and $s^{\mathfrak{A}}\langle z\rangle = s^{\mathfrak{A}} = s'^{\mathfrak{A}}\langle z'\rangle$.

Case iii. $s=f_{n,\alpha}t_0,\ldots,t_{n-1}$ where the t_i's are terms such that $t_i^{\mathfrak{A}}\langle z\rangle = t_i'^{\mathfrak{A}}\langle z'\rangle$. Then $s^{\mathfrak{A}}\langle z\rangle = f_{n,\alpha}^{\mathfrak{A}}t_0^{\mathfrak{A}}\langle z\rangle\cdots t_{n-1}^{\mathfrak{A}}\langle z\rangle = f_{n,\alpha}^{\mathfrak{A}}t_0'^{\mathfrak{A}}\langle z'\rangle\cdots t_{n-1}'^{\mathfrak{A}}\langle z'\rangle = s'^{\mathfrak{A}}\langle z'\rangle$. □

PROOF OF b is by induction on formulas:

Case i. s is an atomic formula. If s is $t_0\approx t_1$, then s' is $t_0'\approx t_1'$ and $\mathfrak{A}\vDash s\langle z\rangle$ iff $t_0^{\mathfrak{A}}\langle z\rangle = t_1^{\mathfrak{A}}\langle z\rangle$ iff (by clause a of the lemma) $t_0'^{\mathfrak{A}}\langle z'\rangle = t_1'^{\mathfrak{A}}\langle z'\rangle$ iff $\mathfrak{A}\vDash s'\langle z'\rangle$. A similar argument applies when s is $R_{n,\alpha}t_0,\ldots,t_{n-1}$.

Case ii. s is $\neg\varphi$, and the lemma is true for φ. Thus s' is $\neg\varphi'$ and

$$\mathfrak{A}\vDash s\langle z\rangle \quad \text{iff} \quad \mathfrak{A}\nvDash\varphi\langle z\rangle \quad \text{iff} \quad \mathfrak{A}\nvDash\varphi'\langle z'\rangle \quad \text{iff} \quad \mathfrak{A}\vDash s'\langle z'\rangle.$$

Case iii. s is $\varphi_1\wedge\varphi_2$, and the lemma holds for φ_1 and φ_2. Then $s'=\varphi_1'\wedge\varphi_2'$ and

$$\mathfrak{A}\vDash s\langle z\rangle \quad \text{iff} \quad \mathfrak{A}\vDash\varphi_1\langle z\rangle \text{ and } \mathfrak{A}\vDash\varphi_2\langle z\rangle \quad \text{iff}$$

$$\mathfrak{A}\vDash\varphi_1'\langle z'\rangle \text{ and } \mathfrak{A}\vDash\varphi_2'\langle z'\rangle \quad \text{iff} \quad \mathfrak{A}\vDash s'\langle z'\rangle.$$

Case iv. s is $\exists v_i\varphi$, and the lemma holds for φ. Then s' is either $\exists u\varphi'$ or $\exists v_i\varphi'$ according as $u=v_i$ or not. Hence the following are equivalent:

$$\mathfrak{A}\vDash s\langle z\rangle.$$

3.5 Normal Forms

There is some $a \in |\mathfrak{A}|$ such that
$$\mathfrak{A} \models \varphi \left\langle z \binom{v_i}{a} \right\rangle.$$

There is some $a \in |\mathfrak{A}|$ such that
$$\mathfrak{A} \models \varphi' \left\langle \binom{u}{z(v)} \binom{v_i}{a} \right\rangle.$$

Noting that
$$\left(z \binom{u}{z(v)} \right) \binom{v_i}{a} = z' \binom{v_i}{a},$$

we see that the last line is equivalent to $\mathfrak{A} \models s' \langle z' \rangle$. This concludes the proof of the lemma. □

Lemma 5.4. *Let u be a variable that does not occur in φ, and let φ' be the result of replacing each bound occurrence of v by u. Then $\models \varphi \leftrightarrow \varphi'$.*

PROOF: The proof is by induction on formulas. We consider only the case where φ has the form $\exists v_i \psi$ under the assumption that the lemma holds for ψ, leaving the remaining easier cases as an exercise.

If $v_i \neq v$, then φ' is $\exists v_i \psi'$, where ψ' is the result of replacing each bound occurrence of v by u in φ. By assumption, $\models \psi \leftrightarrow \psi'$. Hence by Lemma 5.3, $\models \varphi \leftrightarrow \varphi'$.

If $v_i = v$, then φ' is $\exists u \psi''$, where ψ'' is the result of replacing every occurrence of v by u in ψ. By the preceding lemma,
$$\mathfrak{A} \models \psi \left\langle z \binom{v}{a} \right\rangle \quad \text{iff} \quad \mathfrak{A} \models \psi'' \left\langle z \binom{v}{a} \right\rangle.$$

Hence, $\mathfrak{A} \models \varphi \langle z \rangle$ iff for some $a \in |\mathfrak{A}|$,
$$\mathfrak{A} \models \psi \left\langle z \binom{v}{a} \right\rangle$$

iff

for some $a \in |\mathfrak{A}|$, $\mathfrak{A} \models \psi'' \left\langle z \binom{v}{a} \right\rangle$

iff $\mathfrak{A} \models \psi'$. Thus $\models \varphi \leftrightarrow \varphi'$. □

Lemma 5.5.

i. $\models \neg \exists v \varphi \leftrightarrow \forall v \neg \varphi$.
ii. $\models \neg \forall v \varphi \leftrightarrow \exists v \neg \varphi$.
iii. *If u is not free in ψ, then* $\models [\psi \wedge \exists u \varphi] \leftrightarrow \exists u [\psi \wedge \varphi]$ *and* $[\psi \wedge \forall u \varphi] \leftrightarrow \forall u [\psi \wedge \varphi]$.
iv. *If u is not free in ψ, then* $\models [\psi \vee \exists u \varphi] \leftrightarrow \exists u [\psi \vee \varphi]$ *and* $\models [\psi \vee \exists u \varphi] \leftrightarrow \exists u [\psi \vee \varphi]$.

PROOF: The first two clauses follow easily from Definition 4.3. To prove clause iii suppose $\mathfrak{A} \models [\psi \wedge \exists u \varphi] \langle z \rangle$. Then $\mathfrak{A} \models \psi \langle z \rangle$, and for some $a \in |\mathfrak{A}|$,
$$\mathfrak{A} \models \varphi \left\langle z \binom{u}{a} \right\rangle.$$

By Theorem 4.5,
$$\mathfrak{A} \vDash \psi \left\langle z \begin{pmatrix} u \\ a \end{pmatrix} \right\rangle,$$
since u is not free in ψ. Hence
$$\mathfrak{A} \vDash [\psi \wedge \varphi] \left\langle z \begin{pmatrix} u \\ a \end{pmatrix} \right\rangle,$$
and so $\mathfrak{A} \vDash \exists u[\psi \wedge \varphi]\langle z \rangle$. This shows that $\vDash [\psi \wedge \exists u\varphi] \to \exists u[\psi \wedge \varphi]$. Since $\vDash \exists u[\psi \wedge \varphi] \to [\psi \wedge \exists u\varphi]$ is obvious by Definition 4.3, we have iii. The remaining half of clause iii has a similar proof, as does clause iv. □

Definition 5.6.
 i. A formula is *open* if no quantifier occurs in it.
 ii. A *prenex normal form* formula is a formula $Q\varphi$ where φ is open and Q is a sequence $Q_0 u_0 \cdots Q_{n-1} u_{n-1}$ with each $Q_i \in \{\forall, \exists\}$ and each u_i a variable. Q is called the *prefix* of $Q\varphi$, and φ the *matrix* of $Q\varphi$.

EXAMPLE. $\forall v_2 \exists v_0 \forall v_1 [Rv_2 v_0 \wedge \neg Sv_1 v_2]$ is a prenex normal form formula with prefix $\forall v_2 \exists v_0 \forall v_1$ and matrix $Rv_2 v_0 \wedge \neg Sv_1 v_2$. The formula $[Rv_2 v_0 \wedge \neg Sv_1 v_2]$ is open.

Theorem 5.7. *Every formula is logically equivalent to a prenex normal form formula having exactly the same free variables.*

PROOF: We use induction on formulas.

Case i. If ψ is atomic, then ψ is a prenex normal form formula.

Now suppose that φ and φ' are logically equivalent to prenex normal form formulas $Q\rho$ and $Q'\rho'$ respectively, where Q is $Q_0 u_0 \cdots Q_{n-1} u_{n-1}$, Q' is $Q'_0 u'_0 \cdots Q'_{n'-1} u'_{n'-1}$ and ρ and ρ' are open.

Case ii. ψ is $\neg \varphi$. Then $\vDash \psi \leftrightarrow \neg Q\rho$ by Theorem 5.2. Let $Q_i^{\#}$ be $\forall$ if Q_i is $\exists$, and let $Q_i^{\#}$ be $\exists$ otherwise. Then repeated use of Lemma 5.5i and ii gives $\vDash \neg Q\rho \leftrightarrow Q_0^{\#} u_0 \cdots Q_{n-1}^{\#} u_{n-1} \neg \rho$.

Case iii. $\psi = \varphi \wedge \varphi'$. By Lemma 5.4, we can suppose that no u_i occurs in ρ' and no u'_j occurs in ρ. Iterating Lemma 5.5iii and iv, we have $\vDash [\varphi \wedge \varphi'] \leftrightarrow QQ'[\rho \wedge \rho']$. This gives $\vDash \psi \leftrightarrow QQ'[\rho \wedge \rho']$.

Case iv. ψ is $\exists v \varphi$. By Theorem 5.2, $\vDash \psi \leftrightarrow \exists v Q\rho$, as needed. This completes the proof. □

In what follows we usually ignore the fact that a formula φ has many prenex normal form formulas equivalent to it, and speak of "the" prenex normal form of φ. On the other hand, it is easy to see that our proof of the existence of a prenex normal form formula equivalent to φ can be modified to give an algorithm that provides a unique such formula.

3.5 Normal Forms

Definition 5.8. A formula is a *disjunctive normal form* formula if it has the form

$$(\varphi_{1,1} \wedge \varphi_{1,2} \wedge \cdots \wedge \varphi_{1,n_1}) \vee (\varphi_{2,1} \wedge \varphi_{2,2} \wedge \cdots \wedge \varphi_{2,n_2}) \vee \cdots$$
$$\vee (\varphi_{k,1} \wedge \varphi_{k,2} \wedge \cdots \wedge \varphi_{k,n_k}),$$

where each $\varphi_{i,j}$ is an atomic formula or the negation of an atomic formula. A *conjunctive normal form* formula is defined analogously except that the symbols $\wedge$ and $\vee$ are interchanged.

Theorem 5.9. Let $\psi_1, \psi_2, \ldots, \psi_n$ be an enumeration of the atomic subformulas occurring in the open formula ψ. Then ψ is equivalent to a disjunctive normal form formula

$$(\varphi_{1,1} \wedge \varphi_{1,2} \wedge \cdots \wedge \varphi_{1,n}) \vee (\varphi_{2,1} \wedge \varphi_{2,2} \wedge \cdots \wedge \varphi_{2,n}) \vee \cdots$$
$$\vee (\varphi_{k,1} \wedge \varphi_{k,2} \wedge \cdots \wedge \varphi_{k,n}),$$

where $\varphi_{i,j} \in \{\psi_j, \neg\psi_j\}$. The analogous statement for conjunctive normal forms is also true.

PROOF: Let φ_i ($i \leq k$) be an enumeration of all those formulas of the form

$$\varphi_{i,1} \wedge \varphi_{i,2} \wedge \cdots \wedge \varphi_{i,n}, \qquad \varphi_{i,j} \in \{\psi_j, \neg\psi_j\},$$

such that

$$\vDash \varphi_{i,1} \wedge \varphi_{i,2} \wedge \cdots \wedge \varphi_{i,n} \to \psi.$$

We claim that

$$\vDash \varphi_1 \vee \cdots \vee \varphi_k \leftrightarrow \psi.$$

If not, then there is some $\mathfrak{A}$ and $z \in {}^{\mathrm{Vbl}}|\mathfrak{A}|$ such that

$$\mathfrak{A} \vDash \neg [\varphi_1 \vee \cdots \vee \varphi_k]\langle z \rangle \quad \text{and} \quad \mathfrak{A} \vDash \psi \langle z \rangle.$$

Let $\varphi^* = \varphi_1^* \wedge \varphi_2^* \wedge \cdots \wedge \varphi_n^*$, where φ_j^* is ψ_j if $\mathfrak{A} \vDash \psi_j \langle z \rangle$ and is $\neg\psi_j$ otherwise. Then $\mathfrak{A} \vDash \varphi^* \langle z \rangle$. Moreover, since ψ is open, the truth value of ψ in any model with any assignment depends only on the truth values of the ψ_i. Hence $\vDash \varphi^* \to \psi$. But then φ^* is one of the φ_i's and $\mathfrak{A} \vDash \neg [\varphi_1 \vee \cdots \vee \varphi_k]$— a contradiction. Hence $\vDash \varphi_1 \vee \cdots \vee \varphi_k \leftrightarrow \psi$. □

The last normal form we shall discuss is one of the most useful. A formula in this normal form is universal.

Definition 5.10. Say that $\forall u_1, \ldots, u_n \varphi(f(u_1, \ldots, u_n'', w_i, \ldots, w_k))$ is a *one step Skolemization* of ψ if $\psi = \forall u_1, \ldots, u_n \exists u \varphi$, and f is a function symbol not occurring in φ, and $w_1, \ldots, w_k$ is a list of the free variables of ψ. If no variable is free in ψ, and ψ has the form $\exists u \varphi$, and c is a constant symbol not occurring in ψ, then $\varphi(\frac{u}{c})$ is a one step Skolemization.

For example, $\forall v_1[[v_1<f(v_1,v_3)]\wedge[f(v_1,v_3)<v_3]]$ is a one step Skolemization of $\forall v_1\exists v_2[[v_1<v_2]\wedge[v_2<v_3]]$.

Notice that iterating one step Skolemizations will lead to a universal formula provided that we begin with a prenex normal form formula.

Definition 5.11. Let φ be a prenex normal form of ψ, and let $\varphi_1,\varphi_2,\ldots,\varphi_n$ be a sequence of formulas such that

i. $\varphi_1=\psi$,
ii. φ_n is universal,
iii. φ_{i+1} is a one step Skolemization of φ_i for each $i=1,\ldots,n-1$.

Then φ_n is said to be a *Skolem normal form* of ψ, which we shall often abbreviate $\varphi_n\in\text{Sk}(\psi)$.

For example, if ψ is $\forall v_1v_2[v_1<v_2\to\exists v_3[[v_1<v_3]\wedge[v_3<v_2]]]$, then $\forall v_1v_2\exists v_3[[v_1<v_2]\to[[v_1<v_3]\wedge[v_3<v_2]]]$ is a prenex normal form of ψ and $\forall v_1v_2[[v_1<v_2]\to[[v_1<f(v_1,v_2)]\wedge[f(v_1,v_2)<v_2]]]$ is a Skolem normal form of ψ.

Theorem 5.12. *Let $\varphi\in\text{Sk}(\psi)$. Then for every $\mathfrak{A}$ there is an expansion $\mathfrak{B}$ such that for all $z\in{}^{\text{Vbl}}|\mathfrak{A}|$*

$$\mathfrak{A}\vDash\psi\langle z\rangle \quad\text{iff}\quad \mathfrak{B}\vDash\varphi\langle z\rangle.$$

Furthermore, if $z\in{}^{\text{Vbl}}|\mathfrak{C}|$ and $\mathfrak{C}\vDash\varphi\langle z\rangle$, then $\mathfrak{C}\vDash\psi\langle z\rangle$.

PROOF: Clearly it is enough to show that the conclusion of the theorem holds when φ is a one step Skolemization of ψ. Let $w_1,w_2,\ldots,w_k$ be the variables that occur free in ψ. Then φ has the form $\forall u_1,\ldots,u_n\xi(f(u_1^u,\ldots,u_n^u,w_1,\ldots,w_k))$, where f does not occur in ψ. For each $\bar{a}\in{}^{n+k}|\mathfrak{A}|$ let

$$X_{\bar{a}}=\left\{d:\mathfrak{A}\vDash\xi\binom{u_1,\ldots,u_n,w_1,\ldots,w_k,u}{a_1,\ldots,a_n,a_{n+1},\ldots,a_{n+k},d}\right\}.$$

Let a^* be some fixed element of $|\mathfrak{A}|$. Now we let $\mathfrak{B}=(\mathfrak{A},f^{\mathfrak{B}})$, where $f^{\mathfrak{B}}\bar{a}\in X_{\bar{a}}$ if $X_{\bar{a}}\neq\varnothing$, and $f^{\mathfrak{B}}\bar{a}=a^*$ otherwise. (Notice that the existence of such an f requires the axiom of choice; see Exercise 10.) Clearly, for any assignment z to $\mathfrak{A}$

$$\mathfrak{A}\vDash\psi\langle z\rangle \quad\text{iff}\quad \mathfrak{B}\vDash\varphi\langle z\rangle.$$

The second part of the theorem is obvious. □

Definition 5.13. Let $\forall u_1,\ldots,u_n\xi$ be a Skolem normal form for satisfiability of $\neg\psi$, where ξ is quantifier free. Then $\exists u_1,\ldots,u_n(\neg\xi)$ is a *Skolem normal form for validity* of ψ.

Notice that a Skolem normal form for validity is existential.

3.5 Normal Forms

Theorem 5.14. *Let ψ^* be a Skolem normal form for the validity of ψ. Then ψ is valid iff ψ^* is valid.*

PROOF: Let ψ^* be $\exists u_1,\ldots,u_n(\neg\xi)$ with ξ quantifier free. The following statements are equivalent:

ψ is not valid.
$\mathfrak{A}\vDash\neg\psi\langle z\rangle$ for some $\mathfrak{A}$ and $z\in {}^{\mathrm{Vbl}}|\mathfrak{A}|$.
$\mathfrak{B}\vDash\forall u_1,\ldots,u_n\xi\langle z\rangle$ for some $\mathfrak{B}$ and $z\in {}^{\mathrm{Vbl}}|\mathfrak{A}|$.
ψ^* is not valid.

The equivalence of the second and third statement follows from Theorem 5.12. □

EXERCISES FOR §3.5

1. Complete the proof of Lemma 5.4.

2. Show that the stipulation that u is not free in ψ is necessary in parts iii and iv of Lemma 5.5.

3. Complete the proof of Lemma 5.5.

4. Find a prenex normal form formula that is equivalent to $\neg\forall x[Rxy]\vee[\exists y\, Sxy]$.

5. Prove Theorem 5.9 for conjunctive normal forms. (*Hint:* Apply Theorem 5.9 to $\neg\psi$.)

6. Give the disjunctive normal form of $[Ruv\wedge Sv]\leftrightarrow[Su\wedge u\not\approx v]$.

7. Give a Skolem normal form for satisfiability of $\forall v_1\exists v_2[Rv_1v_2v_3\wedge \exists v_4[f(v_2,v_3)\approx v_4]]$.

8. Give a Skolem normal form for validity of the formula in Exercise 7.

9. Prove the axiom of choice from Theorem 5.12.

10. An atomic formula is *simple* if it has the form

$$Ru_1,u_2,\ldots,u_n$$

or

$$f(u_1,u_2,\ldots,u_n)\approx u_{n+1},$$

where each u_i is either a constant symbol or a variable. Show that every atomic formula is equivalent to an existential formula of the form

$$\exists u_1,\ldots,u_n[\varphi_1\wedge\varphi_2\wedge\cdots\wedge\varphi_k]$$

of the same type and with the same free variables where each φ_i is simple.

3.6 The Compactness Theorem

The most studied language of mathematical logic is the first order predicate calculus. It has considerable expressive power combined with a highly tractable theory. It is easy to find languages with greater expressive power, but usually the theory of such languages is far less rich.

One of the properties that makes the first order predicate calculus so amenable is the finitary character of satisfaction. An infinite set of assertions has a model whenever each finite subset has a model. This fact is known as the compactness theorem (for reasons spelled out in Exercise 8) and is another one of Gödel's extraordinary achievements. The theorem is frequently used to construct models that have a wealth of desirable properties from the knowledge that finite subsets of the properties have models. Sometimes such a set of properties may be so demanding that a model for it may appear paradoxical. In this section we shall present several of these examples. Here we state the compactness theorem, but the proof will not be given until 3.7.

Theorem 6.1 (Compactness Theorem). *Let Σ be a set of assertions of L. If each finite subset of Σ has a model, then Σ has a model; in fact Σ has a model of cardinality $\leq \omega + c\Sigma$.*

Here we shall examine several consequences of this theorem. In §3.8 several of the examples discussed here will be sharpened and generalized.

EXAMPLE 6.2. A model of arithmetic with an infinite number: Let $\mathfrak{A} = (\mathbf{N}, +, \cdot, <, 0, 1, 2, \ldots)$, where $n = c_n^{\mathfrak{A}}$ and $+, \cdot, <$ are the symbols denoting $+, \cdot, <$ respectively. Let $\Sigma = \text{Th}\,\mathfrak{A} \cup \{c_n < c_\omega : n \in \omega\}$. We claim that every finite subset Σ' of Σ has a model. Indeed, if $n^* = \max\{n : c_n \text{ occurs in } \Sigma'\}$, then $\mathfrak{B} \in \text{Mod}\,\Sigma'$, where $\mathfrak{B} = (\mathfrak{A}, c_\omega^{\mathfrak{B}})$ and $n^* + 1 = c_\omega^{\mathfrak{B}}$. For clearly, $\mathfrak{B}$ satisfies those assertions of Σ' of the form $c_n < c_\omega$, and since the other members of Σ' belong to $\text{Th}\,\mathfrak{A}$, $\mathfrak{B}$ satisfies them also.

Thus by compactness, there is a model $\mathfrak{C}$ of Σ (and in fact a countable model). By Theorem 4.17 and Theorem 4.18 we can assume $\mathfrak{A} \subseteq \mathfrak{C} \upharpoonright \tau \mathfrak{A}$ with $c_n^{\mathfrak{C}} = n$. $\mathfrak{C}$ has an infinite element $c_\omega^{\mathfrak{C}}$, infinite in the sense that $n <^{\mathfrak{C}} c_\omega^{\mathfrak{C}}$ for all $n \in \mathbf{N}$. In fact $\mathfrak{C}$ has an infinity of infinite elements. Indeed, since $\mathfrak{C} \in \text{Mod}\,\text{Th}\,\mathfrak{A}$, we have $\mathfrak{C} \vDash \forall v_0 v_1 v_2 [v_0 < v_1 \rightarrow v_1 < v_1 + v_2]$. Thus $m < c_\omega + m$ for all $m, n \in \mathbf{N}$, and so all elements of $\mathfrak{C}$ of the form $c_\omega + n$ are infinite. Moreover, if $l \neq k$, then $c_\omega + l \not\approx c_\omega + k$, since $\forall v_0 v_1 v_2 [v_0 + v_1 = v_0 + v_2 \rightarrow v_1 = v_2]$ is in $\text{Th}\,\mathfrak{A}$ and so in $\text{Th}\,\mathfrak{C}$.

Since $\mathfrak{A} \equiv \mathfrak{C} \upharpoonright \tau \mathfrak{A}$ but $\mathfrak{A} \not\approx \mathfrak{C} \upharpoonright \tau \mathfrak{A}$, we see that the class of structures that are isomorphic to $\mathfrak{A}$ is not elementary; hence the converse of Theorem 4.19 fails.

3.6 The Compactness Theorem   159

EXAMPLE 6.3. A model of the reals with infinite numbers and infinitesimals: Let $\mathfrak{A}$ be the ordered field of real numbers $(\mathbf{R},+,\cdot,<,r)_{r\in\mathbf{R}}$ where every real is represented by a constant, say $r=c_r^{\mathfrak{A}}$, and $+,\cdot,<$ denote $+,\cdot,<$ respectively. Let $\Sigma=\text{Th}\,\mathfrak{A}\cup\{c_r<c:r\in\mathbf{R}\}$, where c is a constant symbol different from the c_r's. We claim that every finite subset of Σ has a model. For if $\Sigma'\subseteq\Sigma$ and Σ' is finite, then only finitely many constant symbols c_r occur in Σ', say $c_{r_1},c_{r_2},\ldots,c_{r_n}$. Let $r^*=1+\max\{r_1,r_2,\ldots,r_n\}$. Take $\mathfrak{B}$ to be $(\mathfrak{A},c^{\mathfrak{B}})$, where $c^{\mathfrak{B}}=r^*$. Clearly $\mathfrak{B}\in\text{Mod}\,\Sigma'$, thus proving the claim. Hence by compactness, Σ has a model $\mathfrak{C}$.

By Theorems 4.17 and 4.18 we can assume that $\mathfrak{A}\subseteq\mathfrak{C}\!\restriction\!\tau\mathfrak{A}$, so that $c_r^{\mathfrak{C}}=r$. Also $\mathfrak{A}\equiv\mathfrak{C}\!\restriction\!\tau\mathfrak{A}$. However, $\mathfrak{A}\not\cong\mathfrak{C}$, and in fact $c^{\mathfrak{C}}$ is an infinite number in $\mathfrak{C}$ in the sense that $c>cr$ for all $r\in\mathbf{R}$. Since $\mathfrak{C}$ is a field, every element x in $|\mathfrak{C}|$ other than 0 has a multiplicative inverse, which we call $1/x$. Since $0<r<c$ for all $r\in\mathbf{R}$, it follows that

$$0<\frac{1}{c}<\frac{1}{n+1}$$

for all $n\in\mathbf{N}$ in this sense: $1/c^{\mathfrak{C}}$ is infinitesimal.

EXAMPLE 6.4. Let Γ be a set of sentences such that for every $n\in\omega$ there is an $\mathfrak{A}\in\text{Mod}\,\Gamma$ whose universe has cardinality $\geqslant n$. Then there is a $\mathfrak{B}\in\text{Mod}\,\Gamma$ whose universe is infinite. For let $\Sigma=\Gamma\cup\{d_m\not\approx d_n:m<n<\omega\}$, where the d's are distinct constant symbols none of which occur in Γ. Let Σ' be a finite subset of Σ with $n^*=\max\{n:d_n$ occurs in $\Sigma'\}$. By assumption there is an $\mathfrak{A}\in\text{Mod}\,\Gamma$ with $c|\mathfrak{A}|\geqslant n^*$. Let $a_0,\ldots,a_{n-1}$ be distinct elements in $|\mathfrak{A}|$. Expand $\mathfrak{A}$ to $\mathfrak{B}=(\mathfrak{A},d_0^{\mathfrak{B}},d_1^{\mathfrak{B}},\ldots,d_{n^*-1}^{\mathfrak{B}})$, where $d_i^{\mathfrak{B}}=a_i$. Then clearly $\mathfrak{B}\in\text{Mod}\,\Sigma'$. Hence, every finite subset of Σ has a model, and so by compactness Σ has a model. But every model $\mathfrak{C}$ of Σ is infinite, since $\mathfrak{C}\vDash d_m\not\approx d_n$ for all $m<n<\omega$. Moreover, if $\mathfrak{C}\in\text{Mod}\,\Sigma$, then $\mathfrak{C}\!\restriction\!\tau\Gamma\in\text{Mod}\,\Gamma$ as needed.

Hence we see that the class of all finite groups is not an elementary class, and the same is true of the class of all finite rings and the class of all finite fields.

EXAMPLE 6.5. The class of well-ordered structures is not an elementary class. For let Γ be any set of assertions all of which are true in every well-ordered structure. We must show that $\text{Mod}\,\Gamma$ has a member that is not well-ordered. Let $\Sigma=\Gamma\cup\{d_{n+1}<d_n:n\in\omega\}$. Every finite subset Σ' of Σ has a model; indeed, any infinite model of Γ [such as $(\omega,<)$] can obviously be expanded to a model of Σ'. Hence by compactness Σ has a model. But no model $\mathfrak{A}$ of Σ is well ordered, since $\{d_n^{\mathfrak{A}}:n\in\omega\}$ has no least member (with respect to $<^{\mathfrak{A}}$) in $\mathfrak{A}$. Hence $\mathfrak{A}\!\restriction\!\{<\}$ is a model of Γ which is not well ordered.

EXAMPLE 6.6. Let ZF be the set of axioms for Zermelo-Fraenkel set theory (given in Section 1.11). Included in ZF is the axiom of regularity. But in spite of regularity, if ZF has a model, then ZF has a model $\mathfrak{A}=(A,e^{\mathfrak{A}})$ in which there are elements $c_0^{\mathfrak{A}}, c_1^{\mathfrak{A}},\ldots$ such that $c_{n+1}^{\mathfrak{A}} e^{\mathfrak{A}} c_n^{\mathfrak{A}}$ for $n=0,1,2,\ldots$. Indeed, given $n\in\omega$ and a model $\mathfrak{B}$ of ZF, $\mathfrak{B}$ can be expanded to a model of $\text{ZF}\cup\{c_n e c_{n-1}, c_{n-1} e c_1 e c_0\}$. Hence every finite subset of $\text{ZF}\cup\{c_{n+1} e c_n : n\in\omega\}$ has a model. By compactness, $\text{ZF}\cup\{c_{n+1} e c_n : n\in\omega\}$ has a model, say $\mathfrak{A}$. The regularity axiom is true in $\mathfrak{A}$, yet $\mathfrak{A}\vDash c_{n+1} e c_n$ for all $n\in\omega$. Of course this means that $\{c_n^{\mathfrak{A}}: n\in\omega\}$ is not a set in $\mathfrak{A}$, i.e., there is no $a\in|\mathfrak{A}|$ such that for all $b\in|\mathfrak{A}|$, $b e^{\mathfrak{A}} a$ iff $b=c_n^{\mathfrak{A}}$ for some $n\in\omega$.

EXAMPLE 6.7 (A countable model of the reals). Let $\mathfrak{A}$ be the field of reals, $\mathfrak{A}=(R,+,\cdot)$. Th$\mathfrak{A}$ is countable and so has a countable model. Indeed, if $\mathfrak{B}$ is any expansion of $\mathfrak{A}$ with countable type, then Th$\mathfrak{B}$ is countable and so has a countable model.

EXAMPLE 6.8 (Skolem's paradox). If Zermelo-Fraenkel set theory has a model, then it has a countable model. The reason that this fact at first glance seems paradoxical is that one can prove from the Zermelo-Fraenkel axioms the existence of uncountable sets, i.e., sets x such that there is no function mapping x into ω. This can be written as an assertion of L. Hence this assertion is true in any countable model. Yet for any x in a countable model $\mathfrak{A}$ the cardinality of $\{y : y e^{\mathfrak{A}} x\}$ is countable. However, this only means that in $|\mathfrak{A}|$ there are sets x such that for some $y\in|\mathfrak{A}|$, y is a 1-1 function on x onto a proper subset of x, in the sense of $\mathfrak{A}$, but for no $z\in|\mathfrak{A}|$ is z a function on x 1-1 into ω in the sense of $\mathfrak{A}$.

EXERCISES FOR §3.6

1. $\mathfrak{B}$ is a non-standard model of arithmetic if $\mathfrak{B}\equiv(\mathbf{N},+,\cdot)$ but $\mathfrak{B}\not\cong(\mathbf{N},+,\cdot)$. Show that every non-standard model of arithmetic has an infinite number.

2. Let $\mathfrak{C}$ be a non-standard model of arithmetic.
   (a) Show that $\mathfrak{C}$ has an infinite prime number.
   (b) Show that there is a $k\in|\mathfrak{C}|$ such that n divides k for all $n\in\omega$.

3. A famous open question of number theory asks if there are infinitely many primes p such that $p+2$ is also a prime. Show that the answer is yes iff every non-standard model of arithmetic has an infinite prime p such that $p+2$ is a prime.

4. Let $\mathfrak{B}$ be a non-standard model of arithmetic. Show that there are $b_0, b_1, b_2, \ldots$ all in $|\mathfrak{B}|$ such that $b_{i+1}<b_i$ for all $i\in\mathbf{N}$. ($x<y$ means there is a $c\in|\mathfrak{B}|$ such that $y=x+^{\mathfrak{B}} c$.) Hence $\mathfrak{B}$ is not well ordered.

5. If $a,b\in\mathbf{R}$ and $a>0$, then there is an $n\in\mathbf{N}$ such that $an>b$. This is the Archimedian property for the real number field. Let $\mathfrak{B}$ be a non-standard model of the real number field, i.e., let $\mathfrak{B}\equiv\langle\mathbf{R},+,\cdot\rangle$ and $\mathfrak{B}\not\cong\langle\mathbf{R},+,\cdot\rangle$. Then $\mathfrak{B}$ is non-Archimedian.

3.6 The Compactness Theorem   161

6. Let $\mathfrak{B}$ be a non-standard model of $\mathfrak{A}$, where $\mathfrak{A} = \langle R, Q, I, N, +, \cdot \rangle$, with Q the set of rationals and I the set of integers. So $\mathfrak{B} \equiv \mathfrak{A}$ but $\mathfrak{B} \not\cong \mathfrak{A}$. Suppose also that $\mathfrak{B} \supseteq \mathfrak{A}$. Then $Q^{\mathfrak{B}}$ is uncountable. [*Hint:* For every $r, \varepsilon \in |\mathfrak{A}|$, if $\varepsilon > 0$, then there is a $q \in Q$ such that $q \in (r - \varepsilon, r + \varepsilon)$. Consider the meaning of this in $\mathfrak{B}$ when ε is infinitesimal but r is standard, i.e., in $|\mathfrak{A}|$. If $r_1 \neq r_2$ and $q_1 \in (r_1 - \varepsilon, r_1 + \varepsilon)$ and $q_2 \in (r_2 - \varepsilon, r_2 + \varepsilon)$, then $q_1 \neq q_2$.]

7. To the definition of L formulas (Definition 2.1d) add

   v. if φ is a formula and v is a variable, then $Qv\varphi$ is a formula.

   To the definition of satisfaction (Definition 4.3) add

   v. $\varphi = Qv\psi$, and $\{a \in |\mathfrak{A}| : \mathfrak{A} \models \psi(a)\}$ is infinite.

   Show that this new language is not compact.

   Remark: If we replace clause v by

   v*. $\varphi = Qv\psi$, and $c\{a \in |\mathfrak{A}| : \mathfrak{A} \models \psi(a)\} > \omega$,

   then we get a language that is countably compact in the sense that any countable finitely satisfiable set of assertions is satisfiable.

8. In Definition 2.1d change clause iii to

   iii*. if $\alpha \leq \omega$ and $\{\varphi_i : i < \alpha\} \subseteq X$ then $\bigwedge \{\varphi_i : i < \alpha\} \in X$ and $\bigvee \{\varphi_i : i < \alpha\} \in X$.

   In Definition 2.1d change clause iii to

   iii*. $\varphi = \bigwedge \{\varphi_i : i < \alpha\}$ and $\mathfrak{A} \models \varphi_i$ for each $i < \alpha$, or $\varphi = \bigvee \{\varphi_i : i < \alpha\}$ and $\mathfrak{A} \models \varphi_i$ for some $i < \alpha$.

   (a) Show that there is an assertion σ in this new language such that $\mathfrak{A} \models \sigma$ and $\tau \mathfrak{A} = \{<\}$ iff $\mathfrak{A} \cong (N, <)$.

   (b) Conclude that this new language is not compact.

9. This problem is for those who have some knowledge of point set topology. The problem is to show that compactness in the sense of Theorem 6.1 is equivalent to the compactness of some topological space.

   Let s be a similarity type, and let T be the set of all structures of type s and of cardinality $\leq cs + \omega$. For each $\mathfrak{A} \in T$ we let $\mathfrak{A}^* = \{\mathfrak{B} \in T : \mathfrak{A} \equiv \mathfrak{B}\}$. Now let $T^* = \{\mathfrak{A}^* : \mathfrak{A} \in T\}$. For each Σ of type s let $F_\Sigma = \{\mathfrak{A}^* : \mathfrak{A} \in \text{Mod}\,\Sigma\}$.

   (a) Show that $\mathfrak{T} = \{F_\Sigma : \tau \Sigma \subseteq s\}$ is a base for closed sets for a topology on T^*, i.e., show that $\varnothing$ and $T^* \in \mathfrak{T}$, and if $K \subseteq T^*$, then $\bigcap K \in T^*$.
   (b) $\mathfrak{T}$ is a T_2 topology, i.e., if $\mathfrak{A}^* \neq \mathfrak{B}^*$, then there is some X, Y such that $\tilde{X}, \tilde{Y} \in \mathfrak{T}$ and $\mathfrak{A}^* \in X$, $\mathfrak{B}^* \in Y$, and $X \cap Y = \varnothing$.
   (c) Derive the compactness of $\mathfrak{T}$ from Theorem 6.1 and conversely. (To say that $\mathfrak{T}$ is compact means that if $\mathfrak{T}' \subseteq \mathfrak{T}$ and if $\bigcap K \neq \varnothing$ whenever K is a finite subset of K, then $\bigcap \mathfrak{T}' \neq \varnothing$.)

3.7. Proof of the Compactness Theorem

Although this section is devoted to a proof of the compactness theorem, several of the lemmas are of independent interest.

Definition 7.1. Say that Σ is *finitely satisfiable* if every finite subset of Σ has a model.

Lemma 7.2. *Let σ be of type $\tau\Sigma$. If Σ is finitely satisfiable, then either $\Sigma \cup \{\sigma\}$ is finitely satisfiable or $\Sigma \cup \{\neg\sigma\}$ is finitely satisfiable.*

PROOF: Suppose that neither $\Sigma \cup \{\sigma\}$ nor $\Sigma \cup \{\neg\sigma\}$ is finitely satisfiable. Then there are finite subsets Σ_1 and Σ_2 of Σ such that $\Sigma_1 \cup \{\sigma\}$ and $\Sigma_2 \cup \{\neg\sigma\}$ have no models. But then $\Sigma_1 \cup \Sigma_2$ is a finite subset of Σ having no models, since any model of $\Sigma_1 \cup \Sigma_2$ must be a model of σ or of $\neg\sigma$. □

Definition 7.3. We say that Σ is *complete* if for all σ of type $\tau\Sigma$, either $\sigma \in \Sigma$ or $\neg\sigma \in \Sigma$.

In the next lemma we need a well ordering of the set of all assertions of some fixed type. At first glance this would seem to require the use of the well ordering principle or some other version of the axiom of choice. But this can be avoided by identifying the symbols with ordinals (as suggested in Section 2.10) and then well-ordering the class of expressions by defining $r_0 r_1 \ldots r_{n-1} < s_0 \ldots s_{m-1}$ if either $n < m$ or $n = m$ and $r_j \in s_j$, where j is the least k such that $r_k \neq s_k$. Hence, using Theorem 10.9 of part I, we see that any set of expressions can be indexed by an ordinal. The axiom of choice appears implicitly in Lemmas 7.6 and 7.7, but again can be avoided using these devices.

Lemma 7.4. *If Σ is finitely satisfiable, then there is a Γ such that*

   i. $\Gamma \supseteq \Sigma$,
   ii. $\tau\Gamma = \tau\Sigma$,
   iii. Γ *is complete*,
   iv. Γ *is finitely satisfiable.*

PROOF: By the remark above, we can assume that the set of assertions of type $\tau\Sigma$ is well ordered; say $\{\sigma_\alpha : \sigma \in \beta\}$ is this set, where β is some ordinal. Now define

$$\Sigma_0 = \Sigma.$$

$$\Sigma_{\alpha+1} = \begin{cases} \Sigma_\alpha \cup \{\sigma_\alpha\} & \text{if } \mathrm{Mod}(\Sigma \cup \{\sigma_\alpha\}) \neq \emptyset, \\ \Sigma_\alpha \cup \{\neg\sigma_\alpha\} & \text{otherwise}, \end{cases}$$

if $\alpha < \beta$.

$$\Sigma_\delta = \bigcup_{\gamma \in \delta} \Sigma_\gamma \quad \text{if } \delta = \bigcup \delta \leqslant \beta.$$

3.7 Proof of the Compactness Theorem

By induction on α and Lemma 7.2, it is easy to see that each Σ_α satisfies conditions i, ii, and iii of the theorem. Let $\Gamma = \Sigma_\beta$. We need only observe that Γ is complete: Let $\tau\sigma \subseteq \tau\Sigma$. Then $\sigma = \sigma_\alpha$ for some $\alpha < \beta$. By definition of $\Sigma_{\alpha+1}$, $\sigma \in \Sigma_{\alpha+1}$ or $\neg \sigma \in \Sigma_{\alpha+1}$. Since $\Sigma_{\alpha+1} \subseteq \Gamma$, we have $\sigma \in \Gamma$ or $\neg\sigma \in \Gamma$. Hence Γ is complete as needed. $\square$

Notice that this lemma is an immediate consequence of the compactness theorem. For if every finite subset of Σ has a model, then so does Σ. Let $\mathfrak{A} \in \text{Mod}\,\Sigma$. Then $\text{Th}\,\mathfrak{A}$ is complete and $\text{Th}\,\mathfrak{A} \supseteq \Sigma$. However, we want to use this lemma to prove the compactness theorem, and so we need a proof that does not use the compactness theorem.

Definition 7.5. The constant symbol c is a *witness* for $\exists v \varphi$ in Σ if
$$\varphi\binom{v}{c} \in \Sigma.$$

Lemma 7.6. *If Σ is finitely satisfiable, then there is a Ω such that:*

i. $\Sigma \subseteq \Omega$.
ii. *If $\exists v\varphi \in \Omega$, then $\exists v\varphi$ has a witness in Ω.*
iii. $c\Omega = c\Sigma + \omega$.
iv. Ω *is finitely satisfiable.*

PROOF: Suppose that Σ is finitely satisfiable. Let g be a 1-1 function on Σ into the constant symbols not in $\tau\Sigma$. For each $\sigma \in \Sigma$ let
$$\sigma^* = \varphi\binom{v}{g(\sigma)}$$
if $\sigma = \exists v\varphi$ for some v and φ, and let $\sigma^* = \sigma$ otherwise. Let $\Omega = \Sigma \cup \{\sigma^* : \sigma \in \Sigma\}$. Clearly Ω satisfies i, ii, and iii. Let Ω' be a finite subset of Ω, say $\Omega' = \Sigma' \cup \{\sigma_i^* : i < n\}$, where $\Sigma' \subseteq \Sigma$ and $\sigma_i = \exists u_i \varphi_i$ for each $i < n$. Since Σ is finitely satisfiable, $\Sigma' \cup \{\sigma_i : i < n\}$ has a model $\mathfrak{A}$. For each $i < n$ choose $a_i \in |\mathfrak{A}|$ such that $\mathfrak{A} \models \varphi_i(a_i)$. Now let $\mathfrak{B} = (\mathfrak{A}, a_0, \ldots, a_{n-1})$, where $a_i = g(\sigma_i)^{\mathfrak{B}}$. Clearly $\mathfrak{B}$ is a model of Ω'. Hence Ω' is finitely satisfiable. $\square$

Lemma 7.7. *Suppose Σ is complete and finitely satisfiable, and every assertion $\exists v\varphi \in \Sigma$ has a witness in Σ. Then Σ has a model of cardinality $\leq c\Sigma + w$.*

PROOF: Let A' be the set of all terms of type $\tau\Sigma$ in which no variable occurs. Define $t_1 \sim t_2$ if $t_1 \approx t_2 \in \Sigma$.

We first show that '$\sim$' is an equivalence relation on A'.

Case i. $t \sim t$: If not, then $\neg[t \approx t] \in \Sigma$ by completeness; but this contradicts the finite satisfiability of Σ, since $\{\neg[t \approx t]\}$ has no model.

Case ii. $t_1 \sim t_2$ implies $t_2 \sim t_1$: Suppose $t_1 \sim t_2$. If $t_2 \approx t_1 \notin \Sigma$, then $\neg[t_2 \approx t_1] \in \Sigma$ by completeness; but then $\{t_1 \approx t_2, \neg[t_2 \approx t_1]\}$ is a finite subset of Σ without a model—a contradiction. Hence, $t_2 \sim t_1$.

Case iii. If $t_1 \sim t_2$ and $t_2 \sim t_3$, then $t_1 \sim t_3$: Suppose $t_1 \sim t_2$ and $t_2 \sim t_3$ but $t_1 \not\sim t_3$. Then $\neg[t_1 \approx t_3] \in \Sigma$ by completeness, but $\{t_1 \approx t_2, t_2 \approx t_3, \neg[t_1 \approx t_3]\}$ is a finite subset of Σ without a model. This contradicts the finite satisfiability of Σ. Hence if $t_1 \sim t_2$ and $t_2 \sim t_3$, then $t_1 \sim t_3$.

Hence '$\sim$' is an equivalence relation.

We next note that if $t_i \sim t_i'$ for $i < n$, then for all $f_{n,\alpha} \in \tau\Sigma$, then $f_{n,\alpha} t_0, \ldots, t_{n-1} \approx f_{n,\alpha} t_0', \ldots, t_{n-1}' \in \Sigma$. For if not, then by completeness $f_{n,\alpha} t_0, \ldots, t_{n-1} \not\approx f_{n,\alpha} t_0', \ldots, t_{n-1}' \in \Sigma$, contradicting finite satisfiability, since $\{t_i \approx t_i' : i < n\} \cup \{f_{n,\alpha} t_0, \ldots, t_{n-1} \not\approx t_0', \ldots, t_{n-1}'\}$ has no model.

Similarly, one sees that if $t_i \sim t_i'$ for $i < n$, then for any $R_{n,\alpha} \in \tau\Sigma$, $R_{n,\alpha} t_0, \ldots, t_{n-1} \in \Sigma$ iff $R_{n,\alpha} t_0', \ldots, t_{n-1}' \in \Sigma$.

We can now define a model $\mathfrak{A}$ of Σ as follows:

a. Let $|\mathfrak{A}| = A = \{\bar{t} : t \in A'\}$ where $\bar{t} = \{t' : t \sim t'\}$.
b. $c_\alpha^\mathfrak{A} = \bar{c}_\alpha$ for all $c_\alpha \in \tau\Sigma$.
c. $f_{n,\alpha}^\mathfrak{A} \bar{t}_0, \ldots, \bar{t}_{n-1} = \overline{f_{n,\alpha} t_0, \ldots, t_{n-1}}$ for all $f_{n,\alpha} \in \tau\Sigma$.
d. $R_{n,\alpha}^\mathfrak{A} \bar{t}_0, \ldots, \bar{t}_{n-1}$ iff $R_{n,\alpha} t_0, \ldots, t_{n-1} \in \Sigma$ for all $R_{n,\alpha} \in \tau\Sigma$.

As we have shown in the paragraphs preceding the definition, clauses c and d are unambiguous in that they do not depend on which representatives $t_0, \ldots, t_{n-1}$ are chosen from the equivalence classes $\bar{t}_0, \ldots, \bar{t}_{n-1}$.

Notice that $t^\mathfrak{A} = \bar{t}$ for all terms $t \in A'$. Certainly this is true for constant symbols by clause b of the definition of $\mathfrak{A}$, and if true for $t_0, \ldots, t_{n-1}$, then $(f_{n,\alpha} t_0, \ldots, t_{n-1})^\mathfrak{A} = f_{n,\alpha}^\mathfrak{A} t_0^\mathfrak{A} \ldots t_{n-1}^\mathfrak{A} = f_{n,\alpha}^\mathfrak{A} \bar{t}_0 \ldots \bar{t}_{n-1}$ which by clause b is $\overline{f_{n,\alpha} t_0 \ldots t_{n-1}}$. Hence by induction on formulas, we have $t^\mathfrak{A} = \bar{t}$ for all $t \in A'$.

We now show by induction on assertions that for all σ of type $\tau\Sigma$

$$\mathfrak{A} \models \sigma \quad \text{iff} \quad \sigma \in \Sigma. \tag{*}$$

We take advantage of the remark following Theorem 4.4 to reduce the number of cases considered to the four that follow:

Case 1. σ is atomic: Then σ is either of the form $Rt_0 \ldots t_{n-1}$ or of the form $t_0 \approx t_1$. If $\sigma = Rt_0, \ldots, t_{n-1}$, then by clause d of the definition of $\mathfrak{A}$ and the fact that $t^\mathfrak{A} = \bar{t}$ for all $t \in A'$, we have $\mathfrak{A} \models Rt_0, \ldots, t_{n-1}$ iff $R^\mathfrak{A} t_0^\mathfrak{A}, \ldots, t_{n-1}^\mathfrak{A}$ iff $R^\mathfrak{A} \bar{t}_0, \ldots, \bar{t}_{n-1}$ iff $Rt_0, \ldots, t_{n-1} \in \Sigma$. The argument is completely similar when σ is $t_0 \approx t_1$.

Case 2. σ is $\neg\varphi$ and φ satisfies (*): First notice that φ and $\neg\varphi$ cannot both be in Σ, since Σ is finitely satisfiable but $\{\varphi, \neg\varphi\}$ has no models. By completeness either φ or $\neg\varphi$ is in Σ. Hence $\neg\varphi \in \Sigma$ iff $\varphi \notin \Sigma$ iff not $\mathfrak{A} \models \varphi$ iff $\mathfrak{A} \models \neg\varphi$.

Case 3. σ is $\varphi_1 \wedge \varphi_2$, where both φ_1 and φ_2 satisfy (*): By completeness, Σ intersects each of the pairs $\{\varphi_1, \neg\varphi_1\}, \{\varphi_2, \neg\varphi_2\}, \{\varphi_1 \wedge \varphi_2, \neg[\varphi_1 \wedge \varphi_2]\}$. Since no finitely satisfiable set contains $\{\neg\varphi_1, \varphi_1 \wedge \varphi_2\}$, $\{\neg\varphi_2, \varphi_1 \wedge \varphi_2\}$, or $\{\varphi_1, \varphi_2, \neg[\varphi_1 \wedge \varphi_2]\}$, we have $\sigma \in \Sigma$ iff $[\varphi_1 \in \Sigma$ and $\varphi_2 \in \Sigma]$ iff $[\mathfrak{A} \models \varphi_1$ and $\mathfrak{A} \models \varphi_2]$ iff $\mathfrak{A} \models \sigma$.

Case 4. σ is $\exists v \varphi$ and φ satisfies (*): Notice that if $\sigma \in \Sigma$, then σ has a witness, say c, such that $\varphi(c) \in \Sigma$. On the other hand, if $\varphi(c) \in \Sigma$ for some c, then $\sigma \in \Sigma$. For if not, then $\neg\sigma \in \Sigma$ by completeness, but then

3.7 Proof of the Compactness Theorem

$\{\varphi(c), \neg \sigma\}$ has no model, contradicting the finite satisfiability of Σ. Hence $\sigma \in \Sigma$ iff [for some c, $\varphi(c) \in \Sigma$] iff [for some c, $\mathfrak{A} \vDash \varphi(c)$] iff $\mathfrak{A} \vDash \sigma$.

This completes the proof of (*), and so $\mathfrak{A} \in \text{Mod}\,\Sigma$.

To finish the proof of the theorem we need only observe that $c|\mathfrak{A}| \leq cA' \leq c\Sigma$. The last inequality holds because any complete set of sentences is infinite, and only finitely many terms occur in each sentence. □

PROOF OF THE COMPACTNESS THEOREM. For each finitely satisfiable Σ, let $G(\Sigma)$ be the set Γ of Lemma 7.4, and let $H(\Sigma)$ be the set Ω of Lemma 7.6. Now let Δ be a finitely satisfiable set, and define:

$$\Sigma_0 = \Delta,$$
$$\Sigma_{2n+1} = G(\Sigma_{2n}),$$
$$\Sigma_{2n+2} = H(\Sigma_{2n+1}).$$

A trivial induction shows that for all $n \in \omega$:

i. $\Sigma_n \subseteq \Sigma_{n+1}$.
ii. Σ_n is finitely satisfiable.
iii. Σ_{2n+1} is complete.
iv. If $\exists v \varphi \in \Sigma_{2n+2}$, then $\exists v \varphi$ has a witness in Σ_{2n+2}.
v. Σ_n has cardinality $c\Sigma + \omega$.

Now let $\Delta^* = \bigcup_{i \in \omega} \Sigma_n$. We show that Δ^* satisfies the hypotheses of Lemma 7.7.

i. Δ^* is finitely satisfiable: Suppose $\{\sigma_0, \ldots, \sigma_{n-1}\} \subseteq \Delta^*$. Then for each $i < n$ there is a $j(i) \in \omega$ such that $\sigma_i \in \Sigma_{j(i)}$. Let $j = \max\{j(i): i < n\}$. By conclusion i, $\{\sigma_0, \ldots, \sigma_{n-1}\} \subseteq \Sigma_j$. By conclusion ii, Σ_j is finitely satisfiable. Hence $\{\sigma_0, \ldots, \sigma_{n-1}\}$ has a model. Thus Δ^* is finitely satisfiable.
ii. Δ^* is complete: Let σ be of type $\tau \Delta^*$. Then for some $n \in \omega$, σ is of type $\tau \Sigma_n$. By conclusions i and iii, either $\sigma \in \Sigma_{2n+1}$ or $\neg \sigma \in \Sigma_{2n+1}$. Hence $\sigma \in \Delta^*$ or $\neg \sigma \in \Delta^*$, which shows that Δ^* is complete.
iii. Each formula $\exists v \varphi \in \Delta^*$ has a witness in Δ^*: If $\exists v \varphi \in \Delta^*$, then $\exists v \varphi \in \Sigma_n$ for some $n \in \omega$. By conclusion i, $\exists v \varphi \in \Sigma_{2n+2}$. So by conclusion iv, $\exists v \varphi$ has a witness in Σ_{2n+2} and so a witness in Δ^*.

Thus we can apply Lemma 7.7 and conclude that there is a model $\mathfrak{A}$ of Δ^* such that $c|\mathfrak{A}| \leq c\Delta^*$. Since $\Delta \subseteq \Delta^*$, we have $\mathfrak{A} \in \text{Mod}\,\Delta^*$. It remains only to observe that $c\Delta^* = \omega \cdot (c\Delta + \omega) = c\Delta + \omega$. □

EXERCISES FOR §7

Here we outline our alternate proof of the compactness theorem. This proof has a more algebraic flavor and requires the axiom of choice.

Let J be a set, and let $F \subseteq P(J)$, i.e., F is a set of subsets of J. Say that F is a *filter base* on J if

i. $\emptyset \notin F$,
ii. $X \in F$ and $Y \in F$ implies $X \cap Y \in F$.

F is a *filter* on J if in addition

iii. $J \supseteq Y \supseteq X$ and $X \in F$ implies $Y \in F$.

A filter F is an *ultrafilter* if

iv. for each $Y \subseteq J$ either $Y \in F$ or $J - Y \in F$.

1. Let $a \in J$, and let $F = \{X : a \in X \subseteq J\}$. Show that F is an ultrafilter. An ultrafilter of this kind, i.e., containing a singleton, is called a *principal ultrafilter*.

2. Show that if F is a filter and $X \subseteq J$, then either $F \cup \{X\}$ or $F \cup \{J - X\}$ is contained in a filter. Compare with Lemma 7.2.

3. Every filter is contained in an ultrafilter. [*Hint:* Well-order $P(X)$ and use 2 above or use Theorem 9.3 in Part I.]

4. There are non-principal ultrafilters. [*Hint:* Let J be infinite, and let F be the set of all subsets X of J such that $c(J - X) < cJ$. Then F is a filter, and any ultrafilter containing F is non-principal.]

5. Show that there are 2^{2^κ} ultrafilters on J if $cJ = \kappa$.

Let F be an ultrafilter on J, and let $\{\mathfrak{A}_j : j \in J\}$ be a set of structures all of type s. Let $A^{\#}$ be the set of all choice functions on this set, i.e., $A^{\#} = \{g : \text{Dom}\, g = J \text{ and } g(j) \in |\mathfrak{A}_j| \text{ for each } j \in J\}$. Assuming the axiom of choice, $A^{\#} \neq \emptyset$. Write $g \sim h$ if $\{j : g(j) = h(j)\} \in F$.

6. Show that '$\sim$' is an equivalence relation on $A^{\#}$.

Now let $\bar{g} = \{h : g \sim h\}$, and let $A = \{\bar{g} : g \in A^{\#}\}$. We define a structure $\mathfrak{A} = \Pi_F \mathfrak{A}_j$ as follows:

i. $|\mathfrak{A}| = A$;

and for all $c, f, R \in s$ let

ii. $c^{\mathfrak{A}} = \bar{g}$, where $g(j) = c^{\mathfrak{A}_j}$ for all $j \in J$;

iii. $f^{\mathfrak{A}} \bar{g}_1, \ldots, \bar{g}_n = \bar{g}$, where $g(j) = f^{\mathfrak{A}_j} g_1(j), \ldots, g_n(j)$ for all $j \in J$;

iv. $R^{\mathfrak{A}} \bar{g}_1, \ldots, \bar{g}_n$ if $\{j : R^{\mathfrak{A}_j} g_1(j), \ldots, g_n(j)\} \in F$.

7. Show that $f^{\mathfrak{A}}$ is well defined, i.e., if $h_i \in \bar{g}_i$ for $i = 1, 2, \ldots, n$, and if we define $f^* \bar{g}_1, \ldots, \bar{g}_n = \bar{h}$, where $h(j) = f^{\mathfrak{A}_j} h_1(j), \ldots, h_n(j)$, then $f^* = f^{\mathfrak{A}}$.

8. Show that $R^{\mathfrak{A}}$ is well defined, i.e., if $h_i \in \bar{g}_i$ for $i = 1, 2, \ldots, n$, then $\{j : R^{\mathfrak{A}_j} h_1(j), \ldots, h_n(j)\} \in F$ iff $\{j : R^{\mathfrak{A}_j} g_1(j), \ldots, g_n(j)\} \in F$.

The structure $\mathfrak{A}$ is the *ultraproduct* of $\{\mathfrak{A}_j : j \in J\}$ with respect to F. The main theorem of ultraproducts is

Theorem. $\mathfrak{A} \vDash \varphi(\bar{g}_1, \ldots, \bar{g}_n)$ *iff* $\{j : \mathfrak{A}_j \vDash \varphi(g_1(j), \ldots, g_n(j))\} \in F$.

9. Prove this theorem by induction on formulas.

Principal ultrafilters are uninteresting, since:

10. If $F = \{X \subseteq J : j \in X\}$, then $\mathfrak{A} \cong \mathfrak{A}_j$.

However, non-principal ultrafilters can be used to meld the properties of the various $\mathfrak{A}_j$'s.

11. Compactness theorem via ultraproducts: Let Σ be finitely satisfiable. We want to show that Σ is satisfiable. Without loss of generality we can suppose that $\sigma_1 \wedge \cdots \wedge \sigma_n \in \Sigma$ whenever $\sigma_i \in \Sigma$ for each $i = 1, 2, \ldots, n$. For each $\sigma \in \Sigma$ let $\mathfrak{A}_\sigma \in \text{Mod}\,\sigma$. Let $F_\sigma = \{\rho \in \Sigma : \rho \to \sigma\}$.
    (a) Show that $\{F_\sigma : \sigma \in \Sigma\}$ is a filter base and so is contained in an ultrafilter F.
    (b) If F is any ultrafilter containing $\{F_\sigma : \sigma \in \Sigma\}$, then $\Pi_F \{\mathfrak{A}_\sigma : \sigma \in \Sigma\}$ is a model of Σ.

We end this section with a few miscellaneous problems on ultraproducts of a special kind. If each $\mathfrak{A}_j = \mathfrak{B}$, then $\mathfrak{A} = \Pi_F \mathfrak{B}$ is called the *ultrapower* of $\mathfrak{B}$ with respect to F.

12. For each $b \in |\mathfrak{B}|$ let $h_b \in {}^J\{b\}$, the constant function on J with value b. Prove that the function H defined by
$$H(b) = h_b$$
is an elementary embedding of $\mathfrak{B}$ into $\mathfrak{A}$.

13. Let $\mathfrak{B} = \langle \mathbf{N}, +, \cdot \rangle$, and let F be non-principal on ω. Show that $\Pi_F \mathfrak{B}$ is a non-standard model of $\text{Th}\,\mathfrak{B}$, i.e., a model of $\text{Th}\,\mathfrak{B}$ that is not isomorphic to $\mathfrak{B}$. In fact, if $h(n) = n$, then $\bar{h}$ is an infinite element in $\mathfrak{A}$.

14. Let $\mathfrak{B} = \langle \mathbf{R}, +, \cdot \rangle$, and let F be a non-principal ultrafilter on ω. Then $\Pi_F \mathfrak{B}$ is a non-standard model of $\mathfrak{B}$.

3.8 The Löwenheim-Skolem Theorems

In this section we shall consider a more restrictive notion of substructure, namely the notion of elementary substructure. Many of the examples given in §3.6 can be sharpened by requiring the standard model to be an elementary substructure of the non-standard model. But the main interest in the new notion stems from its usefulness in constructing models having specified properties, as we shall see in §3.10.

Definition 8.1. We say that $\mathfrak{A}$ is an *elementary substructure* of $\mathfrak{B}$ or that $\mathfrak{B}$ is an *elementary extension* of $\mathfrak{A}$ and write $\mathfrak{A} \propto \mathfrak{B}$ if

  i. $\mathfrak{A} \subseteq \mathfrak{B}$ and
  ii. for all $z \in {}^{\text{Vbl}}|\mathfrak{A}|$ and all formulas φ, $\mathfrak{A} \vDash \varphi\langle z \rangle$ iff $\mathfrak{B} \vDash \varphi\langle z \rangle$.

EXAMPLE. Let $\mathfrak{A} = \langle \mathbf{N}^+, ' \rangle$, $\mathfrak{B} = \langle \mathbf{N}, ' \rangle$, where $'$ is the successor function $n' = n + 1$. Clearly $\mathfrak{A} \subseteq \mathfrak{B}$. Moreover, $\mathfrak{A} \cong \mathfrak{B}$, and so $\mathfrak{A} \equiv \mathfrak{B}$ by Theorem 6.1. However, $\mathfrak{A} \not\propto \mathfrak{B}$, for if $z \in {}^{\text{Vbl}}|\mathfrak{A}|$ and $z(v_0) = 1$, then z satisfies $\exists v_1 [v_1' \approx v_0]$ in $\mathfrak{B}$ but not in $\mathfrak{A}$.

The next theorem provides necessary and sufficient conditions for $\mathfrak{A}$ to be an elementary substructure of $\mathfrak{B}$. What makes the test so useful is that the condition involves satisfaction in $\mathfrak{B}$ alone.

Theorem 8.2. Let $\mathfrak{A} \subseteq \mathfrak{B}$, and suppose that for every formula φ and every $z \in {}^{\text{Vbl}}|\mathfrak{A}|$ such that $\mathfrak{B} \models \exists v \varphi \langle z \rangle$, there is an $a \in |\mathfrak{A}|$ for which

$$\mathfrak{B} \models \varphi \left\langle z \binom{v}{a} \right\rangle.$$

Then $\mathfrak{A} \propto \mathfrak{B}$.

PROOF: Suppose $\mathfrak{A}$ and $\mathfrak{B}$ are as in the hypotheses of the theorem. We show by induction on formulas that for all φ

(*) $\mathfrak{A} \models \varphi \langle z \rangle$ iff $\mathfrak{B} \models \varphi \langle z \rangle$, for all $z \in {}^{\text{Vbl}}|\mathfrak{A}|$.

By the remark following Theorem 4.4 it is sufficient to consider the following four cases:

i. For atomic formulas this is clear, since $\mathfrak{A} \subseteq \mathfrak{B}$.
ii. Suppose ψ satisfies (*) and $\varphi = \neg \psi$. Then $\mathfrak{A} \models \varphi \langle z \rangle$ iff not $\mathfrak{A} \models \psi \langle z \rangle$ iff not $\mathfrak{B} \models \psi \langle z \rangle$ iff $\mathfrak{B} \models \varphi \langle z \rangle$. Hence φ satisfies (*).
iii. Suppose ψ_1 and ψ_2 satisfies (*) and $\varphi = [\psi_1 \vee \psi_2]$. Then $\mathfrak{A} \models \varphi \langle z \rangle$ iff [either $\mathfrak{A} \models \psi_1 \langle z \rangle$ or $\mathfrak{A} \models \psi_2 \langle z \rangle$] iff [either $\mathfrak{B} \models \psi_1 \langle z \rangle$ or $\mathfrak{B} \models \psi_2 \langle z \rangle$] iff $\mathfrak{B} \models [\psi_1 \vee \psi_2] \langle z \rangle$ iff $\mathfrak{B} \models \varphi \langle z \rangle$.
iv. Suppose ψ satisfies (*) and $\varphi = \exists v \psi$. If $\mathfrak{A} \models \varphi \langle z \rangle$, then for some $a \in |\mathfrak{A}|$, $\mathfrak{A} \models \psi \langle z \binom{v}{a} \rangle$. Hence, since ψ satisfies (*), $\mathfrak{B} \models \psi \langle z \binom{v}{a} \rangle$; so $\mathfrak{B} \models \varphi \langle z \rangle$. Conversely, if $\mathfrak{B} \models \varphi \langle z \rangle$ then for some $a \in |\mathfrak{A}|$, $\mathfrak{B} \models \psi \langle z \binom{v}{a} \rangle$ by the hypotheses of the theorem. Hence, since ψ satisfies (*), $\mathfrak{A} \models \psi \langle z \binom{v}{a} \rangle$; so $\mathfrak{A} \models \varphi \langle z \rangle$. Thus φ satisfies (*). This concludes the induction and the proof of the theorem. □

Corollary 8.3. Let $\mathfrak{A} \subseteq \mathfrak{B}$, and suppose that for every finite subset $\{a_0, \ldots, a_{n-1}\}$ of $|\mathfrak{A}|$, and for every $b \in |\mathfrak{B}|$, there is an automorphism g on $\mathfrak{B}$ such that

i. $g(b) \in |\mathfrak{A}|$,
ii. $g(a_i) = a_i$ for all $i < n$.

Then $\mathfrak{A} \propto \mathfrak{B}$.

PROOF: Under the conditions of the theorem suppose that $\mathfrak{B} \models \exists v \varphi(a_0, \ldots, a_{n-1})$, where $a_i \in |\mathfrak{A}|$ for all $i < n$. Then for some $b \in \mathfrak{B}$, $\mathfrak{B} \models \varphi(b, a_0, \ldots, a_{n-1})$. Let g be an automorphism satisfying conditions i and ii. By Theorem 4.19, $\mathfrak{B} \models \varphi(g(b), g(a_0), \ldots, g(a_{n-1}))$ i.e., $\mathfrak{B} \models \varphi(g(b), a_0, \ldots, a_{n-1})$. Hence by Theorem 8.2, $\mathfrak{A} \propto \mathfrak{B}$. □

EXAMPLE. If $A \subseteq B$ and $cA \geq \omega$, then $(A) \propto (B)$. This follows trivially from the corollary. Hence, if C and D are infinite, then $(C) \equiv (D)$, for by the

3.8 The Löwenheim-Skolem Theorems

axiom of choice, there is a 1-1 map on C onto some $D' \subseteq D$, or a 1-1 map on D onto some $C' \subseteq C$. Since such a map is an isomorphism on C onto D' or on D onto C', we have $(C) \equiv (D') \propto (D)$ or $(D) \equiv (C') \propto (C)$. In either case $C \equiv D$.

EXAMPLE. Let (x,y) be the open interval on the real line between x and y, i.e., the set of all reals greater than x but less than y. Let $\mathfrak{A} = ((0, \tfrac{1}{2}), <)$, $\mathfrak{B} = ((0, 1), <)$ where '$<$' is the usual ordering. Then by Corollary 8.3, we see that $\mathfrak{A} \propto \mathfrak{B}$.

Definition 8.4. Let $\mathfrak{B} = (\mathfrak{A}, c_a^{\mathfrak{B}})_{a \in |\mathfrak{A}|}$, where $c_a^{\mathfrak{B}} = a$. Then $\text{Th} \mathfrak{B}$ is a *complete diagram* of $\mathfrak{B}$.

No structure has a unique complete diagram, but since the complete diagram in question is usually clear or immaterial in most arguments, we shall speak of *the* complete diagram of $\mathfrak{A}$, abbreviated $\mathfrak{D}^c \mathfrak{A}$.

The relation between elementary substructures and complete diagrams is analogous to that between substructures and diagrams, as a comparison of Theorem 4.17 and the following shows:

Theorem 8.5. *$\mathfrak{A}$ is isomorphic to an elementary substructure of $\mathfrak{B}$ iff there is an expansion $\mathfrak{B}^+$ of $\mathfrak{B}$ such that $\mathfrak{B}^+ \in \text{Mod} \, \mathfrak{D}^c \mathfrak{A}$.*

PROOF: Let $\mathfrak{A} \cong_g \mathfrak{B}' \propto \mathfrak{B}$. Let $\mathfrak{A}^+ = (\mathfrak{A}, c_a^{\mathfrak{A}^+})_{a \in |\mathfrak{A}|}$, where $c_a^{\mathfrak{A}^+} = a$ for each $a \in |\mathfrak{A}|$. Now take $\mathfrak{C} = (\mathfrak{B}, c_a^{\mathfrak{C}})$, where $c_a^{\mathfrak{C}} = g(a)$. We show that $\mathfrak{C} \in \text{Mod} \, \mathfrak{D}^c \mathfrak{A}$. For suppose that $\varphi(c_{a_0}, \ldots, c_{a_{n-1}}) \in \text{Mod} \, \mathfrak{D}^c \mathfrak{A}$ with $\tau \varphi(c_{a_0}, \ldots, c_{a_{n-1}}) - \tau \mathfrak{A} = \{c_{a_0}, \ldots, c_{a_{n-1}}\}$. Then $\mathfrak{A} \models \varphi(a_0, \ldots, a_{n-1})$. Hence by Theorem 4.19, $\mathfrak{B}' \models \varphi(a(a_0), \ldots, g(a_{n-1}))$. Since $\mathfrak{B}' \propto \mathfrak{B}$, $\mathfrak{B} \models \varphi(g(a_0), \ldots, g(a_{n-1}))$. Hence $\mathfrak{C} \models \varphi(c_{a_0}, \ldots, c_{a_{n-1}})$. Thus $\mathfrak{C} \in \text{Mod} \, \mathfrak{D}^c \mathfrak{A}$.

Conversely, let $\mathfrak{C}$ be an expansion of $\mathfrak{B}$ such that $\mathfrak{C} \in \text{Th} \mathfrak{A}^+$, where $\mathfrak{A}^+ = (\mathfrak{A}, c_a^{\mathfrak{A}})_{a \in |\mathfrak{A}|}$ and $c_a^{\mathfrak{A}^+} = a$. Define a function g on $|\mathfrak{A}|$ into $|\mathfrak{B}|$ by $g(a) = c_a^{\mathfrak{C}}$. Let $\mathfrak{B}' = \text{Rng} \, g$. Since $\mathfrak{D}^c \mathfrak{A} \supseteq \mathfrak{D} \mathfrak{A}$, we have by Theorem 4.17 that $\mathfrak{A} \cong_g \mathfrak{B}' \subseteq \mathfrak{B}$. Now let φ be a formula whose free variables are $u_0, \ldots, u_{n-1}$, let $z \in {}^{\text{Vbl}}|\mathfrak{B}'|$, and let $g^{-1}(z(u_i)) = a_i$. Then $\mathfrak{B}' \models \varphi \langle z \rangle$ iff $\mathfrak{A} \models \varphi \langle g^{-1} \circ z \rangle$ iff $\mathfrak{A}^+ \models \varphi(c_{a_0}, \ldots, c_{a_{n-1}})$ iff $\mathfrak{C} \models \varphi(c_{a_0}, \ldots, c_{a_{n-1}})$ (since $\mathfrak{C} \in \text{Mod} \, \mathfrak{D}^c \mathfrak{A}$) iff $\mathfrak{B} \models \varphi \langle z \rangle$. Hence $\mathfrak{B}' \propto \mathfrak{B}$. $\square$

We may write $\mathfrak{A} \propto \mathfrak{B}$ when all we mean is that for some $\mathfrak{B}'$, $\mathfrak{A} \cong \mathfrak{B}' \propto \mathfrak{B}$.

Theorem 8.6 (Upward Löwenheim-Skolem Theorem). *Let $c|\mathfrak{A}| \geqslant \omega$, and let $\kappa \geqslant c|\mathfrak{A}| + c\tau \mathfrak{A}$. Then there is a $\mathfrak{B}$ such that $c|\mathfrak{B}| = \kappa$ and $\mathfrak{A} \propto \mathfrak{B}$.*

PROOF: With $\mathfrak{A}$ and κ as in the hypotheses, let $\mathfrak{A}^+ = (\mathfrak{A}, c_a^{\mathfrak{A}^+})_{a \in |\mathfrak{A}|}$, where $c_a^{\mathfrak{A}^+} = a$. Let $\{d_\alpha : \alpha \in \lambda\}$ be a set of constant symbols disjoint from $\tau \mathfrak{A}^+$, such that $d_\alpha \neq d_\beta$ for $\alpha < \beta < \kappa$. Let $\Gamma = \text{Th} \, \mathfrak{A}^+ \cup \{d_\alpha \not\approx d_\beta : \alpha < \beta < \kappa\}$. Clearly, each finite subset of Γ is satisfied in some expansion of $\mathfrak{A}^+$. Hence by compactness, Γ has a model $\mathfrak{C}$ of power κ. By the preceding theorem, $\mathfrak{A} \propto \mathfrak{C} \restriction \tau \mathfrak{A}$. $\square$

EXAMPLE. For every infinite cardinal κ, there is a $\mathfrak{B}$ such that $(\mathbf{N}, +, \cdot, 0, 1, 2, \ldots) \propto \mathfrak{B}$ and $c|\mathfrak{B}| = \kappa$. Hence there are structures in every cardinality that are elementarily equivalent to $(\mathbf{N}, +, \cdot, 0, 1, 2, \ldots)$.

Theorem 8.7 (Downward Löwenheim-Skolem Theorem). *Let $X \subseteq \mathfrak{B}$. Then there is an $\mathfrak{A}$ such that $X \subseteq \mathfrak{A} \propto \mathfrak{B}$ and $c|\mathfrak{A}| \leq cX + c\tau\mathfrak{B} + \omega$.*

PROOF: Let $<$ be a well ordering of $|\mathfrak{B}|$ (here we use the axiom of choice). For each formula $\exists v\varphi$ and each $z \in {}^{\mathrm{Vbl}}|\mathfrak{B}|$ such that $\mathfrak{B} \vDash \exists v\varphi\langle z\rangle$, let $g(\varphi, z)$ be the first b (with respect to $<$) such that

$$\mathfrak{B} \vDash \varphi\left\langle z\binom{b}{v}\right\rangle.$$

Now define

$$A_0 = X,$$
$$A_{n+1} = A_n \cup \{g(\varphi, z) : z \in {}^{\mathrm{Vbl}}A_n \text{ and } \mathfrak{B} \vDash \exists v\varphi\langle z\rangle\}.$$

Now let $A = \bigcup_{n \in \omega} A_n$. We first observe that if $c \in \tau\mathfrak{B}$, then $c^{\mathfrak{B}} \in A$; indeed, since $\mathfrak{B} \vDash \exists v[v \approx c]$, $c^{\mathfrak{B}} \in A_1$. Also, if $a_0, \ldots, a_{n-1} \in A$ and $f_{n,\alpha} \in \tau\mathfrak{B}$, then $f_{n,\alpha}^{\mathfrak{B}} a_0, \ldots, a_{n-1} \in A$; for if $a_i \in A_{j_i}$ and $m = \max\{j_i : i < n\}$, then $a_0, \ldots, a_{n-1} \in A_m$, since $A_k \subseteq A_l$ for all $k \leq l < \omega$. Hence $f_{n,\alpha}^{\mathfrak{B}} a_0, \ldots, a_{n-1} \in A_{m+1}$, since $\mathfrak{B} \vDash \exists v[f_{n,\alpha}(a_0, \ldots, a_{n-1}) \approx v]$. Hence there is a substructure $\mathfrak{A} \subseteq \mathfrak{B}$ with $|\mathfrak{A}| = A$. Clearly $X \subseteq |\mathfrak{A}|$. Moreover, for each $n \in \omega$, $cA_n \leq cX + c\tau\mathfrak{B} + \omega$, and so $cA \leq cX + c\tau\mathfrak{B} + \omega$. It remains to show that $\mathfrak{A} \propto \mathfrak{B}$.

For this we use Theorem 8.2. Suppose φ is a formula with only the variables $v, u_0, \ldots, u_{n-1}$ free. Let $z \in {}^{\mathrm{Vbl}}|\mathfrak{A}|$, and suppose $\mathfrak{B} \vDash \exists v\varphi\langle z\rangle$. Take m large enough so that $z(u_0), \ldots, z(u_{n-1}) \in A_m$. Then $\mathfrak{B} \vDash \varphi(g(\varphi, z), z(u_0), \ldots, z(u_{n-1}))$ and $g(\varphi, z) \in A_{m+1}$, and so $g(\varphi, z) \in A$. Hence by Theorem 8.2, $\mathfrak{A} \propto \mathfrak{B}$, as needed. □

Definition 8.8. Let $(\mathfrak{A}_\alpha)_{\alpha < \kappa}$ be a sequence of structures of type s such that $\mathfrak{A}_\beta \subseteq \mathfrak{A}_\gamma$ whenever $\beta < \gamma < \kappa$. Let $\bigcup_{\alpha < \kappa} \mathfrak{A}_\alpha$ be that structure $\mathfrak{A}$ of type s such that

i. $|\mathfrak{A}| = \bigcup_{\alpha < \kappa} |\mathfrak{A}_\alpha|$,
ii. $c^{\mathfrak{A}} = c^{\mathfrak{A}_0}$ for each $c \in s$,
iii. $f^{\mathfrak{A}} = \bigcup_{\alpha < \kappa} f^{\mathfrak{A}_\alpha}$ for each $f \in s$,
iv. $R^{\mathfrak{A}} = \bigcup_{\alpha < \kappa} R^{\mathfrak{A}_\alpha}$ for each $R \in s$.

Notice that clause iii is equivalent to: $f^{\mathfrak{A}} a_1, \ldots, a_n = b$ iff $f^{\mathfrak{A}_\alpha} a_1, \ldots, a_n = b$ whenever $a_1, \ldots, a_n, b \in |\mathfrak{A}_\alpha|$. Similarly, clause iv is equivalent to: $R^{\mathfrak{A}} a_1, \ldots, a_n$ iff $R^{\mathfrak{A}_\alpha} a_1, \ldots, a_n$ whenever $a_1, \ldots, a_n \in |\mathfrak{A}_\alpha|$.

In model theory one often builds a structure $\mathfrak{A}$ satisfying certain first order requirements by erecting a tower of approximations $\mathfrak{A}_1, \mathfrak{A}_2, \ldots, \mathfrak{A}_\alpha, \ldots$, where $\mathfrak{A}_\beta \subseteq \mathfrak{A}_\gamma$ for $\beta < \gamma$ and $\mathfrak{A} = \bigcup_{\alpha < \kappa} \mathfrak{A}_\alpha$. But the requirement $\mathfrak{A}_\beta \subseteq \mathfrak{A}_\gamma$ for $\beta < \gamma$ is seldom strong enough to determine the first order

3.8 The Löwenheim-Skolem Theorems

properties of $\mathfrak{A}$. For example, if $\mathfrak{A}_i = (\{-i, -i+1, -i+2, \ldots\}, <)$ for each $i = 1, 2, 3, \ldots$, then not only is $\mathfrak{A}_j \subseteq \mathfrak{A}_k$ for $j < k$, but $\mathfrak{A}_j \cong \mathfrak{A}_k$ and so $\mathfrak{A}_j \equiv \mathfrak{A}_k$. Nevertheless, if $\mathfrak{A} = \bigcup_{i \in \omega} \mathfrak{A}_i$, then $\mathfrak{A} = (I, <) \not\equiv \mathfrak{A}_j$. To assure $\mathfrak{A}_j \propto \mathfrak{A}$ for each j, we need the condition $\mathfrak{A}_\gamma \propto \mathfrak{A}_\delta$ whenever $\gamma < \delta$. Such a sequence $(\mathfrak{A}_\alpha)_{\alpha \in \kappa}$ is called an *elementary chain*.

Theorem 8.9. Let $(\mathfrak{A}_\alpha)_{\alpha < \kappa}$ be an elementary chain, i.e., let $\mathfrak{A}_\beta \propto \mathfrak{A}_\gamma$ whenever $\beta < \gamma < \kappa$. Then $\mathfrak{A}_\beta \propto \bigcup_{\alpha < \kappa} \mathfrak{A}_\alpha$ for each $\beta < \kappa$.

PROOF: Let $\mathfrak{A} = \bigcup_{\alpha < \kappa} \mathfrak{A}_\alpha$. Let Γ be the set of formulas φ such that whenever $\beta < \kappa$ and $z \in {}^{\text{Vbl}}|\mathfrak{A}_\beta|$, then $\mathfrak{A}_\beta \vDash \varphi\langle z\rangle$ whenever $\mathfrak{A} \vDash \varphi\langle z\rangle$. We want to show that all formulas of type $\tau\mathfrak{A}$ belong to Γ.

Clearly the atomic formulas belong to Γ (see Definition 8.8, clauses iii and iv). It is easy to see that if $\varphi_1 \in \Gamma$ and $\varphi_2 \in \Gamma$, then $\neg \varphi_2 \in \Gamma$ and $\varphi_1 \vee \varphi_2 \in \Gamma$.

Suppose that $\varphi = \exists v \psi$, where $\psi \in \Gamma$. If $\mathfrak{A}_\beta \vDash \varphi\langle z\rangle$, then $\mathfrak{A}_\beta \vDash \psi\langle z\binom{v}{b}\rangle$ for some $b \in |\mathfrak{A}_\beta|$. Since $\psi \in \Gamma$, $\mathfrak{A} \vDash \psi\langle z\binom{v}{b}\rangle$, and so $\mathfrak{A} \vDash \varphi\langle z\rangle$. Conversely, if $\mathfrak{A} \vDash \varphi\langle z\rangle$, where $z \in {}^{\text{Vbl}}|\mathfrak{A}_\beta|$, then $\mathfrak{A} \vDash \psi\langle z\binom{v}{b}\rangle$, for some $b \in |\mathfrak{A}|$. Since $|\mathfrak{A}| = \bigcup_{\alpha \in \kappa} |\mathfrak{A}_\alpha|$, $b \in |\mathfrak{A}_\gamma|$ for some $\gamma \in \kappa$. Let $\delta = \max\{\beta, \gamma\}$. Then since $z\binom{v}{b} \in {}^{\text{Vbl}}|\mathfrak{A}_\delta|$ and $\psi \in \Gamma$, we have $\mathfrak{A}_\delta \vDash \psi\langle z\binom{v}{b}\rangle$, and so $\mathfrak{A}_\delta \vDash \varphi\langle z\rangle$. But $\mathfrak{A}_\beta \propto \mathfrak{A}_\delta$, and so $\mathfrak{A}_\beta \vDash \varphi\langle z\rangle$. Hence $\varphi \in \Gamma$. This completes the proof that all formulas belong to Γ, which gives Theorem 8.9. □

Definition 8.10. Say that Σ implies σ if $\text{Mod } \Sigma = \text{Mod}(\Sigma \cup \{\sigma\})$, i.e., if every model of Σ is a model of σ.

Next we generalize Definition 7.3 as follows:

Definition 8.11. Σ is *complete* if Σ has a model and for all σ of type $\tau\Sigma$, either Σ implies σ or Σ implies $\neg \sigma$.

Clearly, if Σ is complete in the old sense, then it is complete in the new sense. Conversely, a set Σ complete in the new sense is contained in a unique set Γ that is complete in the old sense and of type $\tau\Sigma$, namely $\Gamma = \{\sigma : \tau\sigma \subseteq \tau\Sigma \text{ and } \Sigma \text{ implies } \sigma\}$.

Definition 8.12. Σ is κ-*categorical* if Σ has models of cardinality κ, and if any two models of cardinality κ are isomorphic.

The following useful criterion for completeness is known as Vaught's test.

Theorem 8.13. If Σ is κ-categorical, where $\kappa \geq c\Sigma + \omega$, and if every model of Σ is infinite, then Σ is complete.

PROOF: Suppose that Σ is as in the hypotheses but is not complete. Then for some σ of type $\tau\Sigma$, both $\Sigma\cup\{\sigma\}$ and $\Sigma\cup\{\neg\sigma\}$ have models. By the compactness theorem, there are models $\mathfrak{A}$ of $\Sigma\cup\{\sigma\}$ and $\mathfrak{B}$ of $\Sigma\cup\{\neg\sigma\}$ such that $c|\mathfrak{A}| = c|\mathfrak{B}| \leq c\Sigma + \omega$. By Theorem 8.6 and our assumption that all the models of Σ are infinite, there is a $\mathfrak{A}' \infty \mathfrak{A}$ and $\mathfrak{B}' \infty \mathfrak{B}$ such that $c|\mathfrak{A}'| = c|\mathfrak{B}'| = \kappa$. Since Σ is κ-categorical, $\mathfrak{A}' \cong \mathfrak{B}'$. By Theorem 4.19, $\mathfrak{A}' \equiv \mathfrak{B}'$. But this is impossible, since $\mathfrak{A}' \vDash \sigma$ and $\mathfrak{B}' \vDash \neg\sigma$. Hence Σ must be complete. $\square$

EXAMPLE. Let $\bigwedge_{i<j<n} v_i \not\approx v_j$ be a conjunction of all formulas $v_i \not\approx v_j$ where $i<j<n$, and let $\exists^{\geq n}$ be the formula $\exists v_0, \ldots, v_{n-1} \bigwedge_{i<j<n} v_i \not\approx v_j$. Clearly $\mathfrak{A} \vDash \exists^{\geq n}$ iff $c|\mathfrak{A}| \geq n$. Let $\Sigma = \{\exists^{\geq n} : n \in \omega\}$. Then $\mathrm{Mod}\,\Sigma$ is the class of all structures $\mathfrak{A} = (A)$ such that $cA \geq \omega$. Since any two such structures of cardinality ω are isomorphic, Σ is ω-categorical, and so by Vaught's test Σ is complete.

A linearly ordered structure $\mathfrak{A}$ (see §1.8) is *densely ordered* if $c|\mathfrak{A}| \geq 2$ and

$$\mathfrak{A} \vDash \forall v_0 v_1 [v_0 < v_1 \rightarrow \exists v_2 [v_0 < v_2 \land v_2 < v_1]].$$

Our next example is based on the following theorem of Cantor.

Theorem 8.14. *If $\mathfrak{A}$ and $\mathfrak{B}$ are countable densely ordered structures without least or greatest elements, then $\mathfrak{A} \cong \mathfrak{B}$.*

PROOF: Let $\mathfrak{A} = (A, <^{\mathfrak{A}})$ and $\mathfrak{B} = (B, <^{\mathfrak{B}})$, where $A = \{a_0, a_1, \ldots\}$ and $B = \{b_0, b_1, \ldots\}$. Suppose that $c_0 <^{\mathfrak{A}} \cdots <^{\mathfrak{A}} c_{n-1}$ and $d_0 <^{\mathfrak{B}} \cdots <^{\mathfrak{B}} d_{n-1}$. Then for every $c \in A$ there is a $d \in B$ such that for all $i < n$, $c_i < c$ iff $d_i < d$, and $c < c_i$ iff $d < d_i$. Indeed, if $c < c_0$ or if $c_{n-1} < c$, then such a d exists, since $\mathfrak{B}$ has no least or greatest element. If for some $j < n-1$, $c_{j-1} < c < c_j$, then such a d exists because the ordering is dense. Now let $g(\{c_i : i < n\}, c, \{d_i : i < n\})$ be the first term in the sequence $b_0, b_1, \ldots$ that can serve as such a d. Reversing the roles of $\mathfrak{A}$ and $\mathfrak{B}$, we get a function $h(\{d_i : i < n\}, d, \{c_i : i < n\})$.

Now define sequences $a'_0, a'_1, \ldots$ and $b'_0, b'_1, \ldots$ as follows:

$a'_0 = a_0,$

$b'_0 = b_0,$

$a'_{2n+1} = a_{2n+1},$

$b'_{2n+1} = g(\{a'_0, \ldots, a'_{2n}\}, a'_{2n+1}, \{b'_0, \ldots, b'_{2n}\}),$

$b'_{2n+2} = b_{2n+2},$

$a'_{2n+2} = h(\{b'_0, \ldots, b'_{2n+1}\}, b'_{2n+2}, \{a'_0, \ldots, a'_{2n+1}\}).$

It is easy to see that the function l defined by $l(a'_i) = b'_i$ is an isomorphism on $\mathfrak{A}$ onto $\mathfrak{B}$, which proves the theorem. $\square$

3.8 The Löwenheim-Skolem Theorems

EXAMPLE. Let $\mathcal{K}$ be the class of densely ordered structures without endpoints. Clearly, $\mathcal{K} = \text{Mod}\,\Sigma$, where Σ is composed of the following assertions:

$$\forall v_0 \exists v_1 v_2 [[v_1 < v_0] \wedge [v_0 < v_2]],$$
$$\forall v_0 v_1 [v_0 < v_1 \rightarrow v_1 \not< v_0],$$
$$\forall v_0 v_1 v_2 [v_0 < v_1 \wedge v_1 < v_2 \rightarrow v_0 < v_2],$$
$$\forall v_0 v_1 [v_0 < v_1 \vee v_1 < v_0 \vee v_0 \approx v_1],$$
$$\forall v_0 v_1 \exists v_2 [v_0 < v_1 \rightarrow [v_0 < v_2 \wedge v_2 < v_1]].$$

It is easy to see that all models of Σ are infinite. Moreover, the above theorem shows that Σ is ω-categorical. Hence Σ is complete by Theorem 8.13.

EXAMPLE. Let $\mathfrak{A} = (\mathbf{I}, f^{\mathfrak{A}})$, where $\mathbf{I}$ is the set of integers and $f(x) = x + 1$ for all $x \in \mathbf{I}$. By recursion on n, define the term $f_n v$ as follows:

$$f_1 v \approx f v,$$
$$f_{n+1} v \approx f(f_n(v)).$$

It is easy to see that the following assertions are true in $\mathfrak{A}$:

i. $\forall v_0 \exists v_1 [f v_1 \approx v_0]$,
ii. $\forall v_0 v_1 [f v_0 \approx f v_1 \rightarrow v_0 \approx v_1]$,
iii. $\forall v_0 [f_n v_0 \not\approx v_0]$ for each $n \in \omega$.

Let Σ be the set of the above sentences. We claim that the set of sentences implied by Σ is $\text{Th}\,\mathfrak{A}$. To prove this it is enough to show that Σ is complete. For this we shall use Theorem 8.13.

Notice, however, that Σ is not ω-categorical. Indeed, if $B = \mathbf{I} \cup \{x + \pi : x \in \mathbf{I}\}$ and if $f^{\mathfrak{B}}(x) = z + 1$ for all $z \in B$, then $\mathfrak{B} = (B, f^{\mathfrak{B}})$ is clearly a countable model of Σ, but $\mathfrak{B} \not\cong \mathfrak{A}$.

But Σ is categorical in every uncountable cardinality: For let $\mathfrak{B}, \mathfrak{C}$ be models of Σ with $c|\mathfrak{B}| = c|\mathfrak{C}| = \kappa > \omega$. Define $b_0 \sim b_1$ if $f_n^{\mathfrak{B}} b_0 = b_1$ or $f_n^{\mathfrak{B}} b_1 = b_0$. It is easy to check that '$\sim$' is an equivalence relation on $\mathfrak{B}$ and each equivalence class is countable. Analogously, define an equivalence relation $c_0 \sim^{\mathfrak{C}} c_1$ on $|\mathfrak{C}|$. There are κ many $\sim^{\mathfrak{B}}$-equivalence classes, say P_α, $\alpha \in \kappa$, and there are κ many $\sim^{\mathfrak{C}}$-equivalence classes, say Q_α, $\alpha \in \kappa$. Choose a point $p_\alpha \in P_\alpha$ and a point $q_\alpha \in Q_\alpha$ for each $\alpha \in \kappa$. Now define an isomorphism F on $\mathfrak{B}$ onto $\mathfrak{C}$ by defining $F(x) = y$ if for some α and some n either

i. $f_n^{\mathfrak{B}}(x) = p_\alpha$ and $f_n^{\mathfrak{C}}(y) = q_\alpha$, or
ii. $f_n^{\mathfrak{B}}(p_\alpha) = x$ and $f_n^{\mathfrak{C}}(q_\alpha) = y$.

(It is easy to check that F is an isomorphism, and the reader is asked to do so in Exercise 4.) Hence Σ is categorical in every infinite power and so is complete.

EXERCISES FOR §3.8

1. Show that if $\mathfrak{A} \prec \mathfrak{C}$ and $\mathfrak{B} \prec \mathfrak{C}$ and $\mathfrak{A} \subseteq \mathfrak{B}$, then $\mathfrak{A} \prec \mathfrak{B}$.

2. Suppose that $\tau\mathfrak{A}_\alpha \subseteq \tau\mathfrak{A}_\beta$ whenever $\alpha < \beta < \kappa$. Also suppose that $\mathfrak{A}_\alpha \prec \mathfrak{A}_\beta \upharpoonright \tau\mathfrak{A}_\alpha$ whenever $\alpha < \beta < \kappa$. Let $\bigcup_{\alpha \in \kappa} \mathfrak{A}_\alpha$ be that structure $\mathfrak{A}$ of type $s = \bigcup_{\alpha \in \kappa} \tau\mathfrak{A}_\alpha$ where

$$|\mathfrak{A}| = \bigcup_{\alpha \in \kappa} |\mathfrak{A}_\alpha|,$$

$c^{\mathfrak{A}} = c^{\mathfrak{A}_\alpha}$ for any α such that $c \in \tau\mathfrak{A}_\alpha$,

$$f^{\mathfrak{A}} = \bigcup_{\alpha \in \omega} f^{\mathfrak{A}_\alpha} \quad \text{for any } f \in s,$$

$$R^{\mathfrak{A}} = \bigcup_{\alpha \in \kappa} R^{\mathfrak{A}_\alpha} \quad \text{for any } R \in s.$$

Show that $\mathfrak{A}_\alpha \prec \mathfrak{A} \upharpoonright \tau\mathfrak{A}_\alpha$ for each $\alpha \in \kappa$.

3. Let $f_n(v)$ be the term defined in the last example of this section. Let Σ be the following set:
    (a) $\forall v_0 \exists v_1 [v_0 \not\approx c \to [f(v_1) \approx v_0]]$,
    (b) $\forall v_0 \forall v_1 [f(v_0) \approx f(v_1) \to v_0 \approx v_1]$,
    (c) $\forall v_0 [f_n(v_0) \not\approx v_0]$,
    (d) $\forall v_0 [f(v_0) \not\approx c]$.
    Prove that Σ is complete and that $\text{Th}(\mathbf{N}, f^N) = \{\sigma | \Sigma \text{ implies } \sigma\}$, where $f^N(n) = n+1$.

4. Let F be the function from $\mathfrak{B}$ to $\mathfrak{C}$ described in the last example of this section. Show that F is an isomorphism.

5. Let $(D, <)$ be a partially ordered structure such that for each d_1 and d_2 in D there is a $d \in D$ for which $d_1 < d$ and $d_2 < d$. Let $\{\mathfrak{A}_d : d \in D\}$ be a set of structures such that $\mathfrak{A}_d \prec \mathfrak{A}_{d'}$ whenever $d < d'$. Define $\bigcup_{d \in D} \mathfrak{A}_d$ as in Exercise 2. Show that $\mathfrak{A}_{d'} \prec \bigcup_{d \in D} \mathfrak{A}_d$ for each $d' \in D$.

6. Let $\Sigma = \text{Th}(\mathbf{R}, <)$. Show that whenever $\mathfrak{A} \subseteq \mathfrak{B}$ and $\mathfrak{A}$ and $\mathfrak{B}$ are models of Σ, then $\mathfrak{A} \prec \mathfrak{B}$.

7. Find $\mathfrak{A}$, $\mathfrak{B}'$, and $\mathfrak{B}$ such that $\mathfrak{A} \prec \mathfrak{B}$ and $\mathfrak{B} \cong \mathfrak{B}' \prec \mathfrak{A}$ but $\mathfrak{A} \not\cong \mathfrak{B}$.

8. Prove the analog of Theorem 4.18 for elementary substructures, i.e., show that if $\mathfrak{A} \cong \mathfrak{B}' \prec \mathfrak{B}$, then there is a $\mathfrak{C} \cong \mathfrak{B}$ such that $\mathfrak{A} \prec \mathfrak{C}$.

3.9 The Prefix Problem

In this section we show that certain algebraic conditions are satisfied by an elementary class K just in case $K = \text{Mod}\,\Gamma$ for some set Γ each member of which has a particular simple form. For example, if each substructure of a member of K is a member of K, then Γ can be taken to be a set of

3.9 The Prefix Problem

assertions each of the form $\forall u_0 \cdots u_{n-1} \varphi$ where no quantifier occurs in φ (Theorem 9.3).

Lemma 9.1. *Suppose that* $\mathfrak{A} \subseteq \mathfrak{B}$, *that* $z \in {}^{\mathrm{Vbl}}|\mathfrak{A}|$, *and that* ψ *is open. Then*

$$\mathfrak{A} \vDash \psi \langle z \rangle \quad \textit{iff} \quad \mathfrak{B} \vDash \psi \langle z \rangle.$$

PROOF: Let K be the class of formulas ψ for which this is true. Then every atomic formula belongs to K by the definition of substructure. Moreover, if $\psi_1, \psi_2 \in K$, then

i. $\mathfrak{A} \vDash \neg \psi_1 \langle z \rangle$ iff $\mathfrak{A} \nvDash \psi_1 \langle z \rangle$ iff $\mathfrak{B} \nvDash \psi_1 \langle z \rangle$ iff $\mathfrak{B} \vDash \neg \psi_1 \langle z \rangle$, and
ii. $\mathfrak{A} \vDash [\psi_1 \vee \psi_2] \langle z \rangle$ iff ($\mathfrak{A} \vDash \psi_1 \langle z \rangle$ or $\mathfrak{A} \vDash \psi_2 \langle z \rangle$) iff ($\mathfrak{B} \vDash \psi_1 \langle z \rangle$ or $\mathfrak{B} \vDash \psi_2 \langle z \rangle$) iff $\mathfrak{B} \vDash [\psi_1 \vee \psi_2] \langle z \rangle$.

Hence $\psi_1, \psi_2 \in K$ implies $\neg \psi_1, \psi_1 \vee \psi_2 \in K$. Thus K contains the open formulas, which proves the lemma. □

Definition 9.2. A formula in prenex normal form is said to be *universal* (*existential*) if the only quantifier symbol occurring in its prefix in $\forall$ ($\exists$). That is, a universal formula is one of the form $\forall u_0 \cdots u_{n-1} \varphi$, where φ is open. A set Γ of formulas is said to be universal if every $\varphi \in \Gamma$ is universal.

Theorem 9.3. *If K is an elementary class, then the following are equivalent*:

i. *Whenever $\mathfrak{A} \subseteq \mathfrak{B}$ and $\mathfrak{B} \in K$, then $\mathfrak{A} \in K$.*
ii. *For some universal set Γ, $K = \mathrm{Mod}\,\Gamma$.*

PROOF: ii implies i: Let $\mathfrak{A} \subseteq \mathfrak{B} \in \mathrm{Mod}\,\Gamma$ with Γ universal. If $\sigma \in \Gamma$, then σ has the form $\forall u_0 \cdots u_{n-1} \psi$, where ψ is open. If $\mathfrak{B} \in \mathrm{Mod}\,\Gamma$, then $\mathfrak{B} \vDash \sigma$ and so $\mathfrak{B} \vDash \psi \langle z \rangle$ for all $z \in {}^{\mathrm{Vbl}}|\mathfrak{B}|$. In particular, $\mathfrak{B} \vDash \psi \langle z \rangle$ for all $z \in {}^{\mathrm{Vbl}}|\mathfrak{A}|$. Hence by Lemma 9.1, $\mathfrak{A} \vDash \psi \langle z \rangle$ for all $z \in {}^{\mathrm{Vbl}}|\mathfrak{A}|$. Hence $\mathfrak{A} \vDash \sigma$ for all $\sigma \in \Gamma$, as was to be shown.

i implies ii: Let K be an elementary class satisfying condition i. Let Γ be the set of all universal sentences in $\mathrm{Th}\,K$. We show that $K = \mathrm{Mod}\,\Gamma$. Clearly, $K \subseteq \mathrm{Mod}\,\Gamma$. Now let $\mathfrak{B} \in \mathrm{Mod}\,\Gamma$. We claim that every finite subset of $\mathcal{D}\,\mathfrak{B}$ has a model that is an expansion of some structure in K. For suppose $\{\psi_0, \ldots, \psi_{n-1}\} \subseteq \mathcal{D}\,\mathfrak{B}$ but has no such model. Let $d_0, \ldots, d_{l-1}$ be the constant symbols in $\tau[\psi_0 \wedge \cdots \wedge \psi_{n-1}] - \tau K$. Let ψ_i' be the result of replacing each d_j in ψ_i by v_j. Then no $\mathfrak{A}$ in K is a model of $\exists v_0 \cdots v_{l-1} [\psi_0' \wedge \cdots \wedge \psi_{n-1}']$. Hence $\forall v_0 \cdots v_{l-1} \neg [\psi_0' \wedge \cdots \wedge \psi_{n-1}'] \in \mathrm{Th}\,K$, and so $\forall v_0 \cdots v_{l-1} \neg [\psi_0' \wedge \cdots \wedge \psi_{n-1}'] \in \Gamma$. But $\mathfrak{B} \in \mathrm{Mod}\,\Gamma$, so $B \vDash \forall v_0 \cdots v_{l-1} \neg [\psi_0' \wedge \cdots \wedge \psi_{n-1}']$—impossible, since $\psi_0 \wedge \cdots \wedge \psi_{n-1} \in \mathcal{D}\,\mathfrak{B}$. Hence for every finite subset Δ' of $\mathcal{D}\,\mathfrak{B}$, there is an $\mathfrak{A}$ such that $\mathfrak{A} \in \mathrm{Mod}\,\Delta' \cup K$.

Now let $\Delta = \mathcal{D}\,\mathfrak{B} \cup \mathrm{Th}\,K$. We have just shown that the compactness theorem applies. Hence Δ has a model $\mathfrak{C}$. By Theorems 4.17 and 4.18, we can suppose $\mathfrak{B} \subseteq \mathfrak{C} \restriction \tau K$. Since $\mathfrak{C} \restriction \tau K \in \mathrm{Mod}\,\mathrm{Th}\,K$ and $\mathrm{Mod}\,\mathrm{Th}\,K =$

K (K is an elementary class), we have $\mathfrak{C}\!\restriction\!\tau K \in K$. And since we are assuming that K satisfies condition i, $\mathfrak{B} \in K$. Thus $K \supseteq \text{Mod}\,\Gamma$ and so $K = \text{Mod}\,\Gamma$. □

Corollary 9.4. *Suppose $K = \text{Mod}\,\rho$ for some ρ. Then the following are equivalent*:

i. $\mathfrak{A} \subseteq \mathfrak{B}$ *and* $\mathfrak{B} \in K$ *implies* $\mathfrak{A} \in K$
ii. $K = \text{Mod}\,\sigma$ *for some universal sentence σ.*

PROOF: We first show that i implies ii. Assuming condition i and applying the theorem, we have $K = \text{Mod}\,\Sigma$ for some set Σ of universal sentences. Since $\Sigma \cup \{\neg\rho\}$ has no models, the compactness theorem implies that there is a finite subset $\{\sigma_0,\ldots,\sigma_{n-1}\}$ of Σ such that $\{\sigma_0,\ldots,\sigma_{n-1}\} \cup \{\neg\rho\}$ has no models. Letting σ be $\sigma_0 \wedge \cdots \wedge \sigma_{n-1}$, this means that $\vDash \sigma \to \rho$. On the other hand, since $\text{Mod}\,\rho = \text{Mod}\,\Sigma$, $\vDash \rho \to \sigma_i$ for each $i < n$, from which it follows that $\vDash \rho \to \sigma$. Hence $\vDash \sigma \leftrightarrow \rho$, i.e., $K = \text{Mod}\,\sigma$, as needed. The proof that ii implies i is immediate from the theorem. □

Corollary 9.5. *Let $K = \text{Mod}\,\rho$. Then the following are equivalent*:

i. $\mathfrak{A} \subseteq \mathfrak{B}$ *and* $\mathfrak{A} \in K$ *implies* $\mathfrak{B} \in K$.
ii. $K = \text{Mod}\,\sigma$ *for some existential assertion σ.*

PROOF: Suppose $K = \text{Mod}\,\rho$ and condition i holds. Then $\tilde{K} = \text{Mod}\,\neg\rho$, and $\tilde{K}$ is closed under substructure. Hence by the preceding corollary, $\tilde{K} = \text{Mod}\,\sigma'$ for some universal σ'; say σ' is $\forall v_0 \cdots v_{n-1}\psi$, where ψ is open. Letting σ be $\exists v_0 \cdots v_{n-1} \neg\psi$, we have $\vDash \neg\sigma' \leftrightarrow \sigma$. Hence $K = \text{Mod}\,\sigma$, and σ is existential, as needed. □

A completely algebraic characterization of elementary classes $\text{Mod}\,\Sigma$ having the property that whenever $\mathfrak{B} \in \text{Mod}\,\Sigma$ and $\mathfrak{B} \subseteq \mathfrak{A}$ then $\mathfrak{A} \in \text{Mod}\,\Sigma$ is considerably more difficult than Theorem 9.3, and will not be discussed. Instead we consider a version of Theorem 9.3 for τK without function symbols in which the proviso requiring K to be an elementary class is dropped in favor of purely algebraic conditions. For this we need several lemmas.

Lemma 9.6. *For every $\mathfrak{A}$ having a finite type and a finite universe, there is an existential assertion $\sigma_\mathfrak{A}$ such that for all $\mathfrak{B}$, $\mathfrak{B} \vDash \sigma_\mathfrak{A}$ iff $\mathfrak{A} \subseteq \mathfrak{B}$.*

PROOF: The finiteness of $|\mathfrak{A}|$ and $\tau\mathfrak{A}$ implies that $\mathcal{D}\,\mathfrak{A}$ is finite; say $\mathcal{D}\,\mathfrak{A} = \{\psi_0,\ldots,\psi_{n-1}\}$. Let $\{k_0,\ldots,k_{i-1}\} = \tau\mathcal{D}\,\mathfrak{A} - \tau\mathfrak{A}$, and let ψ be the result of replacing k_i by v_i in $\psi_0 \wedge \cdots \wedge \psi_{n-1}$. Clearly $\mathfrak{B} \vDash \exists v_0 \cdots v_{i-1}\psi$ if some expansion of $\mathfrak{B}$ is a model of $\mathcal{D}\,\mathfrak{A}$. The lemma follows by an application of Theorem 4.17. □

Lemma 9.7. *Let $\mathcal{C}$ be a set of substructures of $\mathfrak{B}$. Then $\bigcap \{|\mathfrak{A}| : \mathfrak{A} \in \mathcal{C}\}$ is either empty or a substructure of $\mathfrak{B}$.*

3.9 The Prefix Problem

The proof follows easily from the definition of substructure.

Definition 9.8. Let X be a non-empty subset of $\mathfrak{B}$. Let $\mathcal{C} = \{\mathfrak{A} : X \subseteq |\mathfrak{A}|$ and $\mathfrak{A} \subseteq \mathfrak{B}\}$. Define $S(\mathfrak{B}, X)$ to be that substructure of $\mathfrak{B}$ whose universe is $\bigcap \{|\mathfrak{A}| : \mathfrak{A} \in \mathcal{C}\}$.

Of course $X \subseteq S(\mathfrak{B}, X)$, and if $X \subseteq |\mathfrak{C}|$, where $\mathfrak{C} \subseteq \mathfrak{B}$, then $S(\mathfrak{B}, X) \subseteq \mathfrak{C}$.

For the next theorem, we need to observe that if $\tau\mathfrak{B}$ is finite and contains no function symbols and X is a finite subset of $\mathfrak{B}$, then $S(\mathfrak{B}, X)$ is finite. Indeed, $|S(\mathfrak{B}, X)| = X \cup \{c_\alpha^\mathfrak{B} : c_\alpha \in \tau\mathfrak{B}\}$.

Theorem 9.9. *Let K be a class of structures of some fixed finite type not containing function symbols. Then $K = \text{Mod}\,\Sigma$ for some set Σ of universal assertions iff the following conditions are met:*

i. $\mathfrak{A} \cong \mathfrak{B}$ and $\mathfrak{B} \in K$ implies $\mathfrak{A} \in K$.
ii. $\mathfrak{A} \subseteq \mathfrak{B}$ and $\mathfrak{B} \in K$ implies $\mathfrak{A} \in K$.
iii. $\{S(\mathfrak{B}, X) : X \subseteq |\mathfrak{B}|, 0 < cX < \omega\} \subseteq K$ implies $\mathfrak{B} \in K$.

PROOF: Suppose $K = \text{Mod}\,\Sigma$, where Σ is a set of universal sentences. Theorem 4.19 tells us that condition i is satisfied. Condition ii is a consequence of the preceding theorem. Now suppose that $\{S(\mathfrak{B}, X) : X \subseteq |\mathfrak{B}|, 0 < cX < \omega\} \subseteq K$. Let $\sigma \in \Sigma$; say $\sigma = \forall u_0 \cdots u_{n-1} \psi$, where ψ is open. Let $z \in {}^{\text{Vbl}}|\mathfrak{B}|$. Let $\mathfrak{A} = \text{Sub}(\mathfrak{B}, X)$, where $X = \{z(u_0), \ldots, z(u_{n-1})\}$. $\mathfrak{A} \in \text{Mod}\,K$, and so $\mathfrak{A} \vDash \sigma$. Hence $\mathfrak{A} \vDash \psi\langle z \rangle$, and so by Lemma 9.1, $\mathfrak{B} \vDash \psi\langle z \rangle$. This shows that $\mathfrak{B} \in K$ and so condition iii holds.

For the converse, assume i, ii, and iii. Let $\Gamma = \text{Th}\,K$. We claim that $K = \text{Mod}\,\Gamma$. Clearly, $K \subseteq \text{Mod}\,\Gamma$. To prove the reverse inclusion we take $\mathfrak{B} \notin K$ with $\tau\mathfrak{B} = \tau K$. Since K satisfies condition iii, there must be a finite subset X of $|\mathfrak{B}|$ such that $\text{Sub}(\mathfrak{B}, X) \notin K$. Let $\mathfrak{A} = \text{Sub}(\mathfrak{B}, X)$. Since $\tau\mathfrak{A}$ is finite and does not contain function symbols, $\mathfrak{A}$ is finite. By i and ii, no $\mathfrak{C} \in K$ contains a substructure isomorphic to $\mathfrak{A}$. So by Lemma 9.6, $\mathfrak{C} \vDash \neg \sigma_\mathfrak{A}$ for all $\mathfrak{C} \in K$. Hence $\neg\sigma_\mathfrak{A} \in \Gamma$. Since $\mathfrak{B} \vDash \sigma_\mathfrak{A}$ (again using Lemma 9.6), we have $\mathfrak{B} \notin \text{Mod}\,\Gamma$. Hence $K \supseteq \text{Mod}\,\Gamma$, and so $K = \text{Mod}\,\Gamma$. Applying Theorem 9.3, we see that $K = \text{Mod}\,\Sigma$ for some set Σ of universal assertions. $\square$

Lemma 9.10. *Let $c \notin \tau\Sigma \cup \tau\varphi$. Then*

$$\vDash \Sigma \to \forall u \varphi \quad \text{iff} \quad \vDash \Sigma \to \varphi\binom{u}{c}.$$

PROOF: Suppose $\vDash \Sigma \to \forall u\varphi$. Let $\mathfrak{A}^+ \in \text{Mod}\,\Sigma$, where $\tau\mathfrak{A}^+ \supseteq \tau\Sigma \cup \tau\varphi\binom{u}{c}$. Then $\mathfrak{A}^+ = (\mathfrak{A}, c^{\mathfrak{A}^+})$ for some $\mathfrak{A}$ of type $\tau\Sigma \cup \tau\varphi$ and $\mathfrak{A} \in \text{Mod}\,\Sigma$. Hence $\mathfrak{A} \vDash \forall u\varphi$, and so $\mathfrak{A}^+ \vDash \varphi\binom{u}{c}$. This shows that $\vDash \Sigma \to \varphi\binom{u}{c}$.

For the converse, suppose $\vDash \Sigma \to \varphi\binom{u}{c}$ and $\mathfrak{A} \in \text{Mod}\,\Sigma$. Let $a \in \mathfrak{A}$, and form $\mathfrak{A}^+ = (\mathfrak{A} \restriction \tau\Sigma \cup \tau\varphi, c^{\mathfrak{A}^+})$, where $c^{\mathfrak{A}^+} = a$. Then $\mathfrak{A}^+ \in \text{Mod}\,\Sigma$, so $\mathfrak{A}^+ \vDash \varphi\binom{u}{c}$. Hence, $\mathfrak{A} \vDash \varphi\langle z\binom{u}{a}\rangle$ for all a, so $\mathfrak{A} \vDash \forall u\varphi$, as needed. $\square$

Definition 9.11.

i. An $\forall\exists$-formula is a formula of the form $\forall u_0,\ldots,u_{n-1}\exists w_0,\ldots,w_{n-1}\varphi$ where φ is open.

ii. Let K be a class of structures, and define $\mathrm{Th}_{\forall\exists}K=\{\sigma:\sigma\in\mathrm{Th}\,K$ and σ is a $\forall\exists$-formula$\}$.

Definition 9.12. We write $\mathfrak{A}\propto_\Gamma\mathfrak{B}$ if

i. $\mathfrak{A}\subseteq\mathfrak{B}$,
ii. for all $\varphi\in\Gamma$ and all $z\in{}^{\mathrm{Vbl}}|\mathfrak{A}|$, $\mathfrak{A}\vDash\varphi\langle z\rangle$ iff $\mathfrak{B}\vDash\varphi\langle z\rangle$.

Notice that $\mathfrak{A}\subseteq\mathfrak{B}$ iff $\mathfrak{A}\propto_\varnothing\mathfrak{B}$ and that $\mathfrak{A}\prec\mathfrak{B}$ iff $\mathfrak{A}\propto_\Gamma\mathfrak{B}$ where Γ is the set of all formulas of type $\tau\mathfrak{A}$. We write $\mathfrak{A}\propto_{\mathbf{v}}\mathfrak{B}$ for $\mathfrak{A}\propto_\Gamma\mathfrak{B}$ when Γ is the set of universal formulas of type $\tau\mathfrak{A}$.

Lemma 9.13. *If $\mathfrak{A}\propto_{\mathbf{v}}\mathfrak{B}$, then there is a $\mathfrak{C}$ such that $\mathfrak{B}\subseteq\mathfrak{C}$ and $\mathfrak{A}\prec\mathfrak{C}$.*

PROOF: Let $\mathfrak{B}^+=(\mathfrak{B},c_b^{\mathfrak{B}^+})_{b\in|\mathfrak{B}|}$, where $c_b\notin\tau\mathfrak{B}$ and $b=c_b^{\mathfrak{B}^+}$. Let $\mathfrak{A}^+=(\mathfrak{A},c_b^{\mathfrak{A}^+})_{b\in|\mathfrak{A}|}$, where $c_b^{\mathfrak{A}^+}=c_b^{\mathfrak{B}^+}$ for $b\in|\mathfrak{A}|$. Let $\Sigma=\mathrm{Th}\,\mathfrak{A}^+\cup\Delta$, where Δ is the diagram of $\mathfrak{B}$ determined by $\mathfrak{B}^+$. By Theorems 4.17 and 8.5 it is enough to prove that Σ has a model. Suppose not. Then some finite subset has no model, say $\{\sigma_0,\ldots,\sigma_{p-1},\rho_0,\ldots,\rho_{q-1}\}$, where $\sigma_i\in\mathrm{Th}\,\mathfrak{A}^+$ and $\rho_j\in\Delta$ for each $i<p$, $j<q$. Let σ be $\sigma_0\wedge\cdots\wedge\sigma_{p-1}$ and ρ be $\rho_0\wedge\cdots\wedge\rho_{q-1}$. Then $\vDash\sigma\to\neg\rho$. We can write ρ as $\rho(c_{a_0},\ldots,c_{a_{n-1}},c_{b_0},\ldots,c_{b_{m-1}})$, where $a_0,\ldots,a_{n-1}\in|\mathfrak{A}|$, $b_0,\ldots,b_{m-1}\in|\mathfrak{B}|-|\mathfrak{A}|$, and every c_b occurring in ρ is some c_{a_i} or some c_{b_j}. Using Lemma 9.10, we see that $\vDash\sigma\to\forall u_0,\ldots,u_{m-1}\neg\rho(c_{a_0},\ldots,c_{a_{n-1}},u_0,\ldots,u_{m-1})$, where $u_0,\ldots,u_{m-1}$ are distinct variables not occurring in ρ. Since $\mathfrak{A}^+\vDash\sigma$ and $\mathfrak{A}\propto_{\mathbf{v}}\mathfrak{B}$, we have $\mathfrak{B}^+\vDash\forall u_0,\ldots,u_{m-1}\neg\rho(c_{a_0},\ldots,c_{a_{m-1}},u_0,\ldots,u_{m-1})$, contradicting $\mathfrak{B}^+\vDash\rho(c_{a_0},\ldots,c_{a_{n-1}},c_{b_0},\ldots,c_{b_{m-1}})$. Hence Σ is finitely satisfiable and so has a model, as we needed to show. □

Lemma 9.14. *Let K be an elementary class, and let $\mathfrak{A}\in\mathrm{Mod}\,\mathrm{Th}_{\forall\exists}K$. Then there is a $\mathfrak{B}\in K$ such that $\mathfrak{A}\propto_{\mathbf{v}}\mathfrak{B}$.*

PROOF: Let $\mathfrak{A}^+=(\mathfrak{A},c_a^{\mathfrak{A}^+})_{a\in|\mathfrak{A}|}$, where $c_a\notin\tau\mathfrak{A}$ and $c_a^{\mathfrak{A}^+}=a$ for all $a\in|\mathfrak{A}|$. Let $\mathfrak{D}^{\mathbf{v}}\mathfrak{A}$ be the set of all universal formulas true in $\mathfrak{A}^+$. Let $\Sigma=\mathrm{Th}\,K\cup\mathfrak{D}^{\mathbf{v}}\mathfrak{A}$. We claim that every finite subset of Σ has a model. For if not, then some finite subset Σ' of Σ has no model. Let $\Sigma'=\{\sigma_0,\ldots,\sigma_{n-1},\rho_0,\ldots,\rho_{m-1}\}$, where each $\sigma_i\in\mathrm{Th}\,K$, and each $\rho_j\in\mathfrak{D}^A\mathfrak{A}$. Let ρ_i be $\forall u_{i,0}\cdots u_{i,n_i-1}\varphi_i$. Using Lemma 5.4, we can assume that $u_{i,j}\neq u_{k,l}$ when $(i,j)\neq(k,l)$. It is easy to see that $\rho_0\wedge\cdots\wedge\rho_{m-1}$ is equivalent to $\forall u_0,\ldots,u_l\varphi$ where each $u_{i,j}$ is some u_k and φ is $\varphi_0\wedge\cdots\wedge\varphi_{m-1}$. Letting σ be $\sigma_0\wedge\cdots\wedge\sigma_{n-1}$, we have $\vDash\sigma\to\neg\forall u_0,\ldots,u_l\varphi$, or equivalently $\vDash\sigma\to\exists u_0,\ldots,u_{l-1}\neg\varphi$. Let $\{c_{a_0},\ldots,c_{a_{r-1}}\}=\tau\varphi-\tau K$, and let $w_0,\ldots,w_{r-1}$ be variables not occurring in $\exists u_0,\ldots,u_l\neg\varphi$. Let φ' be the result of replacing c_{a_i} by w_i throughout φ for each $i<r$. Applying Lemma 9.10, we have $\vDash\sigma\to\forall w_0\cdots w_{r-1}\exists u_0\cdots u_{l-1}\neg\varphi'$. Since $\sigma\in\mathrm{Th}\,K$,

$\forall w_0 \cdots v_{r-1} \exists u_0 \cdots u_{l-1} \neg \varphi' \in \text{Th}_{\forall\exists} K$ and so is true in $\mathfrak{A}$. But this contradicts $\mathfrak{A}^+ \vDash \forall u_0 \cdots u_{i-1} \varphi$. Hence Σ is finitely satisfiable and so has a model $\mathfrak{B}^+$. Now let $\mathfrak{B} = \mathfrak{B}^+ \upharpoonright \tau \mathfrak{A}$. Then $\mathfrak{A} \propto_{\forall} \mathfrak{B}$ and $\mathfrak{B} \in K$. □

Theorem 9.15. *Let K be an elementary class. Then the following two conditions are equivalent:*

i. *If for each $i \in \omega$, $\mathfrak{A}_i \subseteq \mathfrak{A}_{i+1}$ and $\mathfrak{A}_i \in K$, then $\bigcup_{i \in \omega} \mathfrak{A}_i \in K$.*
ii. *$K = \text{Mod}\,\Gamma$ for some set Γ of $\forall\exists$-assertions.*

PROOF: ii implies i: Let $\mathfrak{A} = \bigcup_{i \in \omega} \mathfrak{A}_i$, where for each $i \in \omega$, $\mathfrak{A}_i \in K$ and $\mathfrak{A}_i \subseteq \mathfrak{A}_{i+1}$. We assume that $K = \text{Mod}\,\Gamma$, where Γ is a set of $\forall\exists$-assertions. Let $\sigma \in \Gamma$; say σ is $\forall u_0 \cdots u_{n-1} \exists w_0 \cdots w_{m-1} \varphi$, where φ is open. Choose $a_0, \ldots, a_{n-1} \in |\mathfrak{A}|$. Then there is an i^* such that $a_0, \ldots, a_{n-1} \in |\mathfrak{A}_{i^*}|$. Since $\mathfrak{A}_{i^*} \in \text{Mod}\,\Gamma$, there are $b_0, \ldots, b_{m-1} \in |\mathfrak{A}_{i^*}|$ such that $\mathfrak{A}_{i^*} \vDash \varphi\langle z \rangle$, where $z(u_j) = a_j$ and $z(w_l) = b_l$ for every $j < n$, $l < m$. Hence $\mathfrak{A} \vDash \varphi\langle z \rangle$ by Lemma 9.1. Thus for every $a_0, \ldots, a_{n-1} \in |\mathfrak{A}|$, there are $b_0, \ldots, b_{m-1} \in |\mathfrak{A}|$ such that $\mathfrak{A} \vDash \varphi(a_0, \ldots, a_{n-1}, b_0, \ldots, b_{m-1})$, i.e., $\mathfrak{A} \vDash \sigma$ for all $\sigma \in \Gamma$, as needed.

We now assume i and prove ii: Let $\Gamma = \text{Th}_{\forall\exists} K$. Clearly, $\text{Mod}\,\Gamma \supseteq K$, and we need only show that $\text{Mod}\,\Gamma \subseteq K$. Let $\mathfrak{A}_0 \in \text{Mod}\,\Gamma$. Using Lemmas 9.13 and 9.14 alternately gives a sequence $\mathfrak{A}_0, \mathfrak{A}_1, \mathfrak{A}_2, \ldots$ such that for all $n \in \omega$,

i. $\mathfrak{A}_{2n} \propto_{\forall} \mathfrak{A}_{2n+1}$ and $\mathfrak{A}_{2n+1} \in K$,
ii. $\mathfrak{A}_{2n+1} \subseteq \mathfrak{A}_{2n+2}$ and $\mathfrak{A}_{2n} \propto \mathfrak{A}_{2n+2}$.

Notice that $\bigcup_{n \in \omega} \mathfrak{A}_n = \bigcup_{n \in \omega} \mathfrak{A}_{2n} = \bigcup_{n \in \omega} \mathfrak{A}_{2n+1}$. Let $\mathfrak{B} = \bigcup_{n \in \omega} \mathfrak{A}_n$. Since we are assuming condition i, we have $\mathfrak{B} \in K$. Also, $\mathfrak{A}_0 \propto \mathfrak{B}$ by Theorem 8.9, and so $\mathfrak{A}_0 \equiv \mathfrak{B}$. Since $\mathfrak{B}$ is an elementary class, this means that $\mathfrak{A}_0 \in K$. Hence $\text{Mod}\,\Gamma \subseteq K$ and so $\text{Mod}\,\Gamma = K$, as we needed to show.

EXERCISES FOR §3.9

1. Let σ be the conjunction of the following assertions:
   i. $\forall uvw[(u \cdot v) \cdot w = u \cdot (v \cdot w)]$,
   ii. $\exists u \forall v[(u \cdot v) = v]$,
   iii. $\forall u \exists v \forall w[(u \cdot v) \cdot w = w]$.
   Then $(G, \cdot) \vDash \sigma$ iff $(G, \cdot)$ is a group with operation '$\cdot$'. Show that there is no universal assertion equivalent to σ.

2. Let ρ be the conjunction of the following:
   i. $\forall uvw[(u \cdot v) \cdot w = u \cdot (v \cdot w)]$,
   ii. $\forall v[e \cdot v = v]$,
   iii. $\forall v[v \cdot v^{-1} = e]$.
   Then $(G, \cdot, {}^{-1}, e)$ is a group iff it is a model of ρ. Clearly ρ is equivalent to a universal assertion. Reconcile this with Exercise 1.

3. Is the class K of discrete orderings a universal class, i.e., is there a universal set of sentences Σ such that $\text{Mod}\,\Sigma = K$?

4. Find two assertions σ and ρ such that
   i. $\mathfrak{A} \vDash \sigma$ iff $\mathfrak{A} \restriction \tau\sigma$ is a field,
   ii. $\mathfrak{B} \vDash \rho$ iff $\mathfrak{B} \restriction \tau\rho$ is a field,
   iii. σ is universal, and
   iv. ρ is not equivalent to a universal assertion.

5. Is the class of discrete orderings an $\forall\exists$-class? Is the class of dense orderings $\forall\exists$?

6. Show that if $K = \text{Mod}\,\sigma$ and if K is closed under the union of chains, then $K - \text{Mod}\,\rho$ for some $\forall\exists$-assertion ρ.

3.10 Interpolation and Definability

Two structures $\mathfrak{A}$ and $\mathfrak{B}$ such that $\mathfrak{A} \restriction s = \mathfrak{B} \restriction s$, where $s = \tau\mathfrak{A} \cap \tau\mathfrak{B}$, can be glued together in the obvious way. The result is a structure $\mathfrak{C}$ such that $\mathfrak{C} \restriction \tau\mathfrak{A} = \mathfrak{A}$ and $\mathfrak{C} \restriction \tau\mathfrak{B} = \mathfrak{B}$. In this section we will show that '$=$' can be weakened to '$\equiv$', i.e., if $\mathfrak{A} \restriction s \equiv \mathfrak{B} \restriction s$, where $s = \tau\mathfrak{A} \cap \tau\mathfrak{B}$, then there is a $\mathfrak{C}$ such that $\mathfrak{C} \restriction \tau\mathfrak{A} \equiv \mathfrak{A}$ and $\mathfrak{C} \restriction \tau\mathfrak{B} \equiv \mathfrak{B}$.

A closely related result states that if $\vDash \sigma \to \rho$ and $s = \tau\sigma \cap \tau\rho$, then there is a δ such that $\tau\delta = s$ and $\sigma \to \delta$ and $\delta \to \rho$. This is the interpolation theorem, which will be discussed here along with a corollary on explicit definitions.

Theorem 10.1 (Consistency Lemma). *Let* $s = \tau\mathfrak{A} \cap \tau\mathfrak{B}$, *and suppose that* $\mathfrak{A} \restriction s \equiv \mathfrak{B} \restriction s$. *Then there is a* $\mathfrak{C}$ *such that*

   i. $\tau\mathfrak{C} \equiv \tau\mathfrak{A} \cup \tau\mathfrak{B}$,
   ii. $\mathfrak{A} \propto \mathfrak{C} \restriction \tau\mathfrak{A}$,
   iii. $\mathfrak{B} \equiv \mathfrak{C} \restriction \tau\mathfrak{B}$.

PROOF: Let $\mathfrak{A}$, $\mathfrak{B}$ and s satisfy the hypothesis of the theorem. We first prove the existence of a structure $\mathfrak{B}_0$ such that

$$\mathfrak{A} \restriction s \propto \mathfrak{B}_0 \restriction s \quad \text{and} \quad \mathfrak{B} \propto \mathfrak{B}_0 \restriction s.$$

Let $\Sigma = \mathfrak{D}^c\mathfrak{B} \cup \mathfrak{D}^c(\mathfrak{A} \restriction s)$ where $\mathfrak{D}^c\mathfrak{B} = \text{Th}\,\mathfrak{B}^+$, $\mathfrak{D}^c(\mathfrak{A} \restriction s) = \text{Th}\,\mathfrak{A}^+$, $\mathfrak{B}^+ = (\mathfrak{B}, c_b^{\mathfrak{B}^+})_{b \in |\mathfrak{B}|}$, $\mathfrak{A}^+ = (\mathfrak{A} \restriction s, c_a^{\mathfrak{A}^+})_{a \in |\mathfrak{A}|}$, $c_b^{\mathfrak{B}^+} = b$, $c_a^{\mathfrak{A}^+} = a$ and $c_b \neq c_a$ for all $a \in |\mathfrak{A}|$, $b \in |\mathfrak{B}|$. If Σ has no model, then, by compactness, there is some $\sigma \in \mathfrak{D}^c\mathfrak{B}$ and some $\rho \in \mathfrak{D}^c(\mathfrak{A} \restriction s)$, such that $\vDash \sigma \to \neg\,\rho$. Using Lemma 9.10 we get that $\vDash \exists \bar{u}\sigma' \to \forall \bar{w}\,(\neg \rho')$ where σ' and ρ' are the result of replacing the constant symbols c_{a_i} and c_{b_j} by variables u_i and w_j not occurring in either σ or ρ. But $\tau\rho' \subseteq s$ and $\mathfrak{B} \vDash \exists \bar{u}\sigma'$. Hence $\mathfrak{B} \vDash \forall \bar{w}\,(\neg \rho')$, and since $\mathfrak{A} \restriction s \equiv \mathfrak{B} \restriction s$, $\mathfrak{A} \vDash \forall \bar{w}\,(\neg \rho')$—a contradiction since $\rho \in \mathfrak{D}^c\mathfrak{A}$. Hence Σ has model $\mathfrak{B}^+$. By Theorem 8.5 we can satisfy (1) by taking $\mathfrak{B}_0$ to be $\mathfrak{B}^+ \restriction \tau\mathfrak{B}$.

We next show that there is an $\mathfrak{A}_0$ such that

$$\mathfrak{A} \propto \mathfrak{A}_0 \quad \text{and} \quad \mathfrak{B}_0 \restriction s \propto \mathfrak{A}_0 \restriction s. \tag{2}$$

Choose expansions $\mathfrak{A}^+ = (\mathfrak{A}, c_a^{\mathfrak{A}^+})_{a \in |\mathfrak{A}|}$ and $\mathfrak{B}^+ = (\mathfrak{B}_0 \restriction s, c_b^{\mathfrak{B}^+})_{b \in |\mathfrak{B}_0|}$ such that $c_b^{\mathfrak{B}^+} = b$ for all $b \in |\mathfrak{B}_0|$ and $c_a^{\mathfrak{A}^+} = a$ for all $a \in |\mathfrak{A}|$. Let $\Sigma = \text{Th}\,\mathfrak{A}^+ \cup \text{Th}\,\mathfrak{B}^+$. We claim that Σ has a model. If not, then by compactness there is a $\sigma \in \text{Th}\,\mathfrak{B}^+$ such that $\vdash \text{Th}\,\mathfrak{A}^+ \to \neg \sigma$. σ is of the form

$$\varphi \begin{pmatrix} u_0 \ldots u_{n-1} w_0 \ldots w_{m-1} \\ c_{a_0} \ldots c_{a_{n-1}} c_{b_0} \ldots c_{b_{m-1}} \end{pmatrix},$$

where $a_i \in |\mathfrak{A}|$ and $b_j \in |\mathfrak{B}_0| - |\mathfrak{A}|$ for $i < n, j < m$. Applying Lemma 9.10, we get that

$$\vdash \text{Th}\,\mathfrak{A}^+ \to \forall w_0 \ldots w_{m-1} \neg \varphi \begin{pmatrix} u_0 \ldots u_{n-1} \\ c_{a_0} \ldots c_{a_{n-1}} \end{pmatrix}.$$

Then $\mathfrak{A} \restriction s \vDash \forall w_0 \ldots w_{m-1} \neg \varphi \langle z \rangle$ for any $z \in \text{Vbl}|\mathfrak{A}|$ such that $z(u_i) = a_i$ for $i < n$. Since $\mathfrak{A} \restriction s \propto \mathfrak{B}_0 \restriction s$, we have $\mathfrak{B}_0 \restriction s \vDash \forall w_0 \ldots v_{m-1} \neg \varphi \langle z \rangle$. But this contradicts the fact that $\sigma \in \text{Th}\,\mathfrak{B}^+$. Hence Σ has a model $\mathfrak{A}'_0$, and by Theorem 8.5 we can take $\mathfrak{A}_0$ to be $\mathfrak{A}'_0 \restriction \tau \mathfrak{A}$.

Using (2) but substituting $\mathfrak{B}_0$ for $\mathfrak{A}$ and $\mathfrak{A}_0$ for $\mathfrak{B}_0$, we get a structure $\mathfrak{B}_1$ such that

$$\mathfrak{B}_0 \propto \mathfrak{B}_1 \quad \text{and} \quad \mathfrak{A}_0 \restriction s \propto \mathfrak{B}_1 \restriction s.$$

Continuing in this way, we generate two sequences $\{\mathfrak{A}_i : i \in \omega\}$, $\{\mathfrak{B}_i : i \in \omega\}$ such that for all $i < \omega$,

i. $\mathfrak{A}_i \propto \mathfrak{A}_{i+1}$,
ii. $\mathfrak{B}_i \propto \mathfrak{B}_{i+1}$,
iii. $\mathfrak{B}_i \restriction s \propto \mathfrak{A}_i \restriction s$ and $\mathfrak{A}_i \restriction s \propto \mathfrak{B}_{i+1} \restriction s$.

Let $\mathfrak{A}^* = \bigcup \{\mathfrak{A}_i : i \in \omega\}$, $\mathfrak{B}^* = \bigcup \{\mathfrak{B}_i : i \in \omega\}$, and $\mathfrak{D} = \bigcup \{\mathfrak{A}_i \restriction s : i \in \omega\} = \bigcup \{\mathfrak{B}_i \restriction s : i \in \omega\}$. By Theorem 8.9, $\mathfrak{A} \propto \mathfrak{A}_0 \propto \mathfrak{A}^*$ and $\mathfrak{B} \propto \mathfrak{B}_0 \propto \mathfrak{B}^*$. Hence $\mathfrak{A} \propto \mathfrak{A}^*$ and $\mathfrak{B} \propto \mathfrak{B}^*$. Moreover, $\mathfrak{A}^*$ and $\mathfrak{B}^*$ are both expansions of the *same* structure $\mathfrak{D}$. Hence we can expand $\mathfrak{A}^*$ to a structure $\mathfrak{C}$ of type $\tau \mathfrak{A} \cup \tau \mathfrak{B}$ such that $\mathfrak{C} \restriction \tau \mathfrak{B} = \mathfrak{B}^*$. This $\mathfrak{C}$ obviously satisfies the conclusion of the theorem. □

Corollary 10.2. *Let Σ_1 and Σ_2 be sets of assertions such that*

i. $\text{Mod}\,\Sigma_1 \neq \emptyset$,
ii. $\text{Mod}\,\Sigma_2 \neq \emptyset$,
iii. $\{\sigma : \vdash \Sigma_1 \cup \Sigma_2 \to \sigma \text{ and } \tau\sigma \subseteq \tau\Sigma_1 \cap \tau\Sigma_2\}$ *is complete.*

Then $\text{Mod}(\Sigma_1 \cup \Sigma_2) \neq \emptyset$.

PROOF: Let $\mathfrak{A} \in \text{Mod}\,\Sigma_1$, and let $\mathfrak{B} \in \text{Mod}\,\Sigma_2$. $\mathfrak{A}$ and $\mathfrak{B}$ satisfy the hypotheses of the theorem, and so there is a $\mathfrak{C}$ such that $\mathfrak{A} \propto \mathfrak{C} \restriction \tau \mathfrak{A}$ and $\mathfrak{B} \propto \mathfrak{C} \restriction \tau \mathfrak{B}$. Hence $\mathfrak{C} \in \text{Mod}(\Sigma_1 \cup \Sigma_2)$. □

Definition 10.3. Say that σ is an *interpolant* for $\sigma_1 \to \sigma_2$ if

i. $\tau\sigma \subseteq \tau\sigma_1 \cap \tau\sigma_2$,
ii. $\vDash \sigma_1 \to \sigma$,
iii. $\vDash \sigma \to \sigma_2$.

Theorem 10.4 (Interpolation Theorem). *If $\vDash \sigma_1 \to \sigma_2$, then $\sigma_1 \to \sigma_2$ has an interpolant.*

PROOF: Suppose that $\sigma_1 \to \sigma_2$ has no interpolant. Let $s = \tau\sigma_1 \cap \tau\sigma_2$. Let $\Gamma = \{\rho : \tau\rho \subseteq s \text{ and } \vDash \sigma_1 \to \rho\}$. We claim that $\Gamma \cup \{\neg \sigma_2\}$ has a model. For if not, compactness implies the existence of a finite subset $\{\rho_0, \ldots, \rho_{n-1}\}$ of Γ such that $\vDash \rho_0 \wedge \cdots \wedge \rho_{n-1} \to \sigma_2$. Since $\vDash \sigma_1 \to \rho_0 \wedge \cdots \wedge \rho_{n-1}$, this means that $\sigma_1 \to \sigma_2$ has an interpolant, contradicting our assumption. Hence $\Gamma \cup \{\neg \sigma_2\}$ has a model, and so there is a complete set of assertions $\Gamma' \supseteq \Gamma \cup \{\neg \sigma_2\}$ such that Γ' has a model. Let $\Sigma = \{\gamma : \gamma \in \Gamma' \text{ and } \tau\gamma = s\}$. We claim that

i. Σ is complete,
ii. $\Sigma \cup \{\neg \sigma_2\}$ has a model,
iii. $\Sigma \cup \{\sigma_1\}$ has a model.

The first two clauses are obvious. If clause iii is false, then there are $\gamma_0, \ldots, \gamma_{n-1} \in \Sigma$ such that $\vDash \sigma_1 \to \neg(\gamma_0 \wedge \cdots \wedge \gamma_{n-1})$. But then Σ has no model, since $\neg(\gamma \wedge \cdots \wedge \gamma_{n-1}) \in \Gamma$ and hence $\{\neg(\gamma_0 \cdots \gamma_{n-1}), \gamma_0, \ldots, \gamma_{n-1}\} \subseteq \Sigma$, a contradiction. Hence clause iii holds also.

Taking Σ_1 to be $\Sigma \cup \{\sigma_1\}$ and Σ_2 to be $\Sigma \cup \{\neg \sigma_2\}$, we apply Corollary 10.2 and conclude that $\Sigma_1 \cup \Sigma_2$ has a model. But this means that $\sigma_1 \wedge \neg \sigma_2$ has a model, and so $\sigma_1 \to \sigma_2$ is not valid. Hence if $\vDash \sigma_1 \to \sigma_2$, then $\sigma_1 \to \sigma_2$ must have an interpolant. □

In other proofs of the interpolation theorem the interpolant σ for $\sigma_1 \to \sigma_2$ is given constructively in terms of σ_1 and σ_2. In other words, if one is explicitly given a valid implication $\sigma_1 \to \sigma_2$, an algorithm may be followed that will explicitly yield an interpolant. Later we shall obtain such an algorithm, but one that is less expedient than that usually given.

Definition 10.5.

i. Let R be a relation symbol. Say that Σ *implicitly defines* R if whenever $\mathfrak{A} = (\mathfrak{C}, R^{\mathfrak{A}})$ and $\mathfrak{B} = (\mathfrak{C}, R^{\mathfrak{B}})$ and $\mathfrak{A}, \mathfrak{B} \in \text{Mod}\,\Sigma$, then $\mathfrak{A} = \mathfrak{B}$; i.e., any structure $\mathfrak{C}$ of type $\tau\sigma - \{R\}$ has at most one expansion in $\text{Mod}\,\Sigma$.
ii. If R is an n-ary relation symbol and φ is a formula of type $\tau\Sigma - \{R\}$, then we say that φ *is an explicit definition of R with respect to Σ* if whenever $\mathfrak{A} \in \text{Mod}\,\Sigma$ and $\mathfrak{A} = (\mathfrak{C}, R^{\mathfrak{A}})$, then $R^{\mathfrak{A}} = \{(a_0, \ldots, a_{n-1}) : \mathfrak{C} \vDash \varphi(a_0, \ldots, a_{n-1})\}$.

EXAMPLE. Let $\Sigma = \text{Th}\,K$, where K is the class of all structures $\mathfrak{A} = (A, R^{\mathfrak{A}}, S^{\mathfrak{A}})$ such that $R^{\mathfrak{A}}$ is a discrete ordering of A and for all $x, y \in A$,

$S^{\mathfrak{A}}xy$ iff y is the immediate successor of x. Clearly Σ implicitly defines S. Moreover, S is explicitly defined with respect to Σ by the formula $v_0 < v_1 \wedge \forall v_2 [v_0 < v_2 \to v_1 \leqslant v_2]$.

Let R and S be n-placed relation symbols. By $\varphi\left(\dfrac{R}{S}\right)$ we mean the formula we get from φ by replacing each atomic formula $Rt_0,\ldots,t_{n-1}$ occurring in it with $St_0,\ldots,t_{n-1}$. For example if σ is $\exists v_0 v_1 [Rf(v_1), v_0 \to \forall v_2 [Rv_2, v_2 \vee v_0 \approx v_2]]$, then $\sigma\left(\dfrac{R}{S}\right)$ is $\exists v_0 v_1 [Sf(v_1), v_0 \to \forall v_2 [Sv_2, v_2 \vee v_0 \approx v_2]]$. By $\Gamma\left(\dfrac{R}{S}\right)$ we mean

$$\left\{\varphi\left(\dfrac{R}{S}\right) : \varphi \in \Gamma\right\}.$$

Lemma 10.6.

i. σ *implicitly defines the n-relation R iff whenever S is an n-relation symbol not occurring in σ, then*

$$\models \sigma \wedge \sigma\left(\dfrac{R_{n,\alpha}}{S_{n,\alpha}}\right)$$
$$\to \forall v_0 \ldots v_{n-1} [Rv_0 \ldots v_{n-1} \leftrightarrow Sv_0 \ldots v_{n-1}].$$

ii. *Let φ be a formula of type $\tau\sigma - \{R\}$. Then φ defines R explicitly with respect to σ iff*

$$\models \sigma \to \forall v_0 \ldots v_{n-1} [\varphi(v_0 \ldots v_{n-1}) \leftrightarrow Rv_0 \ldots v_{n-1}].$$

The proof of this theorem is straightforward and is left as an exercise.

Lemma 10.7.

i. *If Σ implicitly defines R, then some finite subset of Σ implicitly defines R.*
ii. *If φ explicitly defines R with respect to Σ, then φ explicitly defines R with respect to some finite subset of Σ.*

PROOF OF I. Suppose that no finite subset of Σ implicitly defines R. Then by Lemma 10.6, for each finite subset Σ' of Σ there is a model of

$$\Sigma' \cup \Sigma'\left(\dfrac{R}{S}\right) \cup \{\neg \forall u_0 \ldots u_{n-1}(Ru_0 \ldots u_{n-1} \leftrightarrow Su_0 \ldots u_{n-1})\}.$$

So by compactness there is a model $\mathfrak{C}$ of

$$\Sigma \cup \Sigma\left(\dfrac{R}{S}\right) \cup \{\neg \forall u_0 \ldots u_{n-1}(Ru_0 \ldots u_{n-1} \leftrightarrow Su_0 \ldots u_{n-1}).$$

Let $\mathfrak{C}' = \mathfrak{C} \upharpoonright \tau\Sigma - \{R\}$. Then $(\mathfrak{C}', R^{\mathfrak{C}}), (\mathfrak{C}', S^{\mathfrak{C}}) \in \text{Mod } \mathfrak{C}$ and $R^{\mathfrak{C}} \neq S^{\mathfrak{C}}$. Hence R is not implicitly defined by Σ. This proves part i. □

PROOF OF II. If φ does not explicitly define R with respect to some finite subset of Σ, then by Lemma 10.6, every finite subset Σ' of Σ has a model that is also a model of $\neg \forall u_0 \ldots u_{n-1}(\varphi u_0 \ldots u_{n-1} \leftrightarrow Ru_0 \ldots u_{n-1})$. Hence by compactness, Σ has a model of $\neg \forall u_0 \ldots u_{n-1}(\varphi u_0 \ldots u_{n-1} \leftrightarrow Ru_0 \ldots u_{n-1})$, which means that φ does not define R with respect to Σ, thus proving part ii. □

Theorem 10.8 (Definability Theorem). *If Σ implicitly defines R, then R is explicitly definable with respect to Σ.*

PROOF: By Lemma 10.7 it is enough to prove the theorem for $\Sigma = \{\sigma\}$. Suppose that R is implicitly defined by σ. Then by Lemma 10.6i,

$$\vdash \sigma \wedge \sigma\binom{R}{S} \to \forall v_0 \ldots v_{n-1}[Rv_0 \ldots v_{n-1} \leftrightarrow Sv_0 \ldots v_{n-1}].$$

Let $c_0, \ldots, c_{n-1}$ be constant symbols that do not occur in σ. Then

$$\vdash \sigma \wedge \sigma\binom{R}{S} \to [Rc_0, \ldots, c_{n-1} \leftrightarrow Sc_0 \ldots c_{n-1}],$$

and so

$$\vdash \sigma \wedge Rc_0 \ldots c_{n-1} \to \left[\sigma\binom{R}{S} \to Sc_0 \ldots c_{n-1}\right].$$

Let $\varphi(c_{i_0} \ldots c_{i_{n-1}})$ be an interpolant for this implication (an application of Theorem 10.4). Notice that $\tau\varphi(v_0 \ldots v_{n-1}) \subseteq \tau\sigma - \{R, S\}$. We claim that $\varphi(v_0 \ldots v_{n-1})$ is an explicit definition of R.

Since $\vdash \sigma \wedge Rc_{i_0} \ldots c_{i_{n-1}} \to \varphi(c_{i_0} \ldots c_{i_{n-1}})$, we have $\vdash \sigma \to [Rc_{i_0} \ldots c_{i_{n-1}} \to \varphi c_{i_0} \ldots c_{i_{n-1}}]$. From

$$\vdash \varphi c_{i_0} \ldots c_{i_{n-1}} \to \left[\sigma\binom{R}{S} \to Sc_{i_0} \ldots c_{i_{n-1}}\right]$$

we get $\vdash \varphi c_{i_0} \ldots c_{i_{n-1}} \to [\sigma \to Rc_{i_0} \ldots c_{i_{n-1}}]$ (neither R nor S occur in φ). Hence $\vdash \sigma \to [\varphi c_{i_0} \ldots c_{i_{n-1}} \to Rc_{i_0} \ldots c_{i_{n-1}}]$. With the above, this gives $\vdash \sigma \to [Rc_{i_0} \ldots c_{i_{n-1}} \leftrightarrow \varphi c_{i_0} \ldots c_{i_{n-1}}]$. Applying Lemma 9.10, $\vdash \sigma \to \forall v_0 \ldots v_{n-1}[Rv_0 \ldots v_{n-1} \leftrightarrow \varphi v_0 \ldots v_{n-1}]$. So by clause ii of Lemma 10.6, we see that φ defines R explicitly in terms of σ. □

EXERCISES FOR §3.10

1. Prove Lemma 10.6.

2. It is not necessary to prove the interpolation lemma via the consistency lemma. In fact, there are direct proofs of the interpolation lemma that yield considerably more information about the form of the interpolant relative to the form of the valid implication. For this reason it is convenient to have a proof of the consistency lemma assuming the interpolation theorem. Give such a proof.

3. Let $T = \text{Th}\langle \mathbf{N}, + \rangle$. Is there a formula of type τT such that $\langle \mathbf{N}, + \rangle \vDash \varphi(n, m)$ iff $n < m$?

4. Let P be the set of positive integers. Let $T = \text{Th}\langle \mathbf{I}, +, \cdot, P \rangle$. Show that T defines P explicitly. (*Hint*: Every natural number is the sum of four squares.) Do the same for $T = \text{Th}\langle \mathbf{R}, +, \cdot, P \}$, where P is the set of positive numbers.

5. Show that if Σ defines R explicitly, then Σ defines R implicitly.
6. Let $T = \mathrm{Th}\langle \mathbf{I}, +, -, \cdot \rangle$. Let σ be the conjunction of the following assertions:

$$P(v) \lor P(-v),$$
$$\neg [P(v) \land P(-v)],$$
$$P(u) \land P(v) \to P(u \cdot v) \land P(u+v).$$

Does $T \cup \{\sigma\}$ implicitly define P?

7. Answer the same question as in Exercise 6 except that $T = \mathrm{Th}\langle \mathbf{R}, +, -, \cdot \rangle$.

8. Say that Σ is *existentially complete* if whenever σ is an existential assertion, then either $\sigma \in \Sigma$ or $\neg \sigma \in \Sigma$. Show that if Σ is existentially complete and if $\mathfrak{A} \subseteq \mathfrak{B}_1$ and $\mathfrak{A} \subseteq \mathfrak{B}_2$ and $\mathfrak{A}, \mathfrak{B}_1, \mathfrak{B}_2 \in \mathrm{Mod}\,\Sigma$, then there is a $\mathfrak{C} \in \mathrm{Mod}\,\Sigma$ and isomorphisms i and j such that

$$i : \mathfrak{B}_1 \to \mathfrak{C}_1 \subseteq \mathfrak{C},$$
$$j : \mathfrak{B}_2 \to \mathfrak{C}_2 \subseteq \mathfrak{C},$$

and

$$i \restriction \mathfrak{A} = j \restriction \mathfrak{A}.$$

9. Suppose that $\mathfrak{A} = (A, B, C, R, S)$, where B and C are unary and $(B, R) \equiv (C, S)$. Show that there is some $\mathfrak{A}' \equiv (A', B', C', R', S')$ such that $\mathfrak{A} \prec \mathfrak{A}'$ and $(B, R) \cong (B', S')$.

3.11 Herbrand's Theorem

Suppose that $t_1, \ldots, t_n$ are constant terms (i.e., terms without free variables) and that $\mathfrak{A} \vDash \varphi(t_1) \lor \cdots \lor \varphi(t_n)$. Then surely $\mathfrak{A} \vDash \exists v \varphi v$. Much more interesting is the fact that if $\exists v \varphi$ is valid and φ is open and has a constant symbol in its type, then there are constant terms $t_1, \ldots, t_n$ such that $\varphi(t_1) \lor \cdots \lor \varphi(t_n)$ is valid. This is known as Herbrand's theorem and is the main topic of this section. Since there is an effective procedure for checking the validity of open sentences, this leads to an effective method of enumerating the valid sentences of L and an axiom system in the next section that is complete and correct for the valid L-sentences.

Definition 11.1.

i. We call t a *constant term* if no variable occurs in it.
ii. Let $\varphi(u_0 \ldots u_{n-1})$ be an open formula. Say that $\varphi(t_0 \ldots t_{n-1})$ is a *substitution instance* of φ if each t_i is a constant term of type $\tau \varphi$.

Lemma 11.2. *Suppose that some constant symbol belongs to $\tau \mathfrak{A}$. Let $B = \{t^{\mathfrak{A}} : t \text{ is a constant term of type } \tau \mathfrak{A}\}$. Then B is the universe of a substructure of $\mathfrak{A}$.*

PROOF: By Definition 3.8, it is enough to show that whenever $b_0, \ldots, b_{n-1} \in B$ and $f \in \tau\mathfrak{A}$, then $f^{\mathfrak{A}} b_0 \ldots b_{n-1} \in B$. But each b_i is some $t_i^{\mathfrak{A}}$ for t_i a constant term. Hence $f^{\mathfrak{A}} b_0 \ldots b_{n-1} = f^{\mathfrak{A}} t_0^{\mathfrak{A}} \ldots t_{n-1}^{\mathfrak{A}} = [ft_0 \ldots t_n]^{\mathfrak{A}}$, where $ft_0 \ldots t_{n-1}$ is a constant term. Hence $f^{\mathfrak{A}} b_0 \ldots b_{n-1} \in B$. □

Theorem 11.3 (Herbrand's Theorem). *Let φ be an open formula with free variables $u_0, \ldots, u_{n-1}$, and suppose that some constant symbol belongs to $\tau\varphi$. Then $\exists u_0 \ldots u_{n-1} \varphi$ is valid iff there is a finite sequence $\varphi_0, \ldots, \varphi_{k-1}$ of substitution instances of φ such that $\varphi_0 \vee \ldots \vee \varphi_{k-1}$ is valid.*

PROOF: Let $\Sigma = \{\neg \varphi_i : \varphi_i \text{ is a substitution instance of } \varphi\}$. Suppose that no disjunction $\varphi_0 \vee \ldots \vee \varphi_{k-1}$ of substitution instances of φ is valid. Then every finite subset of Σ has a model. Let $\mathfrak{A} \in \text{Mod}\,\Sigma$, and let $\mathfrak{B}$ be that substructure of $\mathfrak{A}$ whose universe is $\{t^{\mathfrak{A}} : t \text{ is a constant term of type } \subseteq \tau\mathfrak{A}\}$ (such as $\mathfrak{B}$ exists by Lemma 11.2). Since $\mathfrak{A} \vDash \neg \varphi t_0 \ldots t_{n-1}$ for all constant terms t_i of type $\subseteq \tau\mathfrak{A}$, and since $\neg \varphi$ is open, we have by Lemma 9.1 that $\mathfrak{B} \vDash \neg \varphi b_0 \ldots b_{n-1}$ for each $b_0, \ldots, b_{n-1} \in \mathfrak{B}$. Hence $\mathfrak{B} \vDash \neg \exists u_0 \ldots u_{n-1} \varphi$, so that $\exists u_0 \ldots u_{n-1} \varphi$ is not valid. So if $\exists u_0 \ldots u_{n-1} \varphi$ is valid, then there is a disjunction $\varphi_0 \vee \ldots \vee \varphi_{n-1}$ of substitution instances of φ that is valid. □

Our next theorem provides us with an algorithm for determining the satisfiability of open assertions. This algorithm also plays an important role in our presentation of an axiom system that is complete for validities.

Theorem 11.4. *Let σ be an open assertion, and let X be the set of all terms occurring in σ. If σ has a model, then σ has a model of power $\leq X$.*

PROOF: Suppose that σ has a model. Then σ has a model $\mathfrak{A}$ of type $\tau\sigma$. Let $B = \{t^{\mathfrak{A}} : t \in X\}$. Since $B \neq \emptyset$, we can choose some $b^* \in B$. Now let f be an n-function symbol in $\tau\sigma$, and let $b_0 \ldots b_{n-1} \in B$. Define

$$f^{\mathfrak{B}} b_0 \ldots b_{n-1} = \begin{cases} f^{\mathfrak{A}} b_0 \ldots b_{n-1} & \text{if } f^{\mathfrak{A}} b_0 \ldots b_{n-1} \in B \\ b^* & \text{otherwise.} \end{cases}$$

For each constant symbol $c \in \tau\sigma$ define $c^{\mathfrak{B}} = c^{\mathfrak{A}}$; for each n-relation symbol R and each $b_0 \ldots b_{n-1} \in B$ define $R^{\mathfrak{B}} b_0 \ldots b_{n-1}$ iff $R^{\mathfrak{A}} b_0 \ldots b_{n-1}$. Now let $\mathfrak{B} = (B, R^{\mathfrak{B}}, f^{\mathfrak{B}}, c^{\mathfrak{B}})_{R, f, c \in \tau\sigma}$. It is easy to see that $\mathfrak{B} \vDash \sigma$; indeed, an easy induction on subformulas φ of σ shows that $\mathfrak{A} \vDash \varphi$ iff $\mathfrak{B} \vDash \varphi$. □

How does Theorem 11.4 yield a test for satisfiability for open assertions? Suppose we are given an open assertion σ. We can then write down explicitly the set X of terms occurring in σ. By the theorem, we know that σ has a model iff it has a model of power $\leq cX$.

Notice that if γ is a 1-1 function on $|\mathfrak{A}|$ onto B, then γ induces an isomorphism on $\mathfrak{A}$ onto $\mathfrak{B}$, where $\mathfrak{B}$ is defined by

$$|\mathfrak{B}| = B,$$
$$c^{\mathfrak{B}} = \gamma c^{\mathfrak{A}},$$
$$f^{\mathfrak{B}} b_0 \ldots b_{n-1} = \gamma(f^{\mathfrak{A}}(\gamma^{-1} b_0) \ldots (\gamma^{-1} b_{n-1})),$$
$$R^{\mathfrak{B}} b_0 \ldots b_{n-1} \quad \text{iff} \quad R^{\mathfrak{A}}(\gamma^{-1} b_0) \ldots (\gamma^{-1} b_{n-1})$$

for all $c, f, R \in \tau\mathfrak{A}$ and all $b_0, \ldots, b_{n-1} \in B$. Hence if σ has a model, then σ has a model whose universe is either $\{0\}$, $\{0,1\}, \ldots,$ or $\{0, 1, 2, \ldots, m-1\}$, where $m = cX$. Next we make a list of all those sturctures of type $\tau\sigma$ having such a universe. We then check to see if σ is satisfied in any of these (finitely many) structures. If not, then we know that σ is not satisfied in any structure.

These considerations also yield a test for validity of open assertions because σ is valid iff $\neg \sigma$ is not satisfiable. Hence the above test for satisfiability of $\neg \sigma$ is tantamount to a test for validity of σ.

EXERCISES FOR §3.11

1. Let $A = \{0, 1\}$. Let $s = \{c_0, f_{1,0}, R_{2,0}\}$. List all structures $\mathfrak{A}$ such that $|\mathfrak{A}| = A$ and $t\mathfrak{A} = s$.

2. Let $A = \{0, 1, \ldots, n\}$, and let $s = \{c_0, c_1, f_{1,0}, f_{2,0}, f_{3,0}, R_{1,0}, R_{2,0}, R_{3,0}\}$. How many structures $\mathfrak{A}$ are there such that $|\mathfrak{A}| = A$ and $t\mathfrak{A} = s$?

3. Complete the proof of Theorem 11.4.

4. Suppose that $\vdash \sigma \to \exists v_0 \ldots v_n \varphi$, where φ is open and σ is universal. Suppose that $\tau\varphi$ contains a constant symbol. Show that $\vdash \sigma \to \psi$, where ψ is some disjunction of substitution instances of φ using constant terms of type $\tau\sigma \cup \tau\varphi$.

3.12 Axiomatizing the Validities of L

The main purpose of this section is to show that the set of valid L-assertions is effectively enumerable, i.e., there is an algorithm for producing a list $\sigma_0, \sigma_1, \sigma_2, \ldots$ of the valid assertions of L.

In order to realize this goal we must restrict our attention to those validities of some fixed, effectively enumerable type s, since if $\{\sigma_0, \sigma_1, \sigma_2, \ldots\}$ is an effectively enumerable set, then so is $\bigcup \{t\sigma_0, t\sigma_1, t\sigma_2, \ldots\}$. Here we take s to be $\{c_i : i \in \omega\} \cup \{f_{i+1,j} : i,j \in \omega\} \cup \{R_{i+1,j} : i,j \in \omega\}$.

Using the techniques of §2.12 it is easy to effectively assign a number $\#\sigma$ to each σ of type s. Hence our assertion that the validities of type s are effectively enumerable has a more formal counterpart, namely, $\{\#\sigma : t\sigma \subseteq s$ and $\vdash \sigma\}$ is a machine enumerable set. Here we shall be content with the less formal version, although the passage from "effectively enumerable" to "machine enumerable" is routine (but tedious).

Let $V = \{\sigma : t\sigma \subseteq s$ and $\vdash \sigma\}$. We shall give an axiomatization S that is effectively given, complete for V (i.e., every $\sigma \in V$ is an S theorem), and correct for V (i.e., every S theorem belongs to V). As argued in §2.13, the existence of such an S implies that V is effectively enumerable.

The Axiom System S.

i. If $\varphi_0 \vee \ldots \vee \varphi_{n-1}$ is a valid disjunction of substitution instances of some formula φ of type s, then $\varphi_0 \vee \ldots \vee \varphi_{n-1}$ is an axiom.
ii. If for each $i < n$, φ_i is a substitution instance of φ and $\tau \varphi_i \subseteq s$, and $\exists u_0 \ldots u_{m-1} \varphi$ is an assertion, then $(\varphi_0 \vee \ldots \vee \varphi_{n-1}, \exists u_0 \ldots u_{m-1} \varphi)$ is a rule of inference.
iii. If σ^* is a Skolem normal form for validity of σ, and $t\sigma^* \subseteq s$, then (σ^*, σ) is a rule of inference.

Theorem 12.1. *S is effectively given, correct, and complete for V.*

PROOF: The discussion at the end of §3.11 shows that the axioms of S are effectively given. Clearly, given an ordered pair (σ, σ'), we can decide whether or not (σ, σ') is a rule of inference of either form ii or iii. Hence S is effectively given.

Each axiom is valid. Moreover, if $\vdash \varphi_0 \vee \ldots \vee \varphi_{n-1}$, and each φ_i is a substitution instance of φ, and $\exists u_0 \ldots u_{m-1} \varphi$ is an assertion, then surely $\vdash \exists u_0 \ldots u_{m-1} \varphi$. Hence the rules in ii preserve validity. By Theorem 5.14 each rule of the form iii also preserves validity. It follows by an easy induction on the length of proofs that S is correct for V.

To prove completeness we must show that every valid σ of type s has an S proof, i.e., that $\vdash \sigma$ implies $\vdash_\mathsf{S} \sigma$. Suppose $\vdash \sigma$. Then $\vdash \sigma^*$, where σ^* is any Skolem normal form for validity of type s (by Theorem 5.14, σ^* is of the form $\exists u_0 \ldots u_{m-1} \varphi$, where φ is open). By Theorem 11.3, there is a valid disjunction of substitution instances of φ, say $\varphi_0 \vee \ldots \vee \varphi_{n-1}$. Hence $(\varphi_0 \vee \ldots \vee \varphi_{n-1}, \sigma^*, \sigma)$ is an S-proof of σ; so $\vdash_\mathsf{S} \sigma$. This gives the completeness of S and concludes the proof of Theorem 12.1. □

There are many other axiom systems that are effectively given, complete, and correct for the validities of L. Some of these appeal because they seem to be natural deductive systems. Their rules of deduction seem more like those we use informally when reasoning in mathematics; so formal proofs in these systems are organized in a way that appears to mimic our usual informal mathematical arguments. Other axiom systems allow a fine analysis of formal proofs that can yield more information. This study of various axiom systems has grown into a branch of logic called proof

3.12 Axiomatizing the Validities of L

theory. In our presentation, we contented ourselves with a not too natural, not too useful axiomatization, but one that gave us a quick proof of the completeness theorem: the validities of L are effectively enumerable. This is yet another of Gödel's achievements.

EXERCISES FOR §3.12

1. Suppose that L^* is a language that extends L in the sense that every formula of L is also a formula of L^*. Suppose that σ is an assertion of L^* such that $\mathfrak{A} \cong (\mathbf{N}, +, \cdot)$ iff some expansion of $\mathfrak{A}$ is a model of σ. Show that L^* is not axiomatizable, in the sense that there is no effectively given axiom system S^* such that the S^*-theorems are exactly the valid L^*-assertions. (*Hint:* See §2.12).

2. Suppose that L^* is a language that extends L, and that γ is an L^*-assertion such that A is finite iff (A) can be expanded to a model of γ. Show that L^* is not axiomatizable. (*Hint:* Produce a σ satisfying the conditions of Exercise 1.)

3. Again assume L^* extends L, but now suppose there is an L^*-assertion δ such that $\mathfrak{A} \cong (\mathbf{N}, <)$ iff $\mathfrak{A}$ can be extended to a model of δ. Show that L^* is not axiomatizable.

4. Prove that any language L^* satisfying the conditions of any of the above exercises is incompact.

5. Extend L to a language L^* (the monadic second order calculus) as follows: Add second order variables $X_1, X_2, \ldots$ (to range over sets). Extend the definition of atomic formula to include $v_j \epsilon X_i$ whenever $i, j \in \mathbf{N}^+$ and to include $\exists X_i \varphi$ whenever φ is a formula. An assignment z to A is now a function z such that $z(v) \in A$ if v is a (first order) variable as before and for each i, $z(X_i) \subseteq A$. The definition of satisfaction is extended in the obvious way by adding the clauses

$\mathfrak{A} \vDash v \epsilon X \langle z \rangle$ iff $z(v) \in z(X)$, and
$\mathfrak{A} \vDash \exists X \varphi \langle z \rangle$ iff there is a subset u of $|\mathfrak{A}|$ such that $\mathfrak{A} \vDash \varphi \langle z \binom{X}{u} \rangle$.

Show that there is an assertion σ of L^* such that $\mathfrak{A} \cong (\mathbf{N}, +, \cdot)$ iff some expansion of $\mathfrak{A}$ is a model of σ. Hence L^* is neither compact nor axiomatizable.

6. Let L^* be an extension of L, and let Σ be a countable set of L^*-assertions such that each finite subset of Σ has a model but Σ does not. Show that there is a set Γ of L^*-assertions such that $\mathfrak{A} \cong (\mathbf{N}, <)$ iff $\mathfrak{A}$ has an expansion that is a model of Γ.

7. The propositional calculus is that language $\mathscr{P}$ whose symbols are the following.
   Propositional variables: $P_1, P_2, P_3, \ldots,$
   Sentential connectives: $\neg, \wedge,$
   Parenthesis: [,].
   The set of formulas is the least set X that contains the propositional variables and is such that whenever $\varphi, \psi \in X$, then $[\neg \varphi] \in X$ and $[\varphi \wedge \psi] \in X$. An assignment z is a function from $\{P_i : i = 1, 2, 3, \ldots\}$ into $\{t, f\}$ (read t as true and f as false). $\varphi \langle z \rangle$ is defined inductively as follows:
   i. If $\varphi = P_i$, then $\varphi \langle z \rangle$ is $z(P_i)$.
   ii. If $\varphi = [\neg \psi]$, then $\varphi \langle z \rangle$ in t if $\psi \langle z \rangle$ is f, and $\varphi \langle z \rangle$ is f if $\psi \langle z \rangle$ is t.

iii. If $\varphi = [\psi_1 \wedge \psi_2]$, then $\varphi\langle z\rangle$ is t if both $\psi_1\langle z\rangle$ and $\psi_2\langle z\rangle$ are t, and $\varphi\langle z\rangle$ is f otherwise.

φ is valid (or a tautology) if $\varphi\langle z\rangle = t$ for all z. Show that the question 'Is φ a tautology?' is decidable. [*Hint:* The values of $\varphi\langle z\rangle$ and $\varphi\langle z'\rangle$ are the same if $z(P) = z'(P)$ for each P occurring in φ.]

8. Let V_L be the set of valid assertions of L. We have seen that V_L is effectively enumerable. However, V_L *is not decidable*. The proof of this is outlined below.
   (a) Let σ be the universal closure of the conjunction of the following:
      i. $Sv \approx Su \to u \approx v$,
      ii. $1 \not\approx Sv$,
      iii. $v \not\approx 1 \to \exists u(v \approx Su)$,
      iv. $v + 1 \approx Sv$,
      v. $v + Su \approx S(v+u)$,
      vi. $v \cdot 1 \approx v$,
      vii. $v \cdot Su \approx v \cdot u + v$.

   Let $\mathfrak{N} = (\mathbf{N}^+, +, \cdot, S, 1)$, where S is the successor function. Show that if $\mathfrak{B} \vDash \sigma$, then $\mathfrak{N} \subseteq \mathfrak{B}$.
   (b) Define the constant term n^* as follows:
   $$1^* = 1,$$
   $$(n+1)^* = S(n^*).$$
   Let $\varphi(v)$ be the formula at the beginning of §2.12 that says "v is the number of a Turing machine that halts on input v." φ is existential. Show that if $\vDash \sigma \to \varphi(n^*)$, then $\mathfrak{N} \vDash \varphi(n^*)$ and conversely.
   (c) Show that V_L is not decidable.

3.13 Some Recent Trends in Model Theory

The first order predicate calculus has a marvelously rich model theory. The cornerstones of this theory are the compactness theorem and the downward Löwenheim–Skolem theorem. In fact, there is a remarkable theorem of Lindström which characterizes the first order predicate calculus L in terms of these properties. Suppose that L^* is a language that extends L in the sense that every L-assertion is an L^*-assertion. Suppose that L^* is compact and that any countable set of L^* assertions that has a model has a countable model. Then L^* is L. In other words, this extraordinary result shows that the first order predicate calculus is the richest language that is compact and also has the downward Löwenheim–Skolem theorem to $\aleph_0$.

Since so much of the model theory of L is built up from its completeness property and the Löwenheim–Skolem theorem, one might take Linström's result as an indication that languages more expressive than L will not have significant model theories.

3.13 Some Recent Trends in Model Theory

On the other hand, the expressive power of the first order predicate calculus is not sufficient to characterize many of the objects and concepts which mathematicians are interested in. For example, the notion of finiteness, or of cardinality κ for κ infinite, cannot be expressed in L. The natural numbers with $<$ cannot be characterized up to isomorphism in L; nor can one characterize $(\mathbf{N}, +, \cdot)$ or $(\mathbf{R}, <)$ or $(\mathbf{R}, +, \cdot)$, and so on. So there is sufficient motivation to investigate languages more expressive than L with the hope that these languages will be relevant to areas of mathematics beyond the scope of L and yet have interesting model theories.

It is quite easy to invent languages with more expressive power than L. For example, the second order predicate calculus has, in addition to the first order variables, variables of second order intended to range over sets. Thus, adding $\forall X \big[[X0 \wedge \forall y [Xy \to X(y+1)]] \to \forall z [Xz] \big]$ to $\mathrm{Th}(\mathbf{N}, +, 0)$ gives a second order theory Σ which determines $(\mathbf{N}, +, 0)$ up to isomorphism, i.e., $\mathfrak{A} \in \mathrm{Mod}\,\Sigma$ iff $\mathfrak{A} \cong (\mathbf{N}, +, \cdot, 0)$. The second order calculus is expressive enough to state the order completeness property of $(\mathbf{R}, +, \cdot, <)$, and hence can describe this structure up to isomorphism. The notion of well ordering can be captured faithfully, and so on. Moreover, one need not stop with the second order. We can consider third order variables ranging over sets of sets, and fourth order, and so on. However, these languages, from the second order calculus on up, have not admitted a model theory as rich or as beautiful as that for L. The game is then to discover languages that have greater expressive power than L but still possess an amenable model theory. Several such languages have been discovered, and we briefly describe a couple of them.

Let κ and λ be infinite cardinals with $\kappa \geqslant \lambda \geqslant \omega$. We describe a language $L_{\kappa, \lambda}$ that differs from the first order predicate calculus in allowing conjunctions and disjunctions over sets of assertions of cardinality less than κ and simultaneous universalization or existentialization of fewer than λ variables. More precisely, the definition of formula, Definition 2.1d, is modified by replacing clause iii with

iii*. if $\Gamma \subseteq X$ and $c\Gamma < \kappa$ then $[\wedge \Gamma] \in X$ and $[\vee \Gamma] \in X$,

and replacing iv with

iv*. if $\varphi \in X$ and u is a set of variables of cardinality less than λ, then $[\exists u \varphi]$ and $[\forall u \varphi] \in X$.

We define $\mathfrak{A} \vDash \varphi \langle z \rangle$ in the obvious way by replacing iii, iii', iv, and iv' of Definition 4.3 with

iii*. $\varphi = [\wedge \Gamma]$, and $\mathfrak{A} \vDash \gamma \langle z \rangle$ for each $\gamma \in \Gamma$.
iii'*. $\varphi = [\vee \Gamma]$, and $\mathfrak{A} \vDash \gamma \langle z \rangle$ for some $\gamma \in \Gamma$.
iv*. $\varphi = [\exists u \psi]$, and for some $z^* \in {}^{\mathrm{Vbl}}|\mathfrak{A}|$ such that $z \langle v \rangle = z^* \langle v \rangle$ whenever $v \notin u$, we have $\mathfrak{A} \vDash \psi \langle z^* \rangle$.
iv'*. $\varphi = [\forall u \psi]$, and for all $z^* \in {}^{\mathrm{Vbl}}|\mathfrak{A}|$ such that $z \langle v \rangle = z^* \langle v \rangle$ whenever $v \notin u$, we have $\mathfrak{A} \vDash \psi \langle z^* \rangle$.

If κ is ω, and λ is arbitrary, then $L_{\kappa,\lambda}$ is just the first order predicate calculus. But if $\kappa > \omega$ and $\lambda \geqslant \omega$, then the increase in expressive power over L is spectacular. For example, $(\mathbf{N}, <)$ can be characterized up to isomorphism. To see this, let σ be the first order assertion '$<$ is a discrete ordering with first element but no last element', and let $\varphi^n(u,v)$ be the first order formula expressing 'u and v are separated by fewer than n elements'. Then $\sigma \wedge \forall uv \bigvee \{\varphi^n : n \in \mathbf{N}^+\}$ is an $L_{\omega_1,\omega}$-sentence characterizing $(\mathbf{N}, <)$ up to isomorphism. Similarly $(\mathbf{N}, +, \cdot)$ can be characterized in $L_{\omega_1,\omega}$ as can (α, ε) when α is a countable ordering. Scott has shown that if $\mathfrak{A}$ is countable and $\tau\mathfrak{A}$ is countable, then there is a single assertion $\sigma \in L_{\omega_1,\omega}$ such that for any countable $\mathfrak{B}$ we have $\mathfrak{B} \vDash \sigma$ iff $\mathfrak{B} \cong \mathfrak{A}$. However, $L_{\omega_1,\omega}$ cannot characterize the class of well-ordered structures, or the class of structures of any given infinite cardinal.

Of course $L_{\omega_1,\omega}$ is incompact. On the other hand, there are several analogies between its model theory and that of L. For example, every $\sigma \in L_{\omega_1,\omega}$ that has a model has a countable model, and any sentence σ with models of cardinality $\geqslant \beth_n$ for all $n \in \omega$ has models of arbitrarily high cardinality. The obvious analogs of the interpolation theorem and the definability theorem hold, and there are many more similarities.

When we pass to languages as rich as or richer than L_{ω_1,ω_1}, we gain a great deal of expressive power, but the resulting model theory seems to be relatively meager. As an example of the power of L_{ω_1,ω_1}, let σ be the first order assertion '$<$ is a linear ordering', and let δ be the L_{ω_1,ω_1}-assertion

$$\neg \exists \{v_i : i \in \omega\} [\bigwedge \{v_i > v_{i+1} : i \in \omega\}].$$

Then $\mathfrak{A} \in \mathrm{Mod}(\sigma \wedge \delta)$ iff $>^{\mathfrak{A}}$ well-orders $\mathfrak{A}$.

Another way of extending the first order predicate calculus is to add a new quantifier Q^n ($n = 1, 2, \ldots$) to the symbols of L and add the following clause to Definition 2.1d, the definition of L-formula:

v. if $\varphi \in X$ and $u_1, u_2, \ldots, u_n$ are variables, then $Q^n u_1, \ldots, u_n \varphi \in X$.

For each infinite cardinal κ, we have a κ-interpretation of the quantifier $Q^n : \mathfrak{A} \vDash^\kappa Q^n u_1, \ldots, u_n \varphi \langle z \rangle$ iff there is a κ-powered subset Y of $|\mathfrak{A}|$ such that for all $y_1, \ldots, y_n \in Y$ we have

$$\mathfrak{A} \vDash^\kappa \varphi \left\langle z \begin{pmatrix} u_1, \ldots, u_n \\ y_1, \ldots, y_n \end{pmatrix} \right\rangle.$$

Allowing different Q^n's to appear in the same formula gives the language $L^{<\omega}$.

Clearly, none of these languages are compact in the sense of L, for in the κ-interpretation the following set of assertions has no model, but each finite subset does:

$$\{c_\alpha \not\approx c_\beta : \alpha < \beta < \kappa\} \cup \{\neg Q^n u_1, \ldots, u_n [u_1 \approx u_1]\}.$$

(Note that the L^n-assertion requires that any model have cardinality $<\kappa$.) However, if we restrict our attention to countable sets of sentences, then

3.13 Some Recent Trends in Model Theory

L^1 is compact in all uncountable interpretations, and it is consistent to assume that $L^{<\omega}$ is compact in the ω_1-interpretation and all κ^{++}-interpretations. In the interpretations and under the set theoretic hypothesis in which the languages are known to be compact, they are also known to be axiomatizable. Löwenheim–Skolem theorems have not yet been fully investigated for these languages, but it is easy to prove that if $\sigma \in L^{<\omega}$ and σ has a model in the κ-interpretation, then σ has a model of power at most κ in the κ-interpretation. Research in this area has begun only recently, and many fundamental questions remain open.

Recently, Rubin and others have found languages that are properly more expressive than $L^{<\omega}$ and yet still retain the desirable model theoretic attributes of $L^{<\omega}$, such as axiomatizable and countable compactness. The search for stronger languages is still continuing. The exploration of the model theory of these languages has made heavy use of infinitary combinatorics and set theory, and some of us believe that these languages will be of some use in the study of axiomatic set theory and infinitary combinatorics.

Meanwhile, work continues in many directions. Fragments of the first order predicate calculus may give a rich theory. One such example is equational logic, which deals with sentences of the form $\forall u_0, \ldots, u_n[t \approx t']$ where t and t' are terms. A great many interesting algebraic classes such as groups, rings, and fields, can be presented in this form.

Non-standard analysis develops classical analysis on the basis of non-standard models of the reals. Probability theory, certain branches of topology and algebra, and other topics can be given a non-standard treatment as well.

It is impossible to say where the borders are between set theory, combinatorics, model theory, computable function theory, and the more classical branches of mathematics such as algebra and topology. The cross-fertilization between these areas has given enormous impetus to the development of mathematical logic, and we can look forward to many years of exciting progress.

Subject Index

Absolute 54
Addition of cardinals 37
Addition of ordinals 35
∀∃-formula 178
Algebraic number 20
Arithmetical 118
Assertion 115
Assignment 112, 142
At least as numerous 15
Atomic formula 113, 137
Axiom 124
  of choice 27
  of determinacy 49
  of extensionality 44
  of infinity 45
  of the null set 44
  of pairing 44
  of the power set 45
  of regularity 44
  of replacement 45
  of separation 45
  of union 45
  system 124

Barwise, J. 134
Binary relation 6
Bound variable 115
Bounded quantifiers 75
Branch 23

Cantor–Bernstein theorem 17
Cardinal number 36

Cardinality 37
Cartesian product 6
Categorical 171
Chain 23, 29
Chinese remainder theorem 109
Choice function 27
Church, Alonzo 90
Cohen, Paul 48, 57
Compactness theorem 158
Complement of A in B 4
Complete 125, 162, 171
  diagram 169
Composite of f and g 7
Composition of f with $g_1,\ldots,g_r$ 68
Computable function 63
Computable relation 65
Computable set 65
Congruence relation 141
Conjunctive normal form 155
Consistency lemma 180
Consistent 51
Constant term 185
Continuum hypothesis 48
Correct 125
Countable 10
  axiom of choice 49

Definability theorem 184
Defines 118
Definition by cases 76
Definition by transfinite recursion 41
De Morgan's rules 5

Dense ordering 22
Densely ordered structure 22, 172
Denumerable 10
Diagonal function 127
Diagram 147
Difference of A from B 4
Diophantine equation 131
Discrete 22
Disjoint 4
Disjunctive normal form 155
Domain 6

$\in$-transitive 31
Effectively given 124
Element 1
Elementarily equivalent 144
Elementary chain 171
Elementary class 144
Elementary extension 167
Elementary substructure 167
Equinumerous 9
Equivalent 151
Existential 175
Expansion 140
Explicit definition 182
Exponentiation of cardinals 39
Exponentiation of ordinals 35
Expression 111, 136
Extension of g 7, 140

False 116
Field 6
Filter 166
  base 165
Finite 10
Finitely satisfiable 162
Fischer, M. and Rabin, M. 132
Formula 113, 137
Free 115
Free for 117
Friedberg, R. 133
Function 7
Functional 41

Generalized continuum hypothesis 48
Gödel, Kurt 48, 51, 56, 90, 128, 129
Gödel's incompleteness theorem 128
Greatest element 23

Halting problem 94
Herbrand's theorem 186

Hilbert, David 131
Homomorphism 141
Huge cardinal 49

Immediate predecessor 22
Immediate successor 22
Implicit definition 182
Implies 51
Inaccessible cardinals 48, 55
Incompleteness theorem 128
Independent 51
Infinite 10
Initial pairing 24
Initial segment 23
Injection 140
Input 61, 62
Integral polynomial 20
Interpolant 182
Interpolation theorem 182
Intersection 4
  over X 4
Inverse function 7
Isomorphic 34
Isomorphism 34, 140

Kleene, S. C. 90
König's infinity lemma 29
König's lemma 39
Kreisel, G. 134

Least element 23
Lebesgue measurable 49
Less numerous 15
Limit cardinal 36
Limit ordinal 35
Lindenmeyer, A. 134
Linear ordering 22
Linearly ordered structure 22
Löwenheim–Skolem Theorem 169

Machine enumerable relation 101
Machine enumerable set 100
Marker 60
Martin, T. 46
Matrix 154
Matiasevič, Y. 131
Maximal 23, 29
  element 23
Maximal principle 29
Measurable cardinal 49

Minimal element 23
Model 51, 144
More numerous 15
Mučnik, A. A. 133
Multiplication of cardinals 37
Multiplication of ordinals 35
Mycielski, J. 49

Occurs 114
On A 7
One to one 7
Onto B 7
Open 154
Ord α 33
Order preserving 24
Ordinal 33
Output 61, 62

Pairing 9
Pairwise disjoint sets 4
Partial computation 61
Partial ordering 21
Partially ordered structure 21
Post, E. L. 90, 133
Power set 5
Predecessor 22
Prefix 154
Prenex normal form 154
Prime Factorization Theorem 12
Primitive recursive functions 110
Principal ultrafilter 166
Product of cardinals 37
Product of ordinals 37
Proof 124

Quantifiers 103
Quotient structure 142

Ramsey's theorem 49, 50
Range 6
Recursive definitions 41, 42, 70
Recursive function 107
Reduct 140
Relation 6
Representing function 65
Restriction 140
Restriction of f 7
Rozenberg, G. 134
Rubin, M. 193
Rule of inference 124

Satisfaction 114, 142
Scanned term 60
Second incompleteness theorem 129
Self-halting problem 92
Set 1
Simple 147
Skolem normal form 156
  for validity 156
Skolem's paradox 160
Solovay, R. 49
Souslin's hypothesis 49
Standard 54
State 60
Strong inaccessible 48
Structure 51, 138
Subformula 114
Subsequence 114
Subset 2
Substitution instance 185
Substructure 140
Successor 22
  cardinal 36
  ordinal 35
  tape position 61
Super-huge cardinal 49
Symbols of L 136

Tape 60
Tape position 60
Tarski, Alfred 122
Term 111, 136
Theorem 124
Theory of $\mathfrak{A}$ 144
Transcendental number 20
Transfinite induction 40
Tree 29
Truth 116
Turing machine 59, 60
Turing's thesis 89
Type 136

Ulam, S. 49
Ultrafilter 166
Ultrapower 167
Ultraproduct 166
Undecidable 127
Union 3
  over X 4
Unique factorization theorem 91
Universal 175
  closure 151
  machine 95

Valid 151
Vaught's test 171

Weakly compact cardinal 49
Well ordering 23
Witness 163

Young, P. 132

Zermelo-Fraenkel axiomatization 44, 146

Undergraduate Texts in Mathematics

continued from ii

Martin: The Foundations of Geometry and the Non-Euclidean Plane.

Martin: Transformation Geometry: An Introduction to Symmetry.

Millman/Parker: Geometry: A Metric Approach with Models.

Owen: A First Course in the Mathematical Foundations of Thermodynamics.

Prenowitz/Jantosciak: Join Geometries.

Priestly: Calculus: An Historical Approach.

Protter/Morrey: A First Course in Real Analysis.

Protter/Morrey: Intermediate Calculus.

Ross: Elementary Analysis: The Theory of Calculus.

Scharlau/Opolka: From Fermat to Minkowski.

Sigler: Algebra.

Simmonds: A Brief on Tensor Analysis.

Singer/Thorpe: Lecture Notes on Elementary Topology and Geometry.

Smith: Linear Algebra. Second edition.

Smith: Primer of Modern Analysis.

Thorpe: Elementary Topics in Differential Geometry.

Troutman: Variational Calculus with Elementary Convexity.

Wilson: Much Ado About Calculus.

±c P.48